SMART ELECTRONIC DEVICES AND SYSTEMS FOR BIOMEDICAL AND HEALTHCARE APPLICATIONS

SMART ELECTRONIC DEVICES AND SYSTEMS FOR BIOMEDICAL AND HEALTHCARE APPLICATIONS

Upendra Kumar, PhD
Rajesh K. Kesharwani, PhD
Raj K. Keservani, PhD, MPharm
Editors

AAP APPLE ACADEMIC PRESS

First edition published 2026

Apple Academic Press Inc.
1265 Goldenrod Circle, NE,
Palm Bay, FL 32905 USA

760 Laurentian Drive, Unit 19,
Burlington, ON L7N 0A4, Canada

CRC Press
2385 NW Executive Center Drive,
Suite 320, Boca Raton FL 33431

4 Park Square, Milton Park,
Abingdon, Oxon, OX14 4RN, UK

Apple Academic Press exclusively co-publishes with CRC Press, an imprint of Taylor & Francis Group, LLC

Library and Archives Canada Cataloguing in Publication

Title: Smart electronic devices and systems for biomedical and healthcare applications / Upendra Kumar, PhD, Rajesh K. Kesharwani, PhD, Raj K. Keservani, PhD, MPharm, editors.
Names: Kumar, Upendra, 1979- editor. | Kesharwani, Rajesh Kumar, 1978- editor | Keservani, Raj K., 1981- editor
Series: Advances in hospitality and tourism book series.
Description: First edition. | Includes bibliographical references and index.
Identifiers: Canadiana (print) 20250199548 | Canadiana (ebook) 20250199645 | ISBN 9781774918166 (hardcover) | ISBN 9781003560852 (ebook)
Subjects: LCSH: Medical informatics. | LCSH: Medical care—Information technology. | LCSH: Medical electronics. | LCSH: Bioelectronics. | LCSH: Medical instruments and apparatus—Technological innovations.
Classification: LCC R856 .S63 2026 | DDC 610.285—dc23

Library of Congress Cataloging-in-Publication Data

CIP data on file with US Library of Congress

ISBN: 978-1-77491-816-6 (hbk)
ISBN: 978-1-00356-085-2 (ebk)
DOI: 10.1201/9781003560852

About the Editors

Upendra Kumar, PhD
Assistant Professor, Department of Computer Science and Engineering, Institute of Engineering and Technology, Dr. A.P.J. Abdul Kalam Technical University, Uttar Pradesh, Lucknow, India

Upendra Kumar, PhD, is Assistant Professor in the Department of Computer Science and Engineering at the Institute of Engineering and Technology, Dr. A. P. J. Abdul Kalam Technical University, Uttar Pradesh, Lucknow, India. He has around four years of industrial experience at Tata Consultancy Services, where he worked on many corporate projects as a JAVA/J2EE developer. He also has more than 10 years of teaching and research experience. His areas of research interest include digital image processing, machine learning, medical informatics, biomedical signal processing, fractal and chaos analysis, embedded systems and robotics with embedded C, medical imaging and diagnostic system, big data analytics, etc. He has published many papers in international and national journals and conferences. He has also published books and book chapters with international publishers and filed one Indian patent with the collaboration of IIIT Allahabad, India. He supervises research scholars for their PhD thesis work. Dr. Kumar has delivered expert lectures at various international and national conferences, workshops, and seminars and is a member of several review committees of national and international journals, conferences and professional bodies. Apart from teaching and research work, he has held an administrative role working as Deputy Dean of Students Welfare at the Institute of Engineering and Technology. Dr. Kumar has an MTech in IT from IIIT Allahabad and a PhD in Computer Science and Engineering from Dr. A.P.J. Abdul Kalam Technical University, UP, Lucknow, India.

Rajesh K. Kesharwani, PhD, MTech
Associate Professor, Department of Computer Application, Nehru Gram Bharati (Deemed to be University), Prayagraj, India

Rajesh K. Kesharwani, PhD, MTech, is working as Associate Professor, Department of Computer Application, Nehru Gram Bharati (Deemed to be University), Prayagraj, India. He has more than 11 years of research and nine years of teaching experience at various institutes of India, imparting bioinformatics and biotechnology education. He has received several awards, including the NASI-Swarna Jayanti Puruskar by The National Academy of Sciences of India. He has supervised one PhD and more than 20 undergraduate and graduate students for their research work. Dr. Kesharwani has authored over 49 peer-reviewed articles, 20 book chapters, and 14 edited books with international publishers. He is a co-editor of the book series *AAP Advances in Nutraceuticals.* He has been a member of many scientific communities as well as a reviewer for many international journals. He has presented many papers at various national and international conferences. He has been a recipient of a Ministry of Human Resource Development (India) Fellowship and a Senior Research Fellowship from the Indian Council of Medical Research, India. His research fields of interest are medical informatics, protein structure and function prediction, computer-aided drug designing, structural biology, drug delivery, cancer biology, nanobiotechnology, and biomedical sciences. Dr. Kesharwani received his PhD from the Indian Institute of Information Technology, Allahabad, and worked at NIT Warangal for two semesters.

Raj K. Keservani, PhD, MPharm
Associate Professor, Faculty of B. Pharmacy, CSM Group of Institutions, Prayagraj, India

Raj K. Keservani, PhD, MPharm, is working as Associate Professor in the Faculty of B. Pharmacy at CSM Group of Institutions, Prayagraj, India. He has more than 12 years of academic (teaching) experience from various institutes of India in pharmaceutical education. He has supervised more than 25 UG students for their research work, has published over 30 peer-reviewed papers in the field of pharmaceutical sciences in national and international journals, four patents, 70 book chapters, three co-authored books, and over 40 edited books. He is also active as a reviewer for several international scientific journals. He is a co-editor of the book series *AAP Advances in Nutraceuticals.* Dr. Keservani graduated with a pharmacy degree from the Department of Pharmacy, Kumaun University, Nainital (Uttarakhand), India. He received his Master of Pharmacy (MPharm) (specialization in pharmaceutics) from the School of Pharmaceutical Sciences, Rajiv Gandhi Proudyogiki Vishwavidyalaya, Bhopal, India. He received his PhD degree from IKG Punjab Technical University, Jalandhar, Punjab, India. Dr. Keservani is a life member of the Society of Pharmaceutical Education and Research (SPER). His research interests include nutraceutical and functional foods, novel drug delivery systems (NDDS), transdermal drug delivery/drug delivery, health science, cancer biology, and neurobiology.

Contents

Contributors

Harsh Arora
Department of Computer Science, Keshav Mahavidyalaya, University of Delhi, India

Navya Arya
Institute of Engineering and Technology, Lucknow, India

Krishna Lal Baishnab
Department of Electronics and Communication Engineering, National Institute of Technology Silchar, Assam, India.

Rajnish Bhasker
Department of Electrical Engineering, Uma Nath Singh Institute of Engineering and Technology, VBSPU, Jaunpur, India

Swagata Devi
Assistant Professor, Faculty of Engineering, Assam down town University, Sankar Madhab Path, Gandhi Nagar, Panikhaiti, Guwahati, Assam, India

Namrata Dhanda
Department of Computer Science & Engineering, Amity University, Lucknow, Uttar Pradesh, India

Koushik Guha
Department of Electronics and Communication Engineering, National Institute of Technology Silchar, Assam, India

Kapil Kumar Gupta
Department of Computer Science & Engineering, Amity University, Lucknow, Uttar Pradesh, India

Arihant Kulshrestha
Department of Computer Science, Keshav Mahavidyalaya, University of Delhi, India

Neeraj Kumar
Department of Computer Science & Engineering Shri Ramswaroop Memorial University, Barabanki, India

Saurabh Kumar
Department of Electrical Engineering, Engineering Degree Division, Institute of Engineering and Rural Technology, Prayagraj, India

Sawan Kumar
Junior Engineer, Bharat Sachar Nigam Limited, Hoshiarpur, Punjab, India

Upendra Kumar
Department of Computer Science & Engineering, Institute of Engineering & Technology, Lucknow, India

Ashutosh Mishra
Computer Science and Engineering Department, Thapar Institute of Engineering and Technology, Patiala, India

Tulika Narang
Institute of Engineering and Technology, Lucknow, India

Sourav Nath
Department of Electronics and Communication Engineering, National Institute of Technology Silchar, Assam, India.

Pratibha Pandey
Department of Computer Science and Engineering, Institute of Engineering and Technology, Lucknow, Uttar Pradesh, India

Rachna Prabha
Department of Electronics Engineering, Indian Institute of Technology (ISM), Dhanbad, India

Ishita Rai
Department of Computer Science, Keshav Mahavidyalaya, University of Delhi, India

Himangi Sharma
Department of Computer Science, Keshav Mahavidyalaya, University of Delhi, India

Shalini Sharma
Department of Electronics & Communication Engineering, University of Lucknow, Lucknow, India

Himanshu Kumar Shukla
Department of CSE, F.O.E.T., University of Lucknow, Lucknow, Uttar Pradesh, India

Shudha Shukla
Department of CSE, F.O.E.T., University of Lucknow, Lucknow, Uttar Pradesh, India

Ashutosh Singh
Assistant Professor, Department of Computer Science, Keshav Mahavidyalaya, University of Delhi, India

Chandrabhan Singh
Department of CSE, F.O.E.T., University of Lucknow, Lucknow, Uttar Pradesh, India

Shweta Singh
Department of Electrical Engineering, Uma Nath Singh Institute of Engineering and Technology, VBSPU, Jaunpur, India

Rohit Srivastava
Department of CSE, F.O.E.T., University of Lucknow, Lucknow, Uttar Pradesh, India

Gayatri Tiwari
Department of Electronics & Communication Engineering, Institute of Engineering & Technology, Lucknow, India

Dinesh Kumar Tyagi
Department of Computer Science and Engineering, Malaviya National Institute of Technology, Jaipur, Rajasthan, India

Santosh Kumar Vipparthi
School of Artificial Intelligence and Data Engineering, Indian Institute of Technology, Ropar, Rupnagar, Punjab, India

Prem Shanker Yadav
Department of Computer Science and Engineering, Malaviya National Institute of Technology, Jaipur, Rajasthan, India

Foreword

It gives me immense pleasure to write the foreword for this book, *Smart Electronic Devices/Systems for Biomedical and Healthcare Applications.* It provides detailed information of the research done in the development of electronic devices/systems for biomedical and healthcare applications. With the development of information technology, the concept of smart healthcare has gradually come to the fore. Smart healthcare uses the next generation of information technologies such as the Internet of Things (IoT), big data, cloud computing, and artificial intelligence to transform the traditional medical system in a comprehensive way, making healthcare more efficient, convenient, and personalized. In order to introduce the concept of smart health, this book first introduces the key technologies supporting smart health and presents the current state of smart health in several important areas. Subsequently, the existing problems of smart healthcare have been discussed and tried to propose solutions for them. Finally, the editors look ahead and assess the future prospects of smart healthcare.

The main driving force behind the editing of this book is the transformation of the current healthcare system with an increasing aging population, skyrocketing healthcare costs, and the frequent occurrence of chronic diseases. Now, there is a need to provide healthcare services in more cost-effective and responsive ways. The traditional hospital healthcare system is undergoing transformation to provide personalized advanced healthcare facilities for early diagnosis, treatment, and long-term digital monitoring. This book will be of particular interest to graduate students and researchers in biomedical engineering and medical informatics, medical device designers, and researchers from around the world. All contributing authors demonstrate exceptional research expertise in electronic devices/systems for biomedical and healthcare applications and provide a global perspective on current and future advances in the field.

I congratulate the editors for bringing together experts in the field of research in electronic devices/systems for biomedical and healthcare applications and the authors for their excellent contributions.

—Prof. (Dr.) Neelendra Badal

Director, Rajkiy Engineering College, Bijnor, Uttar Pradesh, India

Professor, KNIT Sultanpur, Uttar Pradesh, India

Preface

The book, *Smart Electronic Devices and Systems for Biomedical and Healthcare Applications,* is a compilation of the advanced empirical knowledge of current e-healthcare system and its use in medical sciences. Electronic systems are ever present in our daily lives. In addition, biomedical applications of electronic devices/systems seem to be most important in the development and management of a quality and efficient healthcare system. For this reason, this field is attracting more and more specialists and medical practitioners. Thanks to the development of electronics, it is now possible to quickly and accurately diagnose a number of diseases, and, therefore, it is helpful in effectively monitoring patients with severe or chronic conditions. The support provided by such electronic devices in handling major health crises should not be ignored, as in current drastic scenario of COVID-19 pandemic. This book is committed to the dissemination of the latest advances and newly developed methods in the field of applications of electronic devices in biomedical.

The main driver of recent days is the lack of coherence between academic researchers and real-time clinical applications. This book consists of 13 chapters that attempt to recognize the current direction in biomedicine and the healthcare system. The main drivers of the transformation of the current healthcare system are the increasing aging of the population, the skyrocketing costs of healthcare, and the frequent occurrence of chronic diseases. They now need to deliver healthcare services in more cost-effective and responsive ways. The traditional hospital-centered healthcare system, which focuses primarily on diagnosis and treatment, is now poised to transform into an individual-cantered healthcare system, which in turn focuses primarily on early detection, early diagnosis, and long-term follow up.

The roles of portable, wearable, and connected electronic health devices in the diagnosis, continuous monitoring, and efficient treatment of diseases have become important and necessary components for future biomedical and healthcare applications. In addition, the development of biomedical and healthcare devices should be done cost-effectively, with affordable transportation and energy consumption as low as possible. Therefore, integrated circuits (ICs) designed on the basis of biomedical devices should be developed in these biomedical and healthcare applications. Designing biomedical ICs has great problems in finding low noise signal, low frequency, and low power signal, and, therefore, there is a trade-off between noise and performance and area of geographic location, cost, and bandwidth because of two reasons: (1) Biomedical signals such as ECG and EEG are generally of low frequency, weak, and noisy in nature, and (2) the embedded biomedical devices/systems should be human wearable practically and portable in healthcare applications.

With the advancements and newly invented methods of information technology (IT), the idea of smart healthcare and diagnosis gradually came to the fore. Smart healthcare uses the new generation of IT such as the big data, Internet of Things (IoT), cloud and grid computing, machine learning, and artificial intelligence (AI) to transform and revolutionized the traditional healthcare system in a comprehensive way, making healthcare more efficient and convenient and more customized. In order to introduce the concept of smart healthcare, this book first lists the key technologies supporting smart healthcare and presents the current state of smart healthcare in several important areas. Further, the existing problems of smart healthcare are discussed, and the authors have tried to propose efficient solutions to them. Finally, after wide discussion on applications of current technologies in healthcare, the book also assesses the future prospects of smart healthcare.

Smartphone-based applications revolutionize biomedical practice with AI. Due to all-time need of presence of experts in cloud and continuous data connectivity to the device, rising demand for high-quality medical services, telemedicine suffers from its inherent limitations. The present book is a compilation of chapters that describe various major design challenges of applying IT-based methodologies and research potential in biomedical systems and covers the fundamental principles underlying biomedical engineering and its advancement in healthcare systems. In the last few decades, extensive research have been witnessed in the field of health services and their technological modernization. To be more specific, various IT-enabled technologies have shown their potential use in connecting various medical devices, internet, sensors, computers, and healthcare workers to provide quality medical services at a remote location. This IT-enabled healthcare system has helped in improving patient safety, thereby reducing healthcare costs, increasing ease of availability of healthcare services to remote locations, and enhancing operational and diagnostic efficiency in healthcare.

This book is an effort to bring significant attention of the general public, scientists, biomedical engineers, medical experts, and analytical engineers by providing various possible research applications of newly invented IT-enabled technologies and making the systems more advanced, efficient, easy to use, widely applicable, and robust. So, it is a great idea and right time to come out with a book as soon as possible.

CHAPTER 1

The Role of Big Data Analytics and IoT in the Healthcare Domain

NAVYA ARYA, TULIKA NARANG, and UPENDRA KUMAR

Institute of Engineering and Technology, Lucknow, India

ABSTRACT

The rising role of big data analytics and the Internet of Things has created many opportunities for improvements in every domain. Both technologies have great potential hidden in them and when combined will create wonders. The IoT device creates massive amounts of data every second that can be collected and used for big data analytics to uncover insights from it. The healthcare industry is an important domain that generates huge amounts of data through wearable fitness bands, patient records, hospital data, medical examinations, and biomedical research. The IoT devices generate heterogeneous raw data and real-time that keeps on increasing per second. So it is essential to properly store this big data for proper processing and analysis otherwise its potential will get wasted. The collaboration of big data analytics and the Internet of Things can help healthcare corporations such as performing predictive analysis for early detection in patients, discovering root causes of any disease, or performing digital diagnosis.

1.1 BIG DATA: AN OVERVIEW

Big data as the name suggests is huge data obtained from various sources of technical devices. Its formal definition would be data that contain high volume, variety, and velocity (keeps on increasing). Normal computation

Smart Electronic Devices and Systems for Biomedical and Healthcare Applications. Upendra Kumar, Rajesh K. Kesharwani, & Raj K. Keservani (Ed.)

DOI: 10.1201/9781003560852-1

processes cannot process such a complex and huge amount of data. So where is all this data coming from? If you use internet data services then you are contributing to the proliferation of big data. Giant companies such as Google also play a major role in the mushrooming of big data. For example, Facebook sets down 10 petabytes (PB) of data per month, and over 100 PB per Google per month (Chahal et al., 2020). There is an idea that big data started existing because of Moore's law. It states that the count of transistors in a dense integrated circuit (IC) becomes double about every two years. So how does this relate to big data?

Moore's law led to the creation of hardware that became small with powerful processors and the ability of software to compute increased. Nowadays even personal computers can handle much larger amounts of data. The storage space shifted from kilobytes to megabytes and now gigabytes. Another reason was how databases were handled earlier and nowadays. Tools like Apache Hadoop made the data handling process easier so more and more data was able to be managed and stored.

1.1.1 CHARACTERISTICS OF BIG DATA

Big data was initially defined as a model of three V's that was volume, variety, and velocity. With the advancement in big data, it became four V's adding veracity, and later with further progress scientists have settled down the characteristics of big data with 10 V's (Chahal et al., 2020).

- Volume—It indicates the amount of data that exists or is obtained from a source. It is the base if data is not big enough then it cannot be considered big data.
- Velocity—It refers to the speed at which data keeps growing so the data flow of big data is large and continuous. For example, we keep generating data from our smartphones every second all the time. In healthcare, real-time data is generated and must be processed and analyzed in real time.
- Variety—This means that data can be ingested from a diversity of sources. That is the reason why IoT can easily integrate with big data as big data accept heterogeneous data. However, data from multiple sources or unstructured data must be organized and cleaned for effective data modeling and applying machine learning models.
- Veracity—It is the accuracy of data. Data that has noise, redundancy, errors, or missing values is not conventional to analyze. It is an

assurance that data is not corrupt. Hence, accuracy and standard contribute to the insights that can be gained from it. For example, in the healthcare industry if there is incomplete information about a patient's medical history he might be given something not suitable for him and would prevent doctors from accurate diagnosis endangering the patient's life.

- Value—It signifies the value or worth of the data provided to that particular organization. The big data collected and analyzed can have many hidden patterns that could be beneficial to companies and for research purposes.
- Validity—It checks for the validity of the data which means how correct the data is for a particle purpose. For example, you need patient record data to draw insights about patients, business data or any other data will not be correct. Using the correct form or type of data is essential to obtain the proper and meaningful conclusions.
- Variability—This refers to the number of inconsistencies present in the data and it can also point to the inconsistent velocity at which data gets loaded into the database or cloud. On a graph, if data points lie far from each other then the variability is more otherwise it is less. It is defined by its scatter rate or dispersion rate.
- Viscosity—It means a measure of resistance in the flow of volume of data to the database. It is the same as velocity but it describes the lag time or delays in the loading of data. This delay can be caused due to late processing of previous data, resistance from data integration, or can also be coming directly from sources. To tackle this problem, we can use better streaming or agile integration buses.
- Virality—The general meaning of virality is how easily information can be spread from people to people's networks like a rumor. Similarly for big data, it means how fast data is shared and spread from one node to another. The rate or time at which this activity takes place is virality.
- Visualization—The representation of data in figures and symbols is called visualization. The big data is huge so it makes the work easier to understand. Visualization plays an important role. When data is analyzed and insights are drawn they are to be presented in a report format then also data visualization can help. For example, in textual data, if we overview it we may not find anything but if we visualize it using a tool, for example, WordCloud then we find frequently occurring words and important words of that corpus. It is useful for explaining things to non-technical audiences.

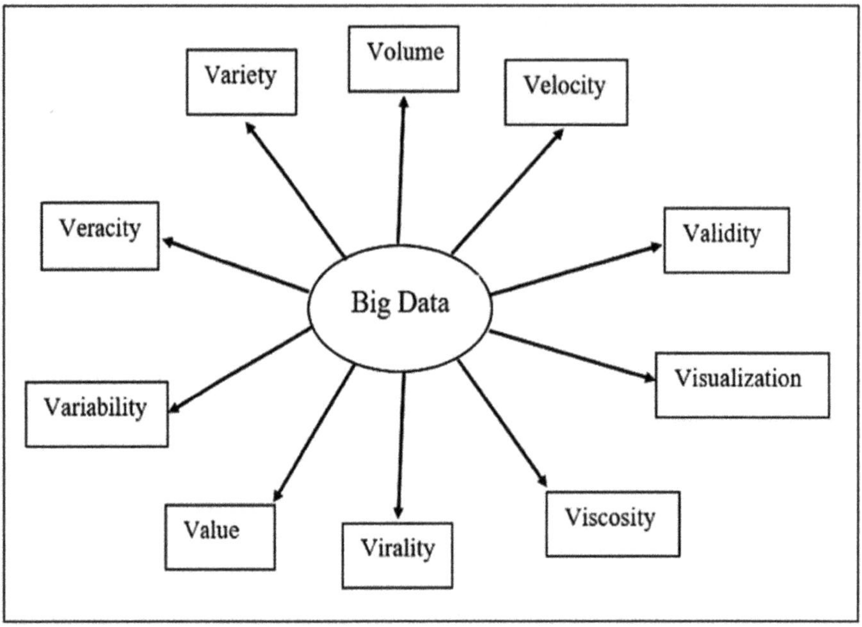

FIGURE 1.1 10 V's of big data.

1.1.2 HISTORY OF BIG DATA

Big data is still emerging and evolving but it came into existence during the 1960s and 70s when relational databases came into existence. Facebook was launched in 2004 and YouTube in 2005, after this time people began realizing how much data is developed through them and other online sites. In 2006, Apache Hadoop came into existence, an open-source platform for storing and solving computational problems of huge data. Gradually a variety of open-source platforms such as Spark, and NoSQL began introducing themselves in the market and big data gained popularity.

1.1.3 APPLICATIONS OF BIG DATA IN HEALTHCARE INDUSTRY

Healthcare analytics with the help of big data helps us reduce the cost of treatments and predictive analysis of diseases which can prevent sudden attacks on health and boost the quality of life. Big data in healthcare is created from patients' records and digital technologies, pay records, and research studies that accumulate patient information such as heart rate, blood pressure, sugar levels, etc. This is a very important aspect that helps in times of pandemics where it gets impossible to even get out of the house. During these times distance diagnosis and predictive analysis may play a major role in helping patients. The National Institute of Health (NIH) has announced "All of Us" which targets to collect all patient's data such as EHR (Electronic Health Records). Its advantage would be that doctors could access the entire history of the patient altogether such as past prescriptions, allergies, family medical history, demographics, or any previous diagnosis. It might also help patients with medical insurance (Dash et al., 2019). Not only it helps patients but can help hospitals manage their staff better.

Big data can help hospitals improve their services. Using time series analysis they can predict at this time of year there are more patients, for example, during rainy seasons dengue and viral fevers are common so hospitals experience more incoming patients and admissions than usual which would require more staff.

Bioinformatics uses big data to perform research studies and draw conclusions.

1.1.4 APPLICATIONS OF BIG DATA IN OTHER DOMAINS

Big data is a collection of huge amounts of data drawn from people, websites, and companies. These data are in raw forms such as the web browsing pattern of a user tweets from Twitter or images or clickbates from YouTube. The obtained is processed and meaningful insights are curated from them. These insights lead to company profits, increased sales and production improved web browsing experience, etc. Some important uses are listed below:

- Big data helps companies understand customer demand by predictive analytics and analyzing sales reports, and customer reviews. It helps them gain insight into products. For example, if a product is not performing well in the market then its production can be inhibited.
- It can also help us predict the maintenance of machinery. Through analyzing log files data that have purchase dates of machinery, faults can be uncovered.
- Fake reviews on websites and products can be eliminated. The same goes for fake news detection and fake profiles on Facebook.
- Big data helps companies make decisions, make plans, and enhance their products with the help of customer feedback and deliver better products and services.

- Companies like Myntra, Flipkart, and Amazon use big data to observe user browsing behavior and then use that to create recommender systems and influence them into buying products.
- Big data is the reason machine learning and deep learning algorithms are flourishing so much. They help us with everything nowadays from predicting the weather, traffic on the roads, and even accidents. It has made our life easier and better.

1.2 INTERNET OF THINGS: AN OVERVIEW

A network or interconnection of objects or things that are planted with sensors, and software and are capable of connecting and exchanging information over the Internet data service is known as the Internet of Things. These things can be mobile phones, smart TVs, Amazon's Alexa, and or a simple kitchen appliance. Over the past years, IoT has innovated and is one of the most important technologies. The development of the Internet of Things has created smart cities, smart hospitals, smart traffic, smart homes, smart cars, etc. IoT makes our lives easier, you can control the switches of your house without even getting up, and everything is at our fingertips.

IoT was first coined by Kevin Ashton of Procter and Gamble and its first literature was published in 2002 in which a tiny wireless chip was designed that enabled stores to have eyes. The growth of IoT has been rapid and research has shown that IoT-connected devices number is 43 billion by 2023.

1.2.1 GENERAL ARCHITECTURE OF INTERNET OF THINGS DEVICES

IoT is a framework of processors, hardware, sensors, and software that enables machine-to-machine communication without human meddling. Its general architecture consists of three layers: Application layer, Network Layer, and Perception Layer. Over the years different IoT architectures have been proposed and hence, there is no standardized architecture.

1.3 THE INTERRELATION BETWEEN IOT AND BIG DATA

Data is the key to many of these technologies. The base of both the Internet of Things and big data is data. Integrating IoT and big data can help solve many research problems in the domain of health, for example, gaining insights on why certain illnesses happen, and in business helps in finding market trends, customer preferences, and many other industries. Although IoT data may differ from other big data as data collected from IoT devices is heterogeneous and has data from sensors, actuators, and other machinery they increase at lightning speed. But here comes big data analytics into the picture that helps in handling such humongous, redundant, and unstructured data.

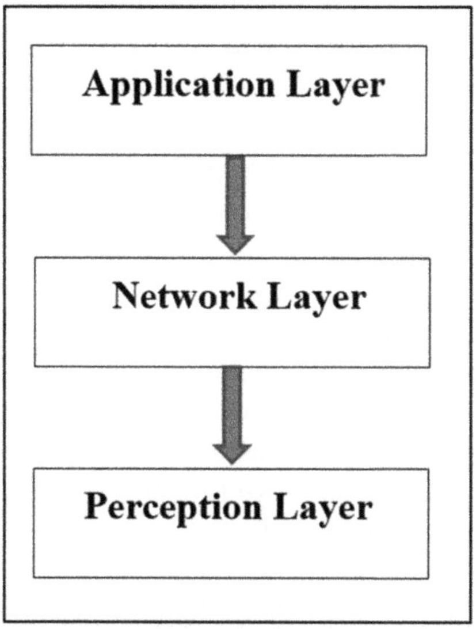

FIGURE 1.2 Architecture of three-layer IoT.

1.3.1 WORKING PROCESS OF IOT AND BIG DATA

A combination of these two technologies works as first data is collected from IoT devices and then stored as big data in cloud storage as other storage might not be able to operate such a gigantic amount of data. Since real-time data is obtained every millisecond, it is better to store it in the cloud and perform cloud computation or fog computation for real-time data processing. Fog computing can also reduce compassion costs and low latency (Elhadad et al., 2022). Later it is analyzed according to need. The Big IoT data has the following features:

- The abundance of terminals in IoT devices creates massive amounts of raw data.
- The raw data obtained can be from multiple sources, sensors, or software so it is heterogeneous and unstructured.

- All the data obtained may not be useful but has to be processed and examined that ensures not skip something essential.

Big data IoT is an evolving technology and can be further integrated with other technologies as well. The big data coming from various IoT devices if not analyzed properly would be wasted. The general methodology that it follows for big IoT data analysis is as follows:

- The first step is to collect the data from all the different sources, i.e., IoT sensors, software, etc. They are raw unstructured heterogeneous data so we must store it usually since most of the IoT data is in real time, it is suitable to use cloud storage for the purpose of local storage.
- Then this raw data has to be pre-processed to make it valuable. Thus, we began by cleaning. Here noise, redundant values, and missing values are handled. Then transformation and normalization can be applied to the data if required.
- Cleaned data is stored and how this data has to be analyzed. We can use fog computing, machine learning, deep learning, and many other techniques to analyze the data to draw relevant conclusions.
- Some open-source platforms such as Apache Hadoop, Spark, MapReduce, or Dryad can be employed for IoT big data analytics. As these platforms provide high-speed, easy-to-use, visualization, some even have real-time analysis.
- Depending on the analysis, the insights drawn are converted into such a format that any user is able to understand it easily.

1.3.2 APPLICATIONS OF IOT AND BIG DATA ANALYTICS IN BIOMEDICAL AND HEALTHCARE INDUSTRY

In today's time where nothing is specific and severe health conditions are common even in toddlers, digital healthcare is not extravagant but a necessity as IoT has helped the healthcare industry to perform better and some of its applications can be seen below.

1.3.2.1 WEARABLE ELECTRONIC DEVICES

They are not just fads but can prevent patients from many unwanted situations. It was published in the news that Apple's smartwatch saved a man's life. The watch monitored heart rate through ECG sensors and gave alerts due to irregular heartbeat. After multiple alerts, he discovered that he suffered from numerous heart conditions when he visited the hospital. Recently in the news, it was published that a teen boy saved his life using the calling feature of his Apple Watch when he fell into a gorge and was left behind during his trek. The watch helped him contact his friends for rescue.

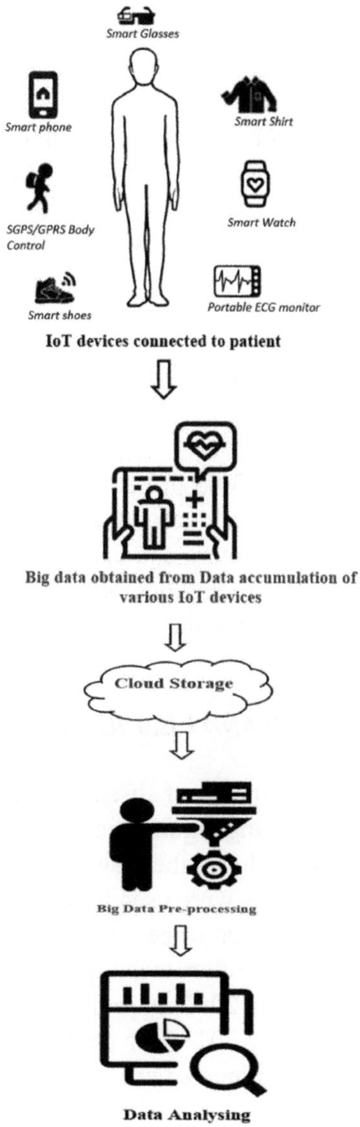

FIGURE 1.3 Working of IoT big data analytics.

1.3.2.2 ALERT SYSTEM IN A MEDICAL EMERGENCY

There are devices or apps present in devices such as smartwatches for mobile phones where one can call an ambulance with a single touch in case of emergencies. In the case of old people, it is crucial as it informs the concerned family members.

1.3.2.3 MEDICAL HOME HEALTH MONITORING DEVICES

Home appliance devices monitor the quality of air inside homes which enables one to protect themselves from respiratory devices. It is beneficial to those who live in urban areas where the air is polluted. There are a plethora of examples of IoT in healthcare. IoT has made being healthy easier with low costs and fewer doctor visits.

1.3.2.4 REMOTE, PERSONALIZED, AND REAL-TIME HEALTH MONITORING

Devices such as fitness bands, blood pressure monitoring, and glucometers are observing patients' vitals every second, and in case of any significant fluctuations, alerts are sent to doctors and concerned family members. This data collected through devices helps doctors design a prescribed tailored regime for a particular patient's health.

1.3.2.5 EARLY PREVENTION AND DIAGNOSIS OF DISEASES

For early prevention and diagnosis of diseases, patient data is used, and machine learning or deep learning algorithms are applied for feature extraction and the predictions are made for the diagnosis. For example, during the Covid-19 pandemic chest X-rays were used to detect coronavirus in patients (Kumar et al., 2019). It has proven to be a potent technique in the accurate diagnosis of disease. There are other techniques too that help radiologists with the machine's opinion. In 2017, a project was developed known as EDUCERF (Martín-Ruíz et al., 2017) for detecting problems with proper child development and help in home education.

1.3.2.6 MEDICAL INSURANCE COMPANY

IoT big data devices create transparency between customers and insurance companies. The data collected will let companies detect fraudsters and help patients with the right pricing and claim the right amount. The insurance companies may also offer incentives to patients for sharing their health-monitored data.

1.3.2.7 MEDICAL IOT BIG DATA IN RESEARCH

The data collected through IoT devices, EHR, and diagnostic labs have proven to be beneficial in research about diseases. The National Cancer Informatics Program (NCIP) is associated with cancer research studies and it makes use of big data for cancer studies. The Cancer Data Science Laboratory (CDSL) in NCIP analyzes and visualizes the big data of patients for cancer welfare research. For example, big data on patients' medical history, family history, population cancer registries, and large-scale genetic sequencing studies helps scientists to curate accurate information from the data (Tsai et al., 2019). So big data can help patients and doctors toward an easier way of having a healthier future.

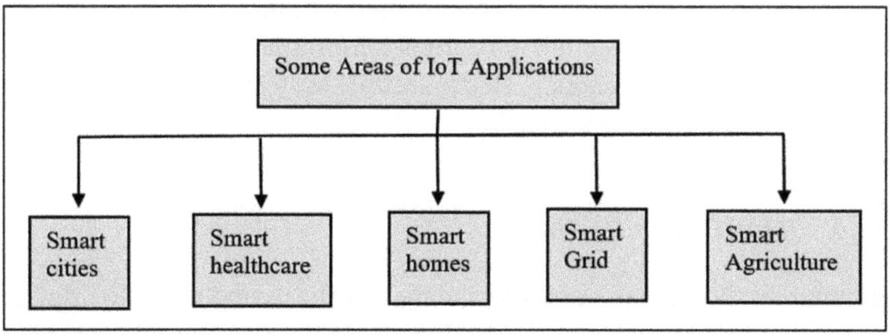

FIGURE 1.4 Areas of IoT applications.

1.3.3 APPLICATION OF IOT AND BIG DATA ANALYTICS IN OTHER DOMAINS

1.3.3.1 DISASTER MANAGEMENT SYSTEM

Disaster management systems have become advanced with the help of technologies like IoT, big data, cloud computing, etc. (Shah et al. 2019). The IoT provides data from various sources and with big data analytics they are immediately processed and results are fetched. In case of danger, for example, landslides or cloudbursts, decision-making is very important. Alerts and rescue teams get prepared beforehand when they observe danger from the data obtained. For example, whenever severe rainfall is about to happen we get a caution notification from the weather department.

1.3.3.2 DATA FOR VARIOUS RESEARCH PURPOSES

Companies like Google and Apple have developed platforms like Apple's Research Kit and Google Fit for gathering data from users about health and fitness statistics. The sensors embedded in the device monitor overall health and the real-time data can be shared with doctors.

1.3.3.3 SMART HOMES

Smart homes or home automation has made people's lives easier and saved lots of time. Lights, fans, room temperature, geysers, entertainment systems, everything could be controlled with your smartphone or by just saying "Alexa turn off the lights" and your command will be followed. IoT is even used for gardening as it can tell the correct temperature if a plant is hydrated or how much sunlight is enough. Better security systems as no one can break in until they are a really good hacker.

1.3.3.4 SMART AGRICULTURE

Agriculture is the basis of our foods and the food basis of our life. IoT devices have the capability to enhance the existing agriculture system. The sensors of the device could detect the moisture levels and elements present in the soil. The knowledge of these things could help the farmer know the deficiencies of soil and which type of crop would be suitable to grow in that soil. The weather predictions also keep farmers alert and knowledge of humidity and temperature prevents the growth of fungus in crops.

1.3.3.5 SMART AGRICULTURE

Sensors and cameras are planted across cities which helps in collecting data and then we use that data to create smart cities. This includes the detection of criminal activities, before taking a route you can check if you will find traffic in that area and you can easily find any place with just a few clicks.

1.3.4 IoT BIG DATA MEDICAL DEVICES USED IN HEALTHCARE INDUSTRY

A medical device that is capable of connecting to the internet and can communicate with other machines is an Internet of Things device. The sensors present in the collection perform various other functions. The big data collected is then cleaned, analyzed, and interpreted. There are numerous IoT big devices present in the healthcare industry and with innovation, new versions and new technologies are launching at great speed. Some of the existing devices are discussed below.

1.3.4.1 THE CONTINUOUS GLUCOSE MONITORING (CGM)

It is a life-saving glucose monitoring kit for patients with Type I diabetes. It keeps monitoring the glucose level of the patient and sends alerts on a mobile or device connected to take insulin when the level glucose level in the blood increases or intake glucose or eat something if the level decreases. It also sends alerts to relatives and tracks the location of patients. Predictive analysis and machine learning can be used for the detection of glucose levels.

A CGM model was proposed using cloud technology and deep learning-based prediction (Nasser et al., 2021). The architecture of the model consisted of sensors, a display, a Message Queuing Telemetry Transport (MQTT) protocol, deep learning model. The IoT device consisted of a display screen for the user, sensors to measure glucose levels, and a microprocessor for establishing connections. The MQTT is an application layer protocol used for creating a link between an IoT device and a cloud environment with the help of Wi-Fi. The cloud environment consisting of deep learning architecture includes a combination of Recurrent Neural Networks (RNN) an artificial neural network and Restricted Boltzmann machine (RBM) a probabilistic learning algorithm. The deep learning algorithm performs all the predictions and decides whether to send alerts or not.

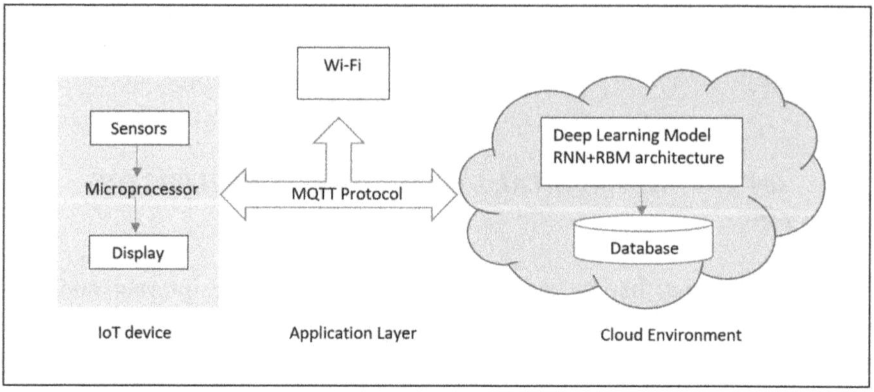

FIGURE 1.5 Architecture of CGM machine.

Source: Adapted from Tiwari et al., 2021.

1.3.4.2 THE POSTURE CORRECTOR DEVICE

It helps you attain and maintain the correct posture throughout the day. It prevents slouched posture, back pain, neck stiffness, and conditions like spondylitis. Whenever you lose the correct posture it will start sending you signals on your phone.

1.3.4.3 SMART TOYS

Toys are an essential part of everyone's childhood. But it would become more advantageous for kids if they were smart toys (Rivera et al., 2016) These toys help children with their mental, physical, cognitive, and sensitive abilities. It also helps in the early detection of any child disabilities or problem with the development of a child by observing the child's responses to stimulation.

1.3.4.4 WEARABLE DEVICES

Smartwatches are the biggest example of IoT in the healthcare domain. We can monitor our calorie burn, heart rate, and oxygen level in our blood and it also helps set healthier goals. Smart shirts have started to be manufactured as well. It has embedded electronic chips. They monitor all your health vitals. By tapping on your sleeves you can control your mobile functions. The growth of healthcare wearable devices is increasing rapidly. Everything is becoming smart these days.

1.4 CASE STUDY: PATIENT BEHAVIOR ANALYSIS

We took a case study of patient behavior analysis in healthcare with IoT that discusses the issue of early detection of stroke-initiated behavioral disorder (Dhiman et al., 2021). A stroke is a medical emergency that can happen due to head injuries or neurological disease or any other underlying health condition. It is a cerebrovascular accident that might lead to abnormal behavior disorders in patients temporarily or permanently. Although early diagnosis and prescriptions can prevent severe impact and permanent damages, the analysis of the patient's behavior may help the doctors or researchers in discovering the fundamental cause of the disease accurately so the valetudinarians can get proper help. The traditional methods have not been flourishing so smart healthcare was required. Automated diagnostic machinery collects bio-signaling, MRI of multiple modalities, single and multi-channel EEG, and bio-imaging data of patients. The data is analyzed and visualized which helps in forecasting the condition of the patient.

1.4.1 IoT HEALTH MONITORING ARCHITECTURE: RASPBERRY PI 3

The case study involves the implementation of Big IoT data technology to create smart healthcare for patients with brain conditions. The architecture of the smart health maintenance monitoring system was made by keeping in mind the requirements of the patients and improving the efficiency of the existing healthcare system. The IoT architecture is named Raspberry Pi 3. It helps in perpetuating the pulse rate sensor, heart sound sensor, Pi camera, and monitor. This device is connected to the internet and the patient's smartphone as well as the doctor's smartphone.

IoT healthcare involves an IoT-enabled healthcare network known as IoThNet. It is a crucial part of smart healthcare. More than 300 clinical leaders and healthcare staff are present in this network. The device Raspberry Pi 3 for patient behavioral analysis stores all the data from sensors in the cloud. The device observed the SpO_2, heart rate, eye blink, and temperature. Capturing element was used to collect data for these and an Arduino UNO was implemented for the processing device. The hardware used is Raspberry Pi 3, ECG sensor, blood pressure sensor, CO_2 sensors, oxygen levels, temperature sensor, and pulse rate sensor. Detection of smoked nicotine is done by MC-135 gas sensors. The data accumulated from these sensors was sent to Raspberry Pi 3. It was processed there and later sent to the IoT network.

The flaw in the system is that it does not support data visualization. Another flaw was that the prototype produced was unsuccessful in capturing the coronary rhythm. Its Arduino Uno processor gets delivery of captured analog sensor data boards. The values captured as data is converted into digital information with the help of an analog-to-digital converter. The data from Arduino was uploaded to the cloud with the use of Wi-Fi and an ethernet shield. The 12C/UART and SPI/SDIO interfaces enabled communication between Wi-Fi and Bluetooth. The susceptibility of detectors used in the architecture can be adjusted with the help of a potentiometer. They used ESP32, a cheap IoT gizmo that provides a Linux system for connecting sensors with actuators.

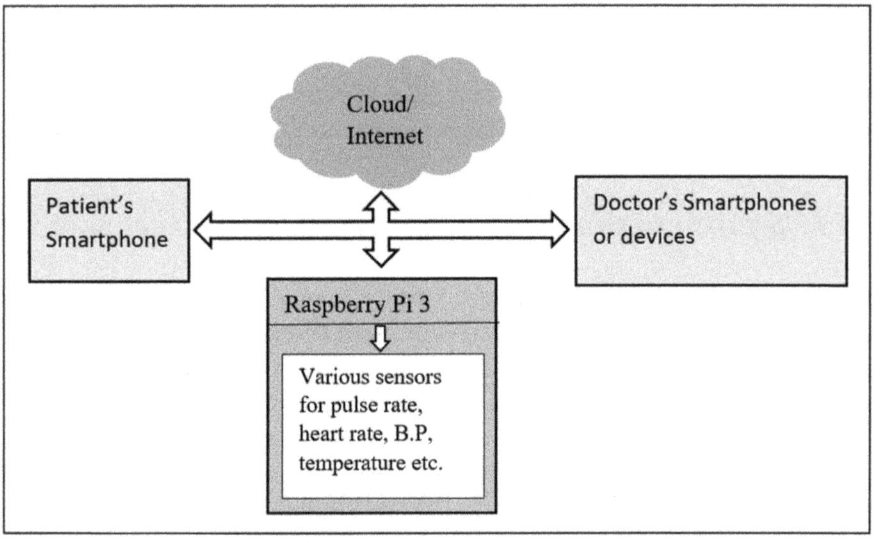

FIGURE 1.6 Architecture of the patient behavior analysis system.

They used the plethysmography principle for designing the heart sound sensor. Although the pulse rate is a more important factor in determining the normal functioning of the heart rate (Jayatilleka et al., 2020). It could also help in determining the change in the blood volume of the patient. When the light gets absorbed by the blood, the blood volume distribution is affected by the heart rate, and signal pulses become equivalent to heart pulses. The researchers have also provided necessary security measures to protect the data of patients and guard the exchange of data between client and server to maintain confidentiality. The data of the patient can be accessed by password and by doctors and authorities as well as concerned parties.

The above smart healthcare system was tested on various patients of different ages and tested under different conditions. There was an error rate of less than 5 percent. The data can be accessed by medical staff or doctors from any distant location. There is a normal level being set, for example, normal range of heartbeat per minute. If data crosses that level then the system must create an alert. Based on this study, the Internet of Things has proven to be crucial in the healthcare department. The monitoring and management become much easier with the involvement of IoT devices and diagnosis and detection with big data analysis.

1.5 CHALLENGES OF IOT AND BIG DATA ANALYTICS IN HEALTHCARE DOMAIN

Medical data is being generated not only from hospitals but also from smartphones, smartwatches, body sensors, etc. This data is used for the early detection of diseases, preventing serious health issues and constant monitoring of patients, and reducing the costs of treatments. This data can be very useful for research purposes to discover the cause of diseases that are still not known. However, there are many posed challenges in analyzing the data obtained from the healthcare domain such as loss of privacy for patients, cybersecurity issues, interoperability, and potential of machine error.

As IoT devices collect data from patients it poses a great risk of the loss of privacy. The data from IoT-enabled devices is susceptible to the risk of data theft and someone might hack your information to sell or buy drugs under your name. Information such as bank details or insurance details can constitute a great threat to patients if leaked. Hence, cybersecurity is a real danger in IoT-enabled devices that must be tackled. Improve security technologies such as blockchain are being used but still, a lot of improvement is yet to be done.

The IoT devices must connect properly with each other before data is transferred so the connection must be good between a device and a hospital or between two devices. This problem is known as interoperability. It means that sometimes integration can become difficult as devices are generally proprietary and use proprietary protocols as well. The same could also lead to hindrances in connecting multiple devices. The IoT devices ecosystem created by different companies may have different consensus due to their proprietary protocols. This non-uniformity causes issues with data aggregation and reduces scalability.

Sometimes machines can make mistakes due to some error in data or if the sensor becomes faulty over time. This can cause major problems in the predictive analysis of disease. Hence, depending on machines can be risky for one's health.

1.6 CONCLUSION

The help of IoT devices and big data analytics has made the process of everything easier in every industry. Especially in the healthcare industry, the inclusion of IoT has improvised traditional methods to a huge extent becoming a boon for both patients and doctors. It is like a Ginnie which has solutions to every problem. Your mood is not pleasant? Do not worry, moldable is an IoT device that can enhance your mood throughout the day. Ingestible food is a holy grail for patients with diabetes as soon as their sugar levels increase it alerts the patients and helps with curbing symptoms. Big data analytics assists in analyzing IoT data, without big data analytics all the massive data coming from IoT ecosystems would have gone to waste. The integration of these two is beneficial for mankind's evolution. Just like anything in the world, IoT and big data has their downsides too but it is nothing compared to the positives it brings to the table. This field has immense potential to transform every aspect of the healthcare industry from diagnosing diseases to research fields, and real-time data sharing of patients with doctors. The healthcare domain must keep up with this rapid advancement in technology.

KEYWORDS

- **Internet of things**
- **big data**
- **big data analytics**
- **healthcare**

REFERENCES

(IoT) Archives of Computational Methods in Engineering. **2021,** 1–9.

6th International Conference on Signal Processing, Computing and Control (ISPCC) 2021 Oct 7(pp. 176-181). IEEE.

Aiswarya, S.; Ramesh, K. In *IoT Based Big Data Analytics in Healthcare: A Survey*, Proceedings of the First International Conference on Advanced Scientific Innovation in Science, Engineering and Technology, Behavioural Neurology. 2022 Apr 11, 2022.

Alarcos, B.; Velasco, J. R. Assessing Developmental Difficulties in Children Through Connected Smart and prospects. *J. Big Data* **2019,** *6* (1), 1–25.

Duan, Y. IoT and Cloud Computing in Health-Care: A New Wearable Device and Cloud-Based Deep Learning Algorithm for Monitoring of Diabetes. *Electronics* **2021,** *10* (21), 2719.

Fawzy, D.; Moussa, M. M.; Badr, N. L. The Internet of Things and Architectures of Big Data Analytics: Challenges of Intersection at Different Domains. *IEEE Access* **2022,** *10,* 4969–4992.

Gulia, P.; Chahal, A. Big Data Analytics for IoT. *Int. J. Adv. Res. Eng. Technol. (IJARET)*. **2020,** *11* (6).

Hong, L.; Mengqi, L.; Ruixue, W.; Peixin, L.; Wei, Lu.; Long, L. Big Data in Healthcare: Applications and Challenges. *Data Inform. Manag.* **2018,** *2* (3), 175–197.

ICASISET. 2020, 16-17 May 2020, Chennai, India 2021 Jan 27.

Kaur, K.; Verma, S.; Bansal, A. IoT Big Data Analytics in Healthcare: Benefits and Challenges. In 2021.

Kaur, M.; Sakhare, S. R.; Wanjale, K.; Akter, F. Early Stroke Prediction Methods for Prevention of Strokes. *Symp. Appl. Comput.* **2017,** 485–491.

Kumar, S.; Tiwari, P.; Zymbler, M. Internet of Things is a Revolutionary Approach for Future Technology Enhancement: A Review. *J. Big Data* **2019,** *6* (1), 1–21.

Laghari, A. A.; Wu, K.; Laghari, R. A.; Ali, M.; Khan, A. A. A Review and State of Art of INTERNET of Things

Nasser, A. R.; Hasan, A. M.; Humaidi, A. J.; Alkhayyat, A.; Alzubaidi, L.; Fadhel, M. A.; Santamaría, J.; Rivera, D.; Martin-Ruiz, M. L.; Cruz-Piris, L.; van der Meulen, K.; García, A.; García, C. S.; Fernández, S.; Sabyasachi, D.; Shakyawar, S. K.; Sharma, M.; Kaushik, S. Big Data in Healthcare: Management, Analysis2021.

Saheb, T.; Izadi, L. Paradigm of IoT big Data Analytics in the Healthcare Industry: A Review of Scientific Literature and Mapping of Research Trends. *Telemat. Inform.* **2019,** *41,* 70–85.

Shah, S. A.; Seker, D. Z.; Hameed, S.; Draheim, D. The Rising Role of Big Data Analytics and IoT in Disaster Management: Recent Advances, Taxonomy and Prospects. *IEEE Access* **2019,** *7,* 54595–54614.

Taherkordi, A.; Eliassen, F., Horn, G. In *From IoT Big Data to IoT Big Services. Proceedings of the Toys*. InYoung Children's Rights in a Digital World 2021. Springer: Cham, pp. 237-253.

Tiwari, A.; Dhiman, V.; Iesa, M. A.; Alsarhan, H.; Mehbodniya, A.; Shabaz, M. Patient Behavioural Analysis WB Saunders.with Smart Healthcare and IoT. *Behav. Neurol.* 2021.

Tsai, C. J.; Riaz, N.; Gomez, S. L. Big Data in Cancer Research: Real-world Resources for Precision Oncology to Improve Cancer Care Delivery. *InSemin. Radiat. Oncol.* **2019,** *29* (4), 306–310.

CHAPTER 2

Machine Learning Techniques Used for Biomedical and Healthcare Systems

KAPIL KUMAR GUPTA[1], NEERAJ KUMAR[2], NAMRATA DHANDA[1], and UPENDRA KUMAR[3]

[1]*Department of Computer Science & Engineering, Amity University, Lucknow, Uttar Pradesh, India*

[2]*Department of Computer Science & Engineering Shri Ramswaroop Memorial University, Barabanki, India*

[3]*Department of Computer Science & Engineering, Institute of Engineering & Technology, Lucknow, India*

ABSTRACT

Machine learning approaches in health care employ increasing volumes of well-being data offered by the Internet of Things to enhance patient outcomes. These approaches have both potential applications and significant obstacles. Purely neuroimaging, natural language processing (NLP) of medical records, and genetic information are the three core domains of machine learning in the healthcare industry. Many of these fields are diagnosed, detected, and predicted. A broad medical device infrastructure currently generates data but often does not provide a supportive infrastructure for the efficient use of such data. The various forms of medical data exist in which data formatting is also challenging and may cause noise. The ability to capture, share, and deliver data is a high priority, given that digitization disrupts every industry, including healthcare. The challenges of the huge amounts of data may be addressed through Big Data, Machine Learning Algorithms, and artificial intelligence (AI). The concept of Machine Learning can also assist health organizations in

Smart Electronic Devices and Systems for Biomedical and Healthcare Applications. Upendra Kumar, Rajesh K. Kesharwani, & Raj K. Keservani (Ed.)

DOI: 10.1201/9781003560852-2

meeting increasing medical needs, improving operations, and decreasing costs. Innovative machine learning can help health practitioners more efficiently and with greater precise and personalized healthcare to detect and treat diseases. The aim of the biological modeling process is to describe data, but ultimately it aims at how systems may be set and how system targets, algorithms, and mechanisms can be better understood. Machine learning enables data to be modeled extremely well, without making strong assumptions about the modeling system, thanks to technical applications. Machine learning usually describes data more clearly than biomedical models, providing an essential benchmark as well as engineering solutions. It can also be an instrument to promote understanding.

2.1 INTRODUCTION

Technology's rapid advancement has an impact on clinical science, including imaging diagnostics. Computed tomography (CT) makes it possible to view internal body organs and tissues non-invasively and also eliminating the need for surgery. This prompts the investigation and creation of novel, improved, and trustworthy diagnostic and therapeutic techniques. In addition to diagnosis, medical imaging, such as biomedical signal collecting, is becoming increasingly important for therapy, monitoring its effects, and rehabilitation. However, as medical diagnostic device data sets grow, it becomes more challenging for clinicians to manually investigate and evaluate data as they have done in the past. Thus, the design and development of autonomous data analysis systems, whose applications encompass a variety of difficulties in the field of healthcare, have recently received an increasing amount of attention. Such systems make it easier to analyze and resolve complicated problems because they can dynamically adapt to changing circumstances. These systems frequently use machine learning techniques. Algorithms provide the best judgments and predictions after being taught to find patterns and correlations in massive data sets. Machine learning algorithms improve over time and become more accurate with the availability of vast data. Deep learning techniques are now frequently applied to medical imaging. Machine learning includes deep learning. Complex artificial neural networks serve as its foundation. Artificial neural networks (ANN) have complex learning processes because of their multiple inputs, outputs, and hidden layers, which are frequently coupled. Deep learning approaches considerably outperform typical machine learning algorithms in terms of medical data identification, classification, and prediction. Because of ML technology, particularly deep learning, healthcare workers, notably doctors, may tackle complex difficulties that would be difficult, time-consuming, and ineffective to resolve on their own (Ash et al., 2019).

2.1.1 POPULAR COMPANIES WORKING ON MACHINE LEARNING TOOL FOR HEALTHCARE SYSTEM

The healthcare sector has a long history of embracing new technologies. Emerging healthcare developments include the creation of innovative medical methods, the administration of personal health data and records, and chronic disease treatments (Thomas, 2022). Machine learning techniques help in identifying the following:

- Disease prediction and treatment
- Offering medical diagnostics and imaging
- Identifying and generating new medications
- Keeping track of medical records
- Cell models, clinical data, and biology and data generated by patients.

Figure 2.1 Explore about the top companies working on advanced machine learning algorithms and biomedical imaging techniques.

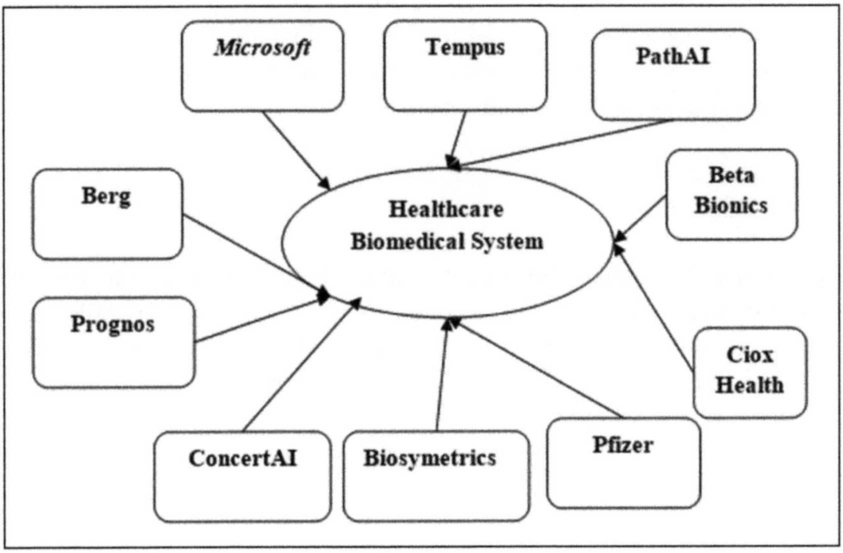

FIGURE 2.1 Top companies working on advanced machine learning algorithms and biomedical imaging techniques.

2.1.1.1 MICROSOFT

In order to help medical professionals plan radiotherapy and surgery, Microsoft's Project InnerEye analyses 3D radiological pictures and employs computer vision and machine learning to discriminate between tumors and healthy anatomy. Microsoft wants to create medicine that is specifically suited to each patient's needs using this AI-based method.

2.1.1.2 TEMPUS

In order to provide patients with individualized therapy, Tempus collects enormous volumes of clinical and medical data in an effort to advance cancer research. Using AI-powered algorithms to analyze its data bank, Tempus supports academic research, clinical trial matching, diagnostic biomarking, and genetic profiling.

2.1.1.3 PATHAI

PathAI's approach uses machine learning to aid pathologists in producing faster and more accurate diagnoses. In addition, the firm provides AI solutions to organize additional tasks for clinical trials and drug research, such as collecting patient data and processing samples. PathAI has the capacity to give patients more effective treatments thanks to a cooperation network of biopharmaceutical companies, labs, and doctors.

2.1.1.4 BETA BIONICS

Beta Bionics is creating the iLet wearable "bionic" pancreas to ease the stress of diabetes patients' lives. In order to relieve the stress of daily blood glucose monitoring for patients with Type 1 diabetes, this gadget, which is currently in the exploratory stage, continuously monitors blood sugar levels.

2.1.1.5 CIOX HEALTH

To give healthcare workers quicker access to patient data, Ciox Health uses machine learning to power its Datavant Switchboard platform. Within the platform, businesses can create customized controls that let employees request particular kinds of data. The technology used by Ciox Health complies with privacy regulations to safeguard patients' electronic medical records.

2.1.1.6 PFIZER

Pfizer conducts immuno-oncology research to better understand how the body's immune system might combat cancer with the aid of IBM's Watson AI technology. Through this collaboration, Pfizer may quickly gain insights about how to create more effective immuno-oncological medicines for patients by analyzing vast amounts of patient data.

2.1.1.7 BIOSYMETRICS

Biosymetrics assists businesses in analyzing vast amounts of raw data to speed up the development of precision medicine through the use of its machine learning platform and contingent AI. The company's platform Elion can give more thorough insights on how to develop drugs thanks to access to millions of electronic health records and human-relevant illness models.

2.1.1.8 CONCERTAI

Oncologists, pharmaceutical firms, payers, and providers may practice precision medicine and health due to ConcertAI's machine learning analysis of oncology data. Healthcare practitioners can fine-tune treatments using patient demographic and clinical data by using the company's product RWD360, which acts as a vast library for tumor clinical data.

2.1.1.9 PROGNOS

Prognos Health collects and analyzes data from prescriptions, health insurance companies, medical reports, and other sources using machine learning to provide healthcare organizations with more thorough patient profiles. Companies may swiftly sell and purchase health data to identify diseases, evaluate insurance coverage, and identify treatment gaps thanks to the company's marketplace Prognos Factor.

2.1.1.10 BERG

Machine learning is used by Berg's Interrogative Biology platform to map diseases and treat them in oncology, neurology, and other rare ailments. Instead of relying on trial-and-error testing, the firm helps healthcare practitioners to be more proactive.

2.1.2 MACHINE LEARNING'S POSSIBLE ADVANTAGES IN THE BIOMEDICAL HEALTHCARE SYSTEM

Machine learning has a wide range of possible applications in clinical care, including the improvement of patient data, diagnosis, and treatment, as well as cost savings and improved patient safety. Just a few advantages that machine learning applications in healthcare can offer medical professionals given in Figure 2.2 are listed below.

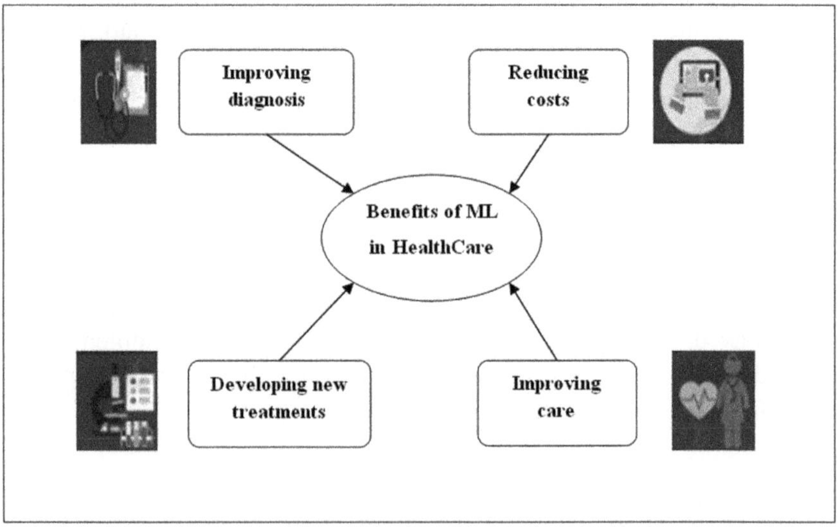

FIGURE 2.2 Potential benefits of machine learning in biomedical healthcare system.

2.1.2.1 IMPROVING DIAGNOSIS

In the healthcare industry, machine learning may be utilized to build more reliable diagnostic tools to assess medical data. A machine learning algorithm, for example, may be used in CT (such as X-rays or MRI scans) to seek patterns that suggest a certain illness using pattern recognition.

2.1.2.2 DEVELOPING NEW TREATMENTS

Deep learning models may also be used to find substantial information in databases that could lead to the identification of novel pharmaceuticals, their creation, and new approaches to cure illnesses. For instance, employing machine learning to analyze clinical trial data can reveal previously unrecognized pharmacological side effects. This could improve patient care, medical treatment effectiveness, and safety.

2.1.2.3 REDUCING COSTS

Machine learning can be used to improve healthcare efficiency, which could lead to lower costs. For example, improved scheduling or patient record management algorithms could be made using machine learning in the healthcare industry. In the healthcare system, there may be less time and money wasted on repetitive tasks.

2.1.2.4 IMPROVING CARE

The healthcare sector may increase the quality of patient care by implementing machine learning. Deep learning algorithms, in particular, have the potential to be utilized to create patient monitoring equipment that informs medical equipment when a patient's status changes. By doing so, service users can receive the appropriate care at the appropriate time (Flam, 2022).

2.1.3 IMPORTANCE OF MACHINE LEARNING FOR HEALTHCARE ORGANIZATIONS

In research and medicine, high-level automation and monitoring devices are frequently employed. MD Anderson data scientists created the first deep neural network algorithm for the health sector to predict neurotoxicity in individuals perceiving radiation treatment for the neck and head tumors. Deep learning in healthcare can offer data that instantly recognizes complicated patterns and provides clinical decision assistance to primary care physicians at the moment of treatment within digital health records.

Large volumes of unprocessed patient records for machine learning account for over 80% of the data stored or "locked" in digital health record systems. They are key data articles or text files containing patient data rather than data components, which previously needed a person to evaluate the medical records. Human language, sometimes known as "natural language," is very intricate, inconsistent, and full of slang, ambiguity, and vagueness. To translate these papers into more useful and analyzable data, machine learning in healthcare usually depends on AI such as NLP technologies. For machine learning, the bulk of NLP-based deep learning technology in the healthcare sector requires some type of healthcare data (Flam, 2022).

2.1.4 REVOLUTIONIZED MEDICINE BY BIOMEDICAL IMAGING

There are still more areas to explore, as well as more chances for imaging to completely transform medicine. Our biological systems are always interacting with different tissues, cells, and proteins in the body as well as with other living, breathing organisms. Without access to all of these mechanisms, our knowledge of the human body and our ability to safeguard it against illness is limited. Take cancer as an illustration. With the technology we now have, we can see tumors inside the body, but not with the depth and resolution needed to see them at the cellular level. We are unable to observe the development of tumors in real time, the interactions of cancer cells with medications or immune system components, or any cancer cells that might endure after a tumor is removed. If we could, cancer would be far better understood and treated.

However, there are a number of problems that the scientific community must address before we can reach that new level of understanding. First, we must be able to view the interior of the body in great detail. MRIs, sonograms, and other procedures can currently offer noninvasive images from deep into our bodies. However, the resolution is insufficient to detect cell interaction. Although we do have alternative high-resolution technologies, they have the opposite issue. We can view high-resolution images of tissues and cells that are outside the body using technologies like fluorescence microscopy, electron microscopy, and others. However, these methods are unable to reveal what is occurring between and inside the cells that are situated deep within our bodies (Zukerberg, 2022)

2.2 EXISTING MEDICAL RESEARCH USING AI AND MACHINE LEARNING TECHNIQUES

In the field of medical image processing, researchers are continually looking for new ideas. Detecting a different disease via medical imaging is a difficult task. In the past 30 years, there have been several studies on brain tumor detection, kidney stone detection, cardiac disease, cephalometric landmark detection, etc. Research has grown at an exponential rate during the previous decade. The details of different machine learning and biomedical imaging techniques in the last decade are listed below:

2.2.1 PROBABILISTIC NEURAL NETWORK-BASED BRAIN TUMOR DETECTION

An image segmentation algorithm developed by Dina Abo Edahab and her colleagues was used to identify brain tumor in MRI images (Edhab et al., 2012). Imaging data such as the location of tumors, other abnormalities, and anatomical structures may be organized using gray-level image segmentation. The anatomical structure of healthy portions is compared to that of sick ones (Shevad, 2005).

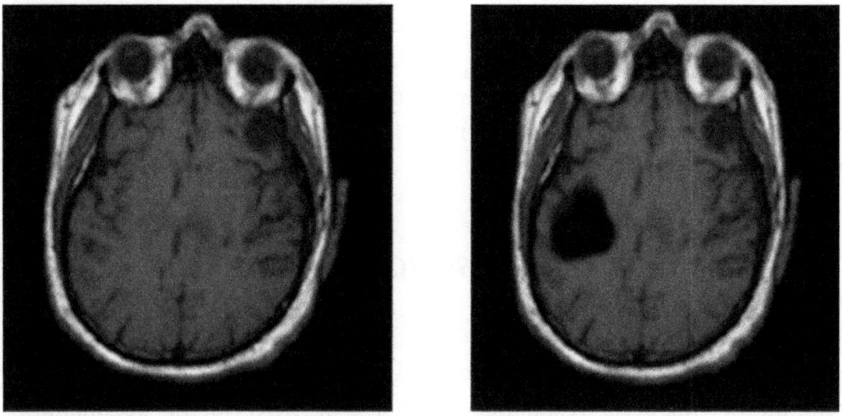

FIGURE 2.3 MRI scans of the standard and disordered brain.

Source: Reprinted from Edhab et al., 2012

They thought that random and additive noise was the primary source of image noise. Prior to any augmentation or smoothing, a spurious signal (m (a, b)) was added to the true pixel value (l (a, b)):

$$I(a, b) = l(a, b) + m(a, b) \tag{2.1}$$

where, I is the resultant image after adding the spurious signal into a genuine pixel value. The spatial domain uses pixels in a small area and displays them as such:

$$h(i, j) = S[f(i, j)] \tag{2.2}$$

where, h(i, j) is the processed image, while f(i, j) is the input image, and S is a function related to f that is specified across a specific area of f. (i, j). The suggested noise reduction technique relies on smoothing and sharpening filters which are included in the spatial filters [10].

Using linear filters for smoothing: In this method, L is considered to be an MN convolution kernel, and B in the image is an odd number less than both M and N. This filtered version of L appears to be given below:

$$LB(a, b) = L*B = \sum_{h-n/2}^{n/2} \sum_{l-m/2}^{m/2} B(f,l)\, L\,(a-f, b-l) \tag{2.3}$$

where, a, and b are each equal to one. L(a, b) is substituted with the weighted sum of L values in the vicinity of (a, b) (Edhab et al., 2012).

The Gaussian filter is used to smooth out the data in the next step. Assume that, G_f. is a nonnegative, real-valued column matrix. :

$$G_f(a,b) = \frac{1}{d}\exp\left[\frac{-[a^2 + b^2]}{2\sigma^2}\right] \tag{2.4}$$

where, d is expressed as $d = \sqrt{2\pi\sigma^2}$. σ is the standard deviation. When compared to linear filters, this form of filter improves the amount of noise reduction.

2.2.2 BRAIN TUMOR MRI CLASSIFICATION USING SUPPORT VECTOR MACHINE (SVM) AND ANN

Rasel Ahmmed and colleagues proposed utilizing a SVM and ANN to classify brain tumor magnetic resonance images. For MRI image segmentation, a combination of temper-based K-means and modified fuzzy C-means (TKFCM) clustering approach is used. SVM and ANN are used to segment, extract features, and classify the image. The TKFCM algorithm is a hybrid of K-means and fuzzy C-means that is obtained after a few changes (Ahmmed et al., 2017). Temper is combined with traditional K-means, distinguished by the intensity of gray levels in brain imaging. Equation 2.5 for the complexity of grey level-based temper and image for coarse image $A(a_i, b_j.)$ is as follows:

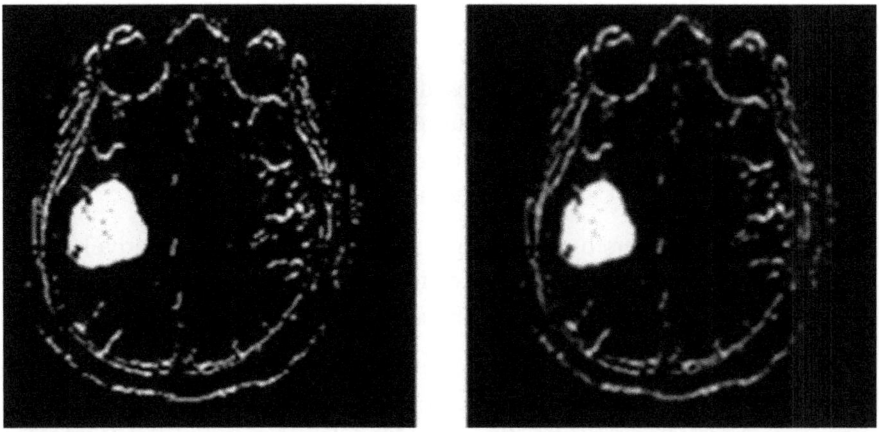

FIGURE 2.4 (a) Application of Gaussian filter, (b) Application of Averaging filter.

Source: Reprinted from Edhab et al., 2012

$$A(ai, bi) = \sum_{i=m+1}^{P+m} \sum_{j=m+1}^{Q+m} R(a_i, b_j)\ S_{PQ(resize)} \tag{2.5}$$

$$S_{PQ} = \sum_{i=m+1}^{P-m} \sum_{j=m+1}^{Q-m} R(a_i, b_j) \tag{2.6}$$

This equation incorporates a temper-based image matrix that includes G number of grey levels and D the number of bins. This image matrix is utilized to identify the temper of the image R (a_i, b_j), Where m is defined as (winsize -1)/2, and n is the number of participants. The required temper may be attained by entering the correct values for nature, row, and column in the appropriate fields. Equations 2.7 and 2.8 illustrated the modified fuzzy C-means clustering, and temper-based K-means clustering:

$$L_g = \sum_{i=1}^{D} \sum_{j=1}^{G} A(a_i, b_j) \| a_i - d_j \|^2 \tag{2.7}$$

$$L_P = \sum_{j=1}^{Q} \sum_{i=1}^{G} (V_{ij}) * (h_{ij})^2 \tag{2.8}$$

where G, Q, and D are the variables that indicate the number of clusters, cluster centroid, and data points, respectively.

The contour, which may be used to locate the tumor location, can be calculated using eqs (2.7) and (2.8):

$$L_{gp} = \oint_d (L_g, L_p) \tag{2.9}$$

P and Q represent the row and column of A(a_i, b_j), respectively, and d represents the contour value.

2.2.3 BRAIN TUMOR DETECTION BASED ON FUZZY C-MEANS

Alexander Zotin and colleagues proposed a novel approach for detecting edges in brain tumors. To tackle the brain tumor detection issue, the author employs the fuzzy C-mean clustering approach (Zotin et al., 2018; Stosic, 2018). The following is how fuzzy C-means clustering works:

- Find the clusters by searching a set of fuzzy clusters iteratively.
- Determine cluster centers of different clusters.
- A number of fuzzy sets to partition an existing collection of points.

A fuzzy membership matrix is defined by the degree to which the kth element of the original set of vectors belongs to the ith cluster. Fuzzy C-means clustering separates the data $X = \{x_1, x_2,..., x_m\}$ into n fuzzy clusters $C = (c_1, c_2, ..., c_n)$.

The objective function is:

$$f(D,C) = \sum_{i=1}^{n} \sum_{k=1}^{m} d \cdot (x_m - c_n)^2 \qquad (2.10)$$

where, D stands for the degree of membership value, x_m for the mth cluster, C for the number of clusters, and c_n for the nth cluster's center. After using the FCM method to split the image into a series of homogenous clusters, the Canny edge detector is used.

2.2.4 AN EFFECTIVE CLASSIFICATION METHOD FOR CARDIAC ARRHYTHMIAS USING AN ENSEMBLE OF DEPTHWISE SEPARABLE CONVOLUTIONAL NEURAL NETWORKS

Placing one's finger on the sensor for roughly 30 seconds makes it simple to capture ECG readings. The ECG signal contains at least two different types of significant data, including those concerning health or biomedicine, and those concerning personal identity or biometrics. Numerous ECG classification algorithms, including manual and machine learning techniques, have been developed as a result of their convenience. The handicraft technique is difficult to apply to non-stationary data such as the ECG, whereas machine learning methods frequently need huge computer resources. Because of its high accuracy, the machine learning approach outperforms the handcraft method. To minimize the processing needs while keeping its high accuracy, a good method is needed (Ihsanto et al., 2020).

When the feature extraction and classification stages are carried out individually, particle swarm optimization (PSO) can be used to enhance SVM

(Melgani et al., 2008). When 500 beats were utilized for supervised training, the model was able to classify 40,438 test beats into five categories with an accuracy of 89.72%, outperforming alternatives such as KNN or radial basis function (RBF). Yet, there is still room to improve the classification's accuracy and/or number of classifications. After feature extraction, principal component analysis (PCA) and discontinuous orthogonal Stockwell transform (DOST), for example, might be used to reduce feature dimensionality. According to (Raj et al., 2017), After 23,996 test beats were supervised, the remaining 86,113 test beats could be categorized into 16 classes with an accuracy of 98.82%. Another intriguing option for increasing efficiency is to combine the phases of feature extraction and classification using Multi-layer Perceptron (MLP) and convolutional neural networks (CNN). As an illustration (Kiranyaz et al., 2015) designed a neural network architecture with three CNN layers and two MLP levels as an example. After the model's input, it can identify five groups with a 95.14% accuracy. Although autoencoder was used in (Nurmaini et al., 2019) with reasonably acceptable results, it is still necessary to fairly compare performance with and without de-noising. Additionally, the deep network setup should be improved further to speed up calculation. Even though there have been several studies, an effective algorithm for classifying cardiac arrhythmias is still needed. Therefore, the goal of this paper was to streamline the overall procedure while keeping high accuracy in order to reduce the computing cost. An updated neural network model was used to categorize 16 different categories. The authors measured the effectiveness of our suggested algorithms using the number of classes, prediction phases, and accuracy.

2.2.5 A MLP NETWORK FOR ESTIMATING PERFUSION PARAMETERS IN DCE-MRI INVESTIGATIONS OF HEALTHY KIDNEY

In contemporary medical diagnostics, keeping track of kidney functions has assumed growing importance. Failures in renal function can affect the body's ability to maintain physiological homeostasis, which can affect how electrolytes are managed, how the acid-base balance is maintained, or how arterial blood pressure is regulated. The kidneys filter blood and remove water-soluble metabolic waste products, excess glucose, and other organic compounds. Acute kidney damage, chronic kidney disease (CKD), and various cancers are examples of kidney diseases (e.g., renal cell carcinoma). Treatment may include kidney removal or transplantation, lifelong dialysis, or both, depending on the pathologic disease. In any instance, accurate information on kidney function is required for the correct diagnosis, therapy, and prognosis afterward. Continuous and precise monitoring of renal function is crucial in all cases, especially in CKD, which can be triggered by long-term hypertension or diabetes mellitus. This is because the illness can be stopped in its tracks before it reaches the end stage if caught early enough. A lifetime's worth of waste removal, humoral communication, and cellular nourishment can all be preserved with tissue perfusion (Vander et al., 2001). As a result, understanding local blood flow and blood filtration in the nephrons is critical for the diagnosis and therapy of the aforementioned diseases. For renal diagnostics, the more reliable iohexol clearance test or the urea and creatinine method are commonly utilized. The findings of these tests allow for the calculation of the rate of glomerular filtration (GFR), which is the major indicator of kidney function. Unfortunately, neither approach distinguishes between the functioning of the left and right kidneys (Stevens and Levey, 2009).

Dynamic contrast-enhanced electromagnetic resonance imaging is another way of assessing renal function (DCE) (Hackstein et al., 2003). During the process, a contrast agent (CA) must be delivered intravenously, and its transit through the kidneys' capillary layer and tubular systems, as well as the abdominal artery system, must be evaluated. Unlike traditional approaches, DCE-MRI can determine GFR for an individual kidney, eventually down to the voxel level, while simultaneously revealing anatomical features of the renal parenchyma. The CA's pharmacokinetic (PK) modeling is essential for the calculation of renal outflow based on DCE-MR images. The CA enters the kidney arteries, the renal parenchyma, as well as the afferent arterioles of the renal corpuscles from the abdominal aorta. It then travels first from the glomerular capillaries through the Bowman's space and the tubular channel before ending up in the renal pelvis. There are numerous PK models, and they differ in the following manner: (1) the quantity of tissue compartments; (2) the extensiveness to which impulse residue processes specific to each compartment are incorporated; (3) the participation of the impact of tracer agent leakage from blood vessels in cancerous cells; and (4) how the peak CA concentration in the renal parenchyma propagates out and delays in relation to the aortic valve are patterned. All contemporary models of the perfectly natural kidney use a two-compartment method (Klepaczko et al., 2020). The kidney's perfusion and filtration, whose effectiveness directly affects a variety of bodily activities. In clinical settings, glomerular filtration rate (GFR) is used to determine kidney function. GFR must be estimated using DCE-MRI data using an organ-specific PK model. Nevertheless, the artery input function (AIF) and starting parameter values have a substantial influence on the derivation of model parameters and, as a result, the characterization of GFR. The authors introduced an MLP network to overcome the shortcomings of current model optimization methodologies that employ non-linear least-squares curve-fitting. The authors employed the trust-region reflective model to statistically optimize the model (Klepaczko et al., 2020). The least-squares curve-fitting technique may be substituted with neural networks, which assure agreement with real-world data. There are two advantages to using a neural network. First, it can compute GFR without knowing the AIF for each individual patient. Second, a reliable estimate may be generated without explicitly specifying the initial model parameters or their limitations. Based on the model, different explicit variables are used to describe the dynamics of the analyte agent.

The transparency product, intervening capacity, plasma volume, and plasma flow are the primary or independent parameters. These properties may be used to compute the extraction percent and volumetric transfer constant, both of which are variable parameters. The latter, denoted as K trans, defines regional renal filtration by quantifying the ratio of plasma flow carried from the glomerular capillary bed into Bowman's space. K trans is a key descriptor in assessing kidney function since it may be used to estimate the GFR by multiplying it by the kidney volume. For PK model parameter estimation, we propose moving out from a non-linear least-squares curve-fitting approach to an MLP network in this paper. The foundation for this idea comes from the possibility that the outcomes of iterative procedures could be local and, generally speaking, poor solutions. As a result, the final result heavily depends on the initial selection of parameter values. Additionally, in order to ensure that the outcome falls within a physiologically acceptable range, restrictions on the expected parameter values must often be established. However, too-narrow bounds could prohibit the fitting algorithm from discovering an optimal. MLP, unlike NLLS, does not necessitate the starting hyperparameters to be provided; additionally, when correctly trained, it guarantees the optimal solution. Although MLP networks could be considered an outdated technique, they are still used and show promise in solving challenging non-linear regression issues (Ohn and Kim, 2019; Liu et al., 2020).

2.2.6 COMPUTER TOOLS TO ANALYZE LUNG CT CHANGES AFTER RADIOTHERAPY

Lung cancer continues to be one of the most important issues in oncology, with more than 2.2 million additional diagnoses and over 1.8 million fatalities in 2020. Even in industrialized nations, five-year survival rates rarely surpass 20%, thus better screening, diagnosis, and treatment are required. An important research field for a long time has been the detection of lung cancer utilizing IT systems (Konkol et al., 2021). Large-scale CT screening systems are now viable thanks to improvements in processing speed and AI technologies. The likelihood of patients receiving effective and aggressive treatment for lung cancer such as surgery, systemic chemotherapy, and immunotherapy, or radiotherapy increases with early detection (RT). Radiation oncology itself requires state-of-the-art IT technology to design and deliver treatments and assess early and late effects as safely as possible.

Definitive radiotherapy is an important part of treatment for all lung cancer patients who cannot undergo definitive surgery due to severity or other medical conditions. In certain situations (positive surgical margins or node-positive), it is also required as adjuvant therapy and essential in palliative care. Notwithstanding the clinical circumstance (organ at risk), the goal is to deliver large radiation dosages to tumor cells while preserving nearby healthy tissue (so-called OAR). There are several examples of modern healthcare advancements that enable us to aptly conform the radiation beam and satisfy the requirements of the OARs' dose constraints. But problems like late and acute toxicity, particularly with regard to lung tissue exist to date.

Pathologic radiation-induced pneumonitis (RIP) and radiation-induced lung fibrosis (RILF), which are identified six to twelve months after conventional RT, respectively, result in late risks in 10 to 30 percent of patients (Konkol et al., 2021). In clinical practice, RILF is evaluated using the Radiation Treatment Oncology Group (RTOG) terminology, LENTSOMA (EORTC) grading, Common Terminology Metrics for Metabolic Abnormalities (CTMMA), and other grading systems. They focus primarily on clinical manifestations, with imaging findings providing some support (Konkol et al., 2021). It appears necessary to employ a precise predictive assessment of CT modifications based on density values because the offered measure does not appear to be enough for radiation oncologists' needs in symptomatic patients. This approach has confirmed both shifts over time and the consequences of high and medium-dose radiation on RILF (Konkol et al., 2021). The research objective is to optimize the workflow of a cutting-edge technique for automated, quick, and thorough radiation comparison of a lung cancer patient's CT images following radiation therapy. Future clinical applications include tracking small changes in tissue density over time in anatomically relevant regions of the lung (expressed in Hounsfield units). Our mature technology supports the investigation of relationships between tissue changes, patient-administered doses, and other dosimetry and clinical parameters. To avoid compromising lung structure and function, examining a large number of real-life patient cases may lead to setting more precise doses and dose limits for future RT. The proposed method (Konkol et al., 2021) was originally intended primarily for RILF analysis.

2.2.7 DEEP-LEARNING-BASED AUTOMATED CEPHALOMETRIC FEATURE IDENTIFICATION ON X-RAY IMAGES

Using skeletal X-ray scans, cephalometric analysis is conducted. This is critical in assisting doctors in identifying orthodontic issues (Song et al., 2020). Locating landmarks in X-ray pictures is the first step in cephalometric analysis. Professionals with experience are required to detect the indications. The slopes and lengths between such spots are measured, which helps with diagnosis and treatment planning. It takes a long time, and the challenge of intra-observer variability emerges because doctors and specialists may employ completely different ways to detect landmarks (Song et al., 2020). In these circumstances, computer-aided detection is highly wanted since it can automatically and impartially identify landmarks. The study of locating anatomical markers has a long history. During the International Symposium on Biomedical Imaging (ISBI) discussed problems with cephalometry landmark localization in 2014 and 2015 (Ching-Wei et al., 2015; Lai, 2016), a wide range of cutting-edge solutions were given. Lindner and Cootes' random-forest technique produced the greatest results in the 2015 challenge, with a 74.84% successful recognition rate for a 2-mm precision range (Lindner et al., 2015) (mistakes within a certain range are permissible because landmark position cannot be precisely the same with handwritten annotation). The permissible range for medical applications is 2 mm. In the 2015 competition, Ibragimov et al. came in second with a recognition rate of 71.7% at a 2-mm proximity. They started with a random-forest regression and a Haar-like feature extraction, and they then added a global-context shape model to make it better. Ibragimov et al. method included CNNs for binary classification in 2017. With an approximate 75.3% range in 2-mm prediction accuracy, they perform somewhat better than Lindner's findings (Lee et al., 2017).

Hansang Lee et al. (2017) introduced a deep learning-based system that produced a slightly reduced image rather than subpar results. Recent interest in deep learning has grown as a result of its accomplishments in computer vision. To address problems like landmark identification (Ramanan and Zhu, 2012;

Belhumeur et al., 2013), picture classification (Simonyan and Andrew, 2014; Claudiu et al., 2011), and image segmentation (Long et al., 2015; Olaf et al., 2015), convolutional neural network models are often used. Often, trained models exceed humans in terms of performance. Because a very non-linear mapping produced by direct regression of multiple landmarks is difficult to learn (Pfister et al., 2015; Tompson et al., 2014).

Authors simply look for one crucial point in a single patch image when using our method. We just need to understand a non-linear mapping function for one key point. Each key point is associated with a non-linear mapping function. As a result, as compared to earlier approaches, we can locate critical locations with more accuracy. The authors extract an ROI patch centered at the coarse landmark position based on the registration results. Next, authors use ResNet50, a cutting-edge convolutional neural network which serves as the foundation for our pre-trained network, to recognize landmarks in the created ROI patches.

2.2.8 DETECTING INTRACRANIAL HEMORRHAGE IN HEAD CT BY EMPLOYING DOUBLE-BRANCH CNN, SUPPORT VECTOR MACHINES, AND RANDOM FORESTS

The medical term for bleeding that occurs within the intracranial vault is intracranial hemorrhage (ICH). A few examples of potential causes include vascular abnormalities, venous infarction, tumors, the effects of trauma, therapeutic anticoagulation, and cerebral aneurysm (Naidech, 2011; Marcolini et al., 2019; Tominari et al., 2015; Yasugi et al., 2018). A hemorrhage is a big concern regardless of the reason. As a result, for the therapeutic procedure to be effective, an accurate and fast diagnosis is required. The people's clinical history, medical examination, and non-contrast CT imaging of the affected area of the brain are used to make the ICH diagnosis. The basic causes of ICH may be identified by CT imaging, and bleeding can be localized (Freeman and Aguilar, 2012). Because of the necessity of the operation, the lengthy and intricate judgment process, the immaturity of new specialists, and the likelihood that most crises occur at night, diagnosing and treating ICH is difficult. As a result, the professional is in desperate need of a computer-aided diagnosing tool. Automatic bleeding detection, however, must be precise enough for medical applications. A number of ICH subtypes can be distinguished depending on where the brain hemorrhage occurred (Figure 2.1).

The epidural subtype (EDH) refers to leakage of blood between both the dura and the bone, whereas the subdural subtype (SDH) refers to leakage of blood between the dura and the arachnoids. Both are common after major injuries. The term "intraparenchymal hemorrhage (IPH)" refers to the leakage of blood within the brain's parenchyma. Figure 2.5 (a–f) shows the locations that are likely to bleed are indicated by yellow arrows. Examples that are more difficult to understand, such as those with numerous ICH subgroups present in Figure 2.5 (g–k) or those with hardly perceptible ICH symptoms, are displayed in the bottom row.

Deep learning techniques for segmenting and categorizing pictures have gained more academic interest in recent years. Due to its dependability and effectiveness, CNNs have grown in popularity and significance in the support of medical diagnostics (Litjens et al., 2017; Hssayeni et al., 2020). The typical 3D structure of a CT scan is made up of stacks of 2D slices. So, although it could require a lot of computation, employing picture voxels is possible. The top row displays slices from a non-contrast CT sample displaying diverse ICH subtypes in Figure 2.5 (a–f). Locations that are likely to bleed are indicated by yellow arrows. Examples that are more difficult to understand, such as those with numerous ICH subtypes present at once in Figure 2.5 (g–k) or those with hardly perceptible ICH symptoms, are displayed in the bottom row (l). Deep learning techniques for segmenting and categorizing pictures have gained more academic interest in recent years. Due to its dependability and effectiveness, CNNs have grown in popularity and significance in the support of medical diagnostics (Litjens et al., 2017; Hssayeni et al., 2020).

The typical 3D structure of a CT scan is made up of stacks of 2D slices. So, although it could require a lot of computation, employing picture voxels is possible. A ResNexT CNN for categorization among five ICH subgroups was created by Danilov et al. (2020). For SAH, EDH, IVH, SDH, and IPH, the accuracy rates were 82.8%, 81.8%, 82.0%, 89.3%, and 83.5%, respectively. Using a hybrid 3D CNN and RNN, Ye et al. (2019) were able to identify ICH and its five subgroups. Leakage of blood accuracy was better than 98. Ker et al. (2019) established a 3-D CNN for the classification of multiple IPH subtypes, including healthy brain, SAH, IPH, ASDH, and BPH. For multi-class analysis and the two-class scenarios, the authors' technique evaluation utilizing the F1 score yielded results of 68.4% and 70.6–95.2%, respectively.

An approach that combines a CNN, autoencoder network, and heat map was proposed by Togaçar et al. (2019) Heatmaps were produced based on each image as part of the dataset analysis using an autoencoder network. The results of the augmentation technique were input into the pre-trained CNN.

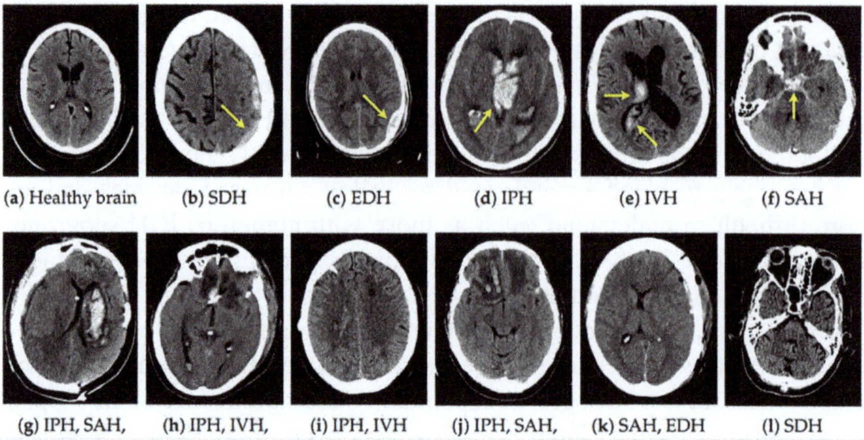

FIGURE 2.5 In the top row, non-contrast CT slices with distinct ICH subtypes are shown (a–f). The regions of bleeding are shown by yellow arrows. Complicated instances are depicted in the bottom row, with combinations of numerous ICH subtypes (g–k) or hardly discernible ICH symptoms.

The SVM was used for the results of the augmentation technique and was input into the pre-trained CNN. The categorization was done using the SVM. Using a dataset of 2101 augmented photos, the researchers achieved 98.6% accuracy, 98.1% sensitivity, and 99.0% specificity for spotting any bleeding. The outcomes of the AlexNet architecture CNN as well as SVM combination and two standard CNN models were studied by Dawud et al. (2019). When compared to conventional structures, the original architecture produced assessment metrics for the identification of ICH over a database of 12,635 photos that were 93% accurate, 95% sensitive, and 90% specific. ResNeXt-101 and bidirectional short-term memory were combined into two neural network architectures by Burduja et al. (2020) to accomplish ICH subtype categorization (BiLSTM). Based on metrics created from annotations received from three radiologists, the model's performance was assessed. The weighted mean log loss for the test dataset was equivalent to 0.04989, and the learned accuracy for ICH subtypes was greater than 96%. However, because the F1 score has lower sensitivity to extremely unequal data, it appears to reflect the system's performance more accurately. Using a double-branch CNN for finding the features and two separate classifiers, we present a method in this study for classifying several ICH subtypes (SDH, IPH, IVH, EDH, and SAH) in brain CT scans. According to the research, multiclass techniques often have worse precision than binary detectors of specific subtypes (Castro et al., 2020; Li et al., 2012; Ye et al., 2019;

Anaya and Beckinghausen, 2020; Cho et al., 2019). Furthermore, a single slice could show many ICH subtypes, making it challenging to use the multiclass architecture to reach a definite conclusion. Therefore, in order to distinguish each ICH subtype specifically, we use examples of the suggested architecture that have been separately trained. Preprocessing is used before feature extraction and categorization, and a method for eliminating skulls is also given. The initial CT slice is converted using three intensity windows. The produced merged grayscale images make up the inputs for the double-branch CNN. The method considers images from various windows in order to extrapolate more information from a single input in the first branch. The adjacent slices are used in the second branch's local 3D context analysis of the slice. In both scenarios, a ResNet-50 architecture is created and trained to extract the necessary characteristics for automatic categorization. SVM and random forest are two techniques used in the classification process (RF). To determine each ICH subtype, we compare the output from the two classifiers. We also assess each ResNet-50 network's performance from both branches. We leverage a public library with 372,556 head CT slices and field labels indicating the existence of SDH, IPH, IVH, EDH, or SAH for training and assessment. Following that, we compare our methodology to the reference approaches indicated in the introduction. Our study's key contributions are the use of two distinct data sources and the use of an outside classifier. A joint vector feeds a classifier with characteristics that are both spatial and intensity-related. Spatial information can be used to detect ICHs, but to our knowledge, no other methods have used a hybrid approach similar to the one shown in this paper.

The two alternate data types used in our work and the use of a third-party classifier are its key contributions. A joint vector feeds a classifier with features that have both spatial and intensity-related relationships. By our research, a hybrid approach like the one utilized in this study has not been used in any of the methods for detecting ICHs that incorporate geographic information.

2.2.9 AUTOMATIC SLEEP STAGE DETECTION USING DEEP LEARNING ALGORITHMS

Sleep is necessary for the upkeep and management of various molecular biological processes that aid in human recovery from mental and physical disease as well as regular daily brain function. There are two forms of sleep: rapid eye motion (REM) and non-rapid eye motion (NREM). After the first four phases of NREM sleep, the REM sleep stage begins. Throughout the sleep cycle, NREM and REM sleep phases exchange in cycles. Sleep issues may result from a lack of sleep phases or an irregular cycle pattern. Too often, sleep problems that result in poor sleep quality are ignored.

Stranges et al. (2012) noted the growing concern for sleep-related problems in terms of global health. Statistics show that 150 million persons, or 16.6% of all adults, have sleep disorders. By 2030, this number will increase to 260 million based on predicted trends. Professionals are now required to manually assess the various stages of sleep. Only a small number of scoring systems for the various sleep stages can be learned by human professionals for different polysomnogram (PSG) recordings, and they are not always able to handle gradual changes in background electroencephalography (EEG) (Stranges et al., 2012). Additionally, human expert evaluations are susceptible to inter- and intra-observer variations that could reduce the accuracy of grading sleep stages (Spriggs, 2014). The patient's comfort level and the diagnostic's price are important variables that affect how the sleep stage is graded. As a result, a sleep lab is a highly regulated environment that requires specialized tools and highly qualified personnel. Patients must, therefore, go to sleep labs, which are frequently found in urban areas, in order to stay there for one or more nights. Due to these factors, patients perceive sleep labs to be inconvenient. Its benefits are better patient accessibility, affordability, and usability. However, It has some disadvantages, such as restricted diagnostic capabilities, device failure, and reliability, overshadowing these advantages (Corral et al., 2013). To improve things, the sleep stage scoring process needs to be completely rewritten. We need machines to take the role of human professionals in the workforce. This can only be done by systems that comprehend the various stages of sleep in a way that is comparable to that of human professionals. Grading sleep stages is one of the many knowledge activities that deep learning (DL) is lauded for automating. Before utilizing and implementing this technology, it is prudent to learn more about the limitations and capabilities of current DL techniques.

2.2.10 THERMAL PICTURE DEEP ENTITY SEGMENTATION OF LAB ANIMALS

It is common practice in research on stress, anxiety, depressive disorders, or neurodegenerative diseases to analyze the behavior of laboratory animals (Lezak et al., 2017). Automating this analysis has a variety of advantages, but objectivity and standardization are two of the most sought ones. Although the parameters for position and motion analysis are straightforward and often used (Franco et al., 2019), it might be difficult to automatically identify actions or behaviors. The quantity or color of the items, for example, places restrictions on the methods currently in use to analyze the behavior of laboratory animals (Franco et al., 2019). The types of behavior analyses that are currently possible are frequently limited to grooming, resting, and exploration. All of these behaviors are detected using straightforward object properties that reflect the position, velocity, direction of motion, or shape parameters. To open up new possibilities, more complex behaviors could be objectively identified and analyzed. An animal's social behavior is a crucial sign of its health and happiness. It is a one-of-a-kind characteristic of a particular person and ensures the survival of the species. Furthermore, social oddities and aberrations may be related to fear, stress, or illness (Lezak et al., 2017). There are three recognized social behavior types: aggressive, defensive, and neutral (Grant and Mackintosh,1963). Forceful behavior includes assaults, bites, and aggressive grooming.

However, meticulous grooming can also be a defensive tactic and is sometimes confused with other behaviors like climbing. This issue can be handled by using camcorders and reviewing the behavior that was captured on film. Another advantage of this technique, which is unquestionably helpful during swift animal movements like fighting, is the capacity to halt or slow down the motion. However, nocturnal animals like rodents find it extremely unpleasant and stressful to need excellent lighting for cameras (Aslani et al., 2014; Roedel et al., 2006). In this case, technology that offers alternative visuals can be useful. The surface temperature can be recorded by thermal cameras even in complete darkness or low light. Additionally, they provide data on the temperature distribution of an object's surface, which has been demonstrated to be useful in the early diagnosis of disease (Manzano-Szalai et al., 2016; Etehadtavakol et al., 2019). In reaction to stress, anxiety, or behavioral changes, the rodent's skin temperature may change (Jang et al., 2015; Tan and Knight, 2018). The authors of the study (Sona et al., 2014) employ the scattering of mouse body temperature for the automated analysis of rodents from video footage to distinguish objects and retrieve tracking algorithm recognition swap after close contact, each animal in a scene must be identified and tracked. Animal tracking visual systems frequently distinguish between the object and the backdrop via color threshold, color matching, or difference between frames (Koniar et al., 2017). A noticeable temperature difference makes foreground-background segmentation in thermal imaging substantially easier. Typically, Otsu thresholding backed by morphology is appropriate (Franco et al., 2019; Fleuret et al., 2016). The outcomes of Otsu's background subtraction on a thermal picture are shown in Figure 2.6 b. It might also be important to tell people apart in a crowded setting or to distinguish an object from other foreground elements (Kim et al., 2017). In (ii), each item is identified as a separate segment, represented, and given a class name. Instance segmentation is a job in computer vision that is particularly difficult. Hariharan et al. (2014) first presented it, and COCO (Lin et al., 2014) popularized it. Two-Stage Methodologies Most two-stage methods (Dai et al., 2015; Chen et al., 2017; He et al., 2017; Huang et al., 2019) view segmentation as a progression of object detection. For instance, they are produced utilizing the detect-then-segment paradigm of segmentation.

The Mask R-CNN model is further developed by the Hybrid Task Cascade (HTC), which this time incorporates a cascade architecture (Chen et al., 2019). It improves the segmentation accuracy by combining cascade and multitasking at each level and investigating additional contextual data from the spatial context. Excellent results are obtained with modern two-stage methods. However, due to the cascade strategy, these solutions frequently need a lot of time and memory. As a result, despite not performing as well as two-stage techniques currently do, single-stage designs for instance identification and segmentation (Yao et al., 2019; Tian et al., 2019) are currently gaining prominence. One-Stage Approaches Single-stage methods (Mazur-Milecka et al., 2020) are easier to implement because the re-pooling step is skipped, but they are frequently unable to produce masks with the same level of accuracy as two-stage methods. Some of those methods work in a way that anticipates every pixel, much like semantic segmentation. Fully Convolutional One-Stage detection (FCOS) (Tian et al., 2019) minimizes time-consuming computations such as calculating boxes that overlapped during training by completely deleting pre-defined anchor boxes from its method. EmbedMask, which was suggested by (Ying et al., 2019), combines segmentation- and proposal-based approaches and gains from both. It is based on recognition systems and has similar detection capabilities to proposal-based approaches. Unlike segmentation-based methods, EmbedMask builds embeddings for pixels and suggestions using additional embedding modules, which enables it to create high-resolution masks without losing features through repolling. TensorMask (Chen et al., 2019) is a method for dense sliding-window instance segmentation. The goal is to express masks across a spatial domain using structured 4-D tensors. In addition to mask prediction, multiclass classification is employed. The authors demonstrate that this approach produces results comparable to Mask R-CNN. This project's major purpose is to recognize each rat in relation to other rats and background items. As a result, we will concentrate on the instance segmentation approach and, in particular, the deep semantic segmentation frameworks represented by the two-stage and one-stage algorithms. We especially addressed the issue of segmenting animals in thermal pictures taken near humans.

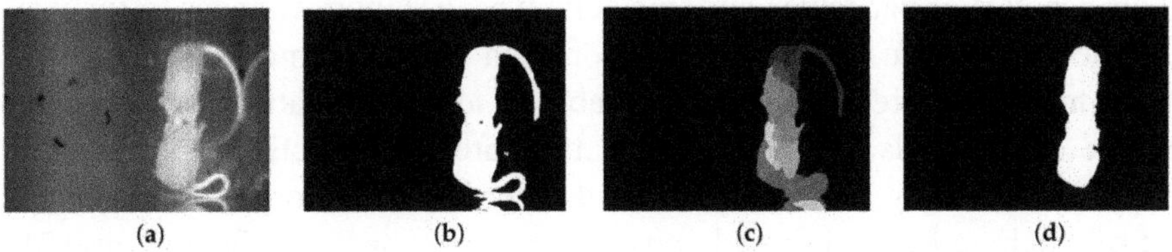

FIGURE 2.6 Results of the most popular algorithms used for segmentation of (a) original image, (b) Otsu's thresholding, (c) watershed with gradient regions markers, and (d) U-Net—semantic segmentation.

2.3 CONCLUSIONS AND FUTURE SCOPE

In recent years, machine learning research has made incredible strides toward automating drug discovery, medical image analysis, genome analysis, disease pre-screening, diagnosis, and prognosis. These achievements have successfully delivered significant performance improvements in comparison to other existing traditional algorithms. It has many practical uses for a variety of everyday problems, especially in biotechnology and medical science. The majority of researchers today believe that within a few years, real-time apps based on machine learning and DL will surpass humans at par, and the automated systems will aid us in carrying out most of the everyday regular chores. This is due to their tremendous performance in the development of classification and analytical systems. Despite everything we are unable to see, one thing is very clear: the imaging experts are setting the pace in resolving the biomedical imaging issues we are facing. Dr. Mikhail Shapiro from the California Institute of Technology, is a leader in the development of a new deep tissue imaging tool to monitor cellular processes deep within the body. He is modifying cells to produce air-filled proteins, which make them show out in ultrasound images and make it easier to monitor them over time. He is already employing this technique to study the behavior of cancer cells (Shapiro, 2022). Dr. Laura Waller, a CZI grantee uses computational microscopy to view previously inaccessible data. She is giving us fresh perspectives into the most fundamental structure of organs like the brain by employing cutting-edge methods like 3D phase microscopy, which is an essential first step toward comprehending neural function (Waller, 2022). These techniques may soon enable us to create individualized treatments for our conditions. For instance, after receiving a diagnosis of cancer, a patient may not require a harsh treatment like chemotherapy. Instead, their doctor may choose to fight the illness by focusing on individual cells depending on their genetics or particular proteins gone wrong. Future changes to the way the medical system operates can be foreseen with even more radiance. The same medical professional could be able to identify a patient's risk factors before cancer cells even appear and take preventive actions to avoid the disease ever taking root. Hospitals and other healthcare organizations use a lot of machine learning-based apps to increase patient satisfaction, provide individualized care, produce precise predictions, and improve overall quality of life. It is also used to expedite the process of creating new medications and doing clinical trials. Techniques for DL and accompanying research, although being an exciting future autonomous systems tool, nevertheless have significant limitations. For starters, the availability of a large amount of high-quality data will affect the performance and trustworthiness of DL models. The pharmaceutical business generates massive volumes of biological data, but it is generally kept as a costly private commercial asset and is not made accessible to the general public. The paucity of full explanations of linked biological pathways is a second issue with DL models. While having demonstrated remarkable generalization ability, DL models remain "black boxes," making it impossible to identify the underlying pathways incorporated in the modeling data. Overall, DL has shown promise for usage in the new big data era of drug development as a recently established machine intelligence technology. DL techniques will soon overtake other approaches in computer-aided drug design (CADD) as more data becomes available and fresh ideas are produced.

ACKNOWLEDGMENT

I would like to kindly acknowledge those authors whose references have been taken in this book chapter. I would like to give sincere thanks to my head of the institution, Prof. Dr. O.P. Singh, Amity University, Lucknow, India, and Pro.VC sir, Prof. Wg. Cdr. (Dr.) Anil Kumar (Retd.), Amity University, Lucknow, India for continuous guidance and encouragement. I would like to give thanks to anonymous reviewers, and all those who were involved directly or indirectly in the preparation of this book chapter.

KEYWORDS

- **biomedical imaging**
- **machine learning**
- **healthcare system**
- **convolutional neural network**
- **deep neural network**
- **deep learning**
- **k-nearest neighbor**

REFERENCES

Abdulla, W. Mask R-CNN for Object Detection and Instance Segmentation on Keras and TensorFlow. 2017. Available online: https://github.com/matterport/Mask_RCNN (accessed on 1 February 2020).

Zotin, A.; Simonov, K.; Kurako, M.;, Hamad, Y.; Kirillova, S.; Edge Detection in MRI Brain Tumor Images Based on fuzzy C-means Clustering, 22nd International Conference on Knowledge-Based and Intelligent Information & Engineering Systems, PP-1262-1270, 2018.

Anaya, E.; Beckinghausen, M. A DL Approach to Classifying Intracranial Hemorrhages. Available online: http://cs230.stanford.edu/projects_fall_2019/reports/26260039.pdf (accessed on 13 August 2020).

Arbabshirani, M.; Fornwalt, B.; Mongelluzzo, G.; Suever, J.; Geise, B.; Patel, A.; Moore, G. Advanced Machine Learning in Action: Identification of Intracranial Hemorrhage on Computed Tomography Scans of the Head with Clinical Workflow Integration. *npj Dig. Med.* **2018,** *1,* 9.

Aslani, S.; Harb, M.; Costa, P.; Almeida, O.; Sousa, N.; Palha, J. Day and Night: Diurnal Phase Influences the Response to Chronic Mild Stress. *Front. Behav. Neurosci.* **2014,** *8,* 82.

Belhumeur, P. N.; David, W.; David, J.; Kriegman, J.; Neeraj Kumar, T. Localizing Parts of Faces Using a Consensus of Exemplars. IEEE Trans. *Pattern Anal. Mach. Intell.* **2013,** *35,* 2930–2940.

Bolya, D.; Zhou, C.; Xiao, F.; Lee, Y. J. In *YOLACT: Real-time Instance Segmentation,* Proceedings of the IEEE International Conference on Computer Vision, Seoul, Korea, 27 October–2 November 2019; pp. 9157–9166.

Breiman, L. Random Forests. *Mach. Learn.* **2001,** *45,* 5–32.

Burduja, M.; Ionescu, R. T.; Verga, N. Accurate and Efficient Intracranial Hemorrhage Detection and Subtype Classification in 3D CT Scans with Convolutional and Long Short-Term Memory Neural Networks. *Sensors* **2020,** *20,* 5611.

Castro, J.; Chabert, S.; Saavedra, C.; Salas, R. In *Convolutional Neural Networks for Detection Intracranial Hemorrhage in CT Images*, Proceedings of the 5th Congress on Robotics and Neuroscience, Valparaíso, Chile, Feb 27–29, 2020; 2564.

Cervantes, J.; García-Lamont, F.; Rodríguez, L.; Lopez-Chau, A. A Comprehensive Survey on Support Vector Machine Classification: Applications, challenges and trends. *Neurocomputing* 2020.

Zukerberg, C. Intiatives, The Past, Present, and Future of Medical Imaging. < https://chan-zuckerberg.com/blog/the-past-present-and-future-of-medical-imaging > (accessed July 25, 2022).

Chang, P.; Kuoy, E.; Grinband, J.; Weinberg, B.; Thompson, M.; Homo, R.; Chen, J.; Abcede, H.; Shafie, M.; Sugrue, L.; et al. Hybrid 3D/2D Convolutional Neural Network for Hemorrhage Evaluation on Head CT. *Am. J. Neuroradiol.* **2018,** *39,* 1609–1616. Appl. Sci. 2020, 10, 7577 12 of 12

Chen, K.; Pang, J.; Wang, J.; Xiong, Y.; Li, X.; Sun, S.; Feng, W.; Liu, Z.; Shi, J.; Ouyang, W.; et al. In *Hybrid Task Cascade for Instance Segmentation*, Proceedings of the IEEE Conference on Computer Vision and Pattern Recognition, Long Beach, CA, USA, June 16–20, 2019; pp. 4974–4983.

Chen, L.; Hermans, A.; Papandreou, G.; Schroff, F.; Wang, P.; Adam, H. MaskLab: Instance Segmentation by Refining Object Detection with Semantic and Direction Features. arXiv 2017, arXiv:1712.04837.

Chen, X.; Girshick, R. B.; He, K.; Dollár, P. In *TensorMask: A Foundation for Dense Object Segmentation*, Proceedings of the IEEE International Conference on Computer Vision, Seoul, Korea, 27 October–2 November 2019; pp. 2061–2069.

Chilamkurthy, S.; Ghosh, R.; Tanamala, S.; Biviji, M.; Campeau, N.; Venugopal, V.; Mahajan, V.; Rao, P.; Warier, P. Deep Learning Algorithms for Detection of Critical Findings in Head CT Scans: A Retrospective Study. *Lancet* **2018,** *392,* 2388–2396.

Ching-Wei, W.; Huang, C.; Hsieh, M.; Li, C.; Chang, S.; Li, W.; Vandaele, R. Evaluation and comparison of Anatomical Landmark Detection Methods for Cephalometric X-ray Images: A Grand Challenge. IEEE Trans. *Biomed. Eng.* **2015,** *53,* 1890–1900. 6. Wang, C. W.; Huang, C. T.; Lee, J. H.; Li, C. H.; Chang, S. W.; Siao, M. J.; Lai, T. M.; Ibragimov, B.; Vrtovec, T.; Ronneberger, O.; et al. A Benchmark for Comparison of Dental Radiography Analysis Algorithms. IEEE Trans. *Biomed. Eng.* **2016,** *31,* 63–76.

Cho, J.; Park, K. S.; Karki, M.; Lee, E.; Ko, S.; Kim, J. K.; Lee, D.; Choe, J.; Son, J.; Kim, M.; et al. Improving Sensitivity on Identification and Delineation of Intracranial Hemorrhage Lesion Using Cascaded Deep Learning Models. *J. Digit. Imaging* **2019,** *32,* 450–461.

Claudiu, C. D.; Meier, U.; Masci, J.; Gambardella, L. M.; Schmidhuber, J. In *Flexible, High Performance Convolutional Neural Networks for Image Classification*, Proceedings of the Twenty-Second International Joint Conference on Artificial Intelligence, Barcelona, Spain, July 16–22, 2011.

Corral, J.; Pepin, J.-L.; Barbé, F. Ambulatory Monitoring in the Diagnosis and Management of Obstructive Sleep Apnoea Syndrome. *Eur. Respir. Rev.* **2013,** *22,* 312–324.

Cortes, C.; Vapnik, V. Support Vector Networks. *Mach. Learn.* **1995,** *20,* 273–297.

Shevad, D, Multi Object Segmentation and Tracking in Image Sequence, Lamar University-Beaumount, 2005.

Dai, J.; He, K.; Sun, J. In *Instance-aware Semantic Segmentation via Multi-task Network Cascades*, Proceedings of the IEEE Conference on Computer Vision and Pattern Recognition, Boston, MA, USA, June 7–12, 2015; pp. 3150–3158.

Danilov, G.; Kotik, K.; Negreeva, A.; Tsukanova, T.; Shifrin, M.; Zakharova, N.; Batalov, A.; Pronin, I.; Potapov, A. Classification of Intracranial Hemorrhage Subtypes Using Deep Learning on CT Scans. *Stud. Health Technol. Inf.* **2020,** *272,* 370–373.

Dash, S.; Shakyawar, S. K.; Sharma, M. et al. Big data in Healthcare: Management, Analysis and Future Prospects. *J. Big. Data* **2019,** *6,* 54. https://doi.org/10.1186/s40537-019-0217-0

Dawud, A. M.; Yurtkan, K.; Oztoprak, H. Application of Deep Learning in Neuroradiology: Brain Haemorrhage Classification Using Transfer Learning. *Comput. Intell. Neurosci.* **2019,** *2019,* 4629859.

Edhab, D. A.; Ghoniemy, S. S. A.; Selim, G. M. Automated Brain Tumor Detection and Identification using Image Processing and Probabilistic Neural Network Techniques. *Int. J. Image Proces. Visual Commun.* **2012,** *1,* 1–8.

Waller, L. Computational Microscopy with Multiple-Scattering Samples, Deep Tissue Imaging, 2022, https://chanzuckerberg.com/science/programs-resources/imaging/frontiers/computational-microscopy-with-multiple-scattering-samples.

Flam, S. CEO, Foresee Medical, Benefits of Machine Learning in Healthcare.https://www.foreseemed.com/blog/machine-learning-in-healthcare#:~:text=Machine%20learning%20in%20healthcare%20can,that%20indicate%20a%20particular%20disease (accessed Nov 1, 2022).

Ihsanto, E.; Ramli, K.; Sudiana, D.; Gunawan, T. S. An Efficient Algorithm for Cardiac Arrhythmia Classification Using Ensemble of Depthwise Separable Convolutional Neural Networks, 2020.

Etehadtavakol, M.; Emrani, Z.; Ng, E. Y. K. Rapid Extraction of the Hottest or Coldest Regions of Medical Thermographic Images. *Med. Biol. Eng. Comput.* **2019,** *57,* 379–388.

Flanders, A.; Prevedello, L.; Shih, G.; Halabi, S.; Kalpathy-Cramer, J.; Ball, R.; Mongan, J.; Stein, A.; Kitamura, F.; Lungren, M.; et al. Construction of a Machine Learning Dataset through Collaboration: The RSNA 2019 Brain CT Hemorrhage Challenge. *Radiol. Artif. Intell.* **2020,** *2,* e190211.

Fleuret, J.; Ouellet, V.; Moura-Rocha, L.; Charbonneau, É.; Saucier, L.; Faucitano, L.; Maldague, X. A Real Time Animal Detection And Segmentation Algorithm For IRT Images In Indoor Environments. Quant. *InfraRed Thermogr*. **2016,** 265–274.

Franco, N. H.; Gerós, A.; Oliveira, L.; Olsson, I. A. S.; Aguiar, P. ThermoLabAnimal—A High-Throughput Analysis Software for Non-Invasive Thermal Assessment of Laboratory Mice. *Physiol. Behav*. **2019,** *207,* 113–121.

Freeman, W. D.; Aguilar, M. I.. Intracranial Hemorrhage: Diagnosis and Management. *Neurol. Clin.* **2012,** *30,* 211–240.

Girshick, R. B. Fast R-CNN. In Proceedings of the IEEE International Conference on Computer Vision, Santiago, Chile, Dec 7–13, 2015; pp. 1440–1448.

Grant, E.; Mackintosh, J. A Comparison of the Social Postures of Some Common Laboratory Rodents. *Behaviour* **1963,** *21,* 246–259.

Hackstein, N.; Heckrodt, J.; Rau, W. S. Measurement of Single-kidney Glomerular Filtration Rate Using a Contrast-enhancedDynamic Gradient-echo Sequence and the Rutland-Patlak Plot Technique. *J. Magn. Reson. Imaging* **2003,** *18,* 714–725

Hariharan, B.; Arbelaez, P.; Girshick, R. B.; Malik, J. Simultaneous Detection and Segmentation. In European Conference on Computer Vision; Springer: Cham, Switzerland, 2014; pp. 297–312.

He, K.; Gkioxari, G.; Dollár, P.; Girshick, R. B. Mask R-CNN. In Proceedings of the IEEE International Conference on Computer Vision, Venice, Italy, Oct 22–29, 2017; pp. 2961–2969.

He, K.; Zhang, X.; Ren, S.; Sun, J. In *Deep Residual Learning for Image Recognition.* Proceedings of the 2016 IEEE Conference on Computer Vision and Pattern Recognition (CVPR), Las Vegas, NV, USA, June 27–30 2016; pp. 770–778.

Hssayeni, M. D.; Croock, M. S.; Salman, A. D.; Al-khafaji, H. F.; Yahya, Z. A.; Ghoraani, B. Intracranial Hemorrhage Segmentation Using Deep Convolutional Model. *Data* **2020,** *5,* 14.

Huang, Z.; Huang, L.; Gong, Y.; Huang, C.; Wang, X. Mask Scoring R-CNN. In Proceedings of the IEEE Conference on Computer Vision and Pattern Recognition, Long Beach, CA, USA, June 16–20, 2019; pp. 6409–6418.

Jang, E.; Park, B.; Park, M.; Kim, S.; Sohn, J. Analysis of Physiological Signals for Recognition of Boredom, Pain, and Surprise Emotions. *J. Phys. Anthropol.* **2015,** *34,* 1–12.

Ker, J.; Singh, S.; Bai, Y.; Rao, J.; Lim, T.; Wang, L. Image Thresholding Improves 3-Dimensional Convolutional Neural Network Diagnosis of Different Acute Brain Hemorrhages on Computed Tomography Scans. *Sensors* **2019,** *19,* 2167.

Kim, W.; Cho, Y. B.; Lee, S. Thermal Sensor-Based Multiple Object Tracking for Intelligent Livestock Breeding. IEEE Access **2017,** *5,* 27453–27463.

Kiranyaz, S.; Ince, T.; Gabbouj, M. Real-time Patient-specific ECG Classification by 1-D Convolutional Neural Networks. *IEEE Trans. Biomed. Eng.* **2015,** *63,* 664–675.

Klepaczko, A.; Strzelecki, M.; Kociołek, M.; Eikefjord, E.; Lundervold, A. A Multi-Layer Perceptron Network for Perfusion Parameter Estimation in DCE-MRI Studies of the Healthy Kidney. *App. Sci.* **2020,** *10* (16), 5525. https://doi.org/10.3390/app10165525

Koniar, D.; Hargaš, L.; Loncová, Z.; Simonová, A.; Ducho ˇn, F.; Be ˇno, P. Visual System-based Object Tracking Using Image Segmentation for Biomedical Applications. *Electr. Eng.* **2017,** *99,* 1349–1366.

Konkol, M.; Śniatała, K.; Śniatała, P.; Wilk, S.; Baczyńska, B.; Milecki, P. Computer Tools to Analyze Lung CT Changes after Radiotherapy. *Appl. Sci.* **2021,** *11,* 1582. https://doi.org/10.3390/app11041582

Kuo, W.; Häne, C.; Mukherjee, P.; Malik, J.; Yuh, E. L. Expert-level Detection of Acute Intracranial Hemorrhage on Head Computed Tomography using deep Learning. *Proc. Natl. Acad. Sci. USA* **2019,** *116,* 22737–22745.

Jain, L. C.; Virvou, M.; Piuri, V.; Balas, V. E. Advances in Bioinformatics, Multimedia, and Electronics Circuits and Signals, Proceedings of GUCON 2019

Lee, H.; Park, M.; Kim, J. In *Cephalometric Landmark Detection in Dental X-ray Images Using Convolutional Neural Networks.* Proceedings of the Medical Imaging 2017: Computer-Aided Diagnosis, Orlando, FL, USA, March 3, 2017.

Lezak, K.; Missig, G.; Carlezon, W. A. Jr. Behavioral Methods to Study Anxiety in Rodents. *Dialogues Clin. Neurosci.* **2017,** *19,* 181–191.

Li, Y.; Wu, J.; Li, H.; Li, D.; Du, X.; Chen, Z.; Jia, F.; Hu, Q. Automatic Detection of the Existence of Subarachnoid Hemorrhage from Clinical CT Images. *J. Med. Syst.* **2012,** *36,* 1259–1270.

Lin, T.; Maire, M.; Belongie, S. J.; Bourdev, L. D.; Girshick, R. B.; Hays, J.; Perona, P.; Ramanan, D.; Dollár, P.; Zitnick, C. L. Microsoft COCO: Common Objects in Context. In European Conference on Computer Vision; Springer: Cham, Switzerland, 2014; pp. 740–755.

Lindner, C.; Tim, F.; Cootes, T. In *Fully Automatic Cephalometric Evaluation Using Random Forest Regression-voting.* Proceedings of the IEEE International Symposium on Biomedical Imaging (ISBI), Brooklyn Bridge, NY, USA, April 16–19, 2015.

Litjens, G.; Kooi, T.; Bejnordi, B. E.; Arindra, A.; Setio, A.; Ciompi, F.; Ghafoorian, M.; van der Laak, J. A. W. M.; van Ginneken, B.; Sánchez, C. I. A Survey on Deep Learning in Medical Image Analysis. *Med. Image Anal.* **2017,** *42,* 60–88.

Liu, L.; Ma, D.; Azar, A. T.; Zhu, Q. Neural Computing Enhanced Parameter Estimation for Multi-Input and Multi-Output Total Non-Linear Dynamic Models. *Entropy* **2020,** *22*, 510

Liu, S.; Qi, L.; Qin, H.; Shi, J.; Jia, J. In *Path Aggregation Network for Instance Segmentation*, Proceedings of the IEEE Conference on Computer Vision and Pattern Recognition, Salt Lake City, UT, USA, June 18–23, 2018; pp. 8759–8768.

Long, J.; Shelhamer, E.; Darrell, T. In *Fully Convolutional Networks for Semantic Segmentation*, Proceedings of the IEEE Conference on Computer Vision and Pattern Recognition (CVPR), Boston, MA, USA, June 7–12, 2015; pp. 3431–3440.

Majumdar, A.; Brattain, L.; Telfer, B.; Farris, C.; Scalera, J. In *Detecting Intracranial Hemorrhage with Deep Learning*, Proceedings of the 2018 40th Annual International Conference of the IEEE Engineering in Medicine and Biology Society (EMBC), Honolulu, HI, USA, 18–21 July 2018; Volume 2018, pp. 583–587.

Manzano-Szalai, K.; Pali-Schöll, I.; Krishnamurthy, D.; Stremnitzer, C.; Flaschberger, I.; Jensen-Jarolim, E. Anaphylaxis Imaging: Non-Invasive Measurement of Surface Body Temperature and Physical Activity in Small Animals. *PLoS ONE* **2016,** *11,* e0150819.

Marcolini, E.; Stretz, C.; DeWitt, K. M. Intracranial Hemorrhage and Intracranial Hypertension. *Emerg. Med. Clin. N. Am.* **2019,** *37,* 529–544.

Mazur-Milecka, M.; Kocejko, T.; Ruminski, J. Deep Instance Segmentation of Laboratory Animals in Thermal Images. *Appl. Sci.* **2020,** *10,* 5979. https://doi.org/10.3390/app10175979

Mazur-Milecka, M.; Ruminski, J. In *Automatic Analysis of the Aggressive Behavior of Laboratory Animals using Thermal Video Processing*, Proceedings of the IEEE Conference of the Engineering in Medicine and Biology Society, EMBC, Seogwipo, Korea, July 11–15, 2017; pp. 3827–3830.

Mazur-Milecka, M.; Ruminski, J. Deep Learning Based Thermal Image Segmentation for Laboratory Animals Tracking. *Quant. InfraRed Thermogr. J.* **2020,** 1–18.

Melgani, F.; Bazi, Y. Classification of Electrocardiogram Signals with Support Vector Machines and Particle Swarm Optimization. *IEEE Trans. Inf. Technol. Biomed.* **2008,** *12,* 667–677

Thomas, M. Bulitin, 16 Machine Learning in Healthcare Examples. < https://builtin.com/artificial-intelligence/machine-learning-healthcare> (accessed Oct 03, 2022)

Shapiro, M. Biomolecular Ultrasound for Deep Tissue Imaging of Cellular Functions, 2022, <https://chanzuckerberg.com/science/programs-resources/imaging/frontiers/biomolecular-ultrasound-for-deep-tissue-imaging-of-cellular-functions/>

Milletari, F.; Navab, N.; Ahmadi, S. In *V-Net: Fully Convolutional Neural Networks for Volumetric Medical Image Segmentation*, Proceedings of the IEEE 2016 Fourth International Conference on 3D Vision (3DV), Stanford, CA, USA, Oct 25–28, 2016; pp. 565–571.

Naidech, A. M. Intracranial Hemorrhage. *Am. J. Respir. Crit. Care Med.* **2011,** *184,* 998–1006.

Nguyen, N.; Tran, D.; Nguyen, N.; Nguyen, H. A CNN-LSTM Architecture for Detection of Intracranial Hemorrhage on CT scans. Med. Imaging Deep Learn. 2020.

Nurmaini, S.; Partan, R. U.; Caesarendra, W.; Dewi, T.; Rahmatullah, M. N.; Darmawahyuni, A.; Bhayyu, V.;Firdaus, F. An Automated ECG Beat Classification System Using Deep Neural Networks with an Unsupervised Feature Extraction Technique. *Appl. Sci.* **2019,** *9,* 2921

Ohn, I.; Kim, Y. Smooth Function Approximation by Deep Neural Networks with General ActivationFunctions. *Entropy* **2019,** *21*, 627

Olaf, R.; Fischer, P.; Brox, T. In *U-net: Convolutional Networks for Biomedical Image Segmentation*, Proceedings of the International Conference on Medical Image Computing and Computer-assisted Intervention, Munich, Germany, Oct 5–9, 2015; pp. 234–241.

Otsu, N. A Threshold Selection Method from Gray-Level Histograms. *IEEE Trans. Syst. Man Cybern.* **1979,** *9,* 62–66.

Pfister, T.; Charles, J.; Zisserman, A. In *Flowing Convnets for Human Pose Estimation in Videos*, Proceedings of the IEEE International Conference on Computer Vision (ICCV), Santiago, Chile, Dec 13–16 2015.

Raj, S.; Ray, K. C. ECG Signal Analysis using DCT-based DOST and PSO Optimized SVM. *IEEE Trans. Instrum. Meas.* **2017,** *66,* 470–478

Ramanan, D.; Zhu, T. In *Face Detection, Pose Estimation, and Landmark Localization in the Wild*, Proceedings of the 2012 IEEE Conference on Computer Vision and Pattern Recognition (CVPR), Providence, RI, USA, June 16–21, 2012; pp. 2879–2886.

Ahmmed R.; Swakshar, A. S.; Hossain, M. F.;, Rafiq, M. A. Classification of Tumors and It Stages in Brain MRI Using Support Vector Machine and Artificial Neural Network, In Proc. IEEE International Conference on Electrical, Computer and Communication Engineering (ECCE), Bangladesh, 229–237, Feb 2017.

Ren, S.; He, K.; Girshick, R.; Sun, J. Faster R-CNN: Towards Real-Time Object Detection with Region Proposal Networks. arXiv 2015, arXiv:1506.01497.

Roedel, A.; Storch, C.; Holsboer, F.; Ohl, F. Effects of light or Dark Phase Testing on Behavioural and Cognitive Performance in DBA Mice. *Lab. Anim.* **2006,** *40,* 371–381.

Ronneberger, O.; Fischer, P.; Brox, T. U-Net: Convolutional Networks for Biomedical Image Segmentation. In International Conference on Medical Image Computing and Computer-Assisted Intervention; Springer: Cham, Switzerland, 2015; pp. 234–241.

Simonyan, K.; Andrew, Z. T. Very Deep Convolutional Networks for Large-scale Image Recognition. arXiv 2014, arXiv:1409.1556

Sona, D.; Zanotto, M.; Papaleo, F.; Murino, V. In *Automated Discovery of Behavioural Patterns in Rodents*, Proceedings of the 9th International Conference on Methods and Techniques in Behavioral Research, Wageningen, The Netherlands, Aug 27–29, 2014.

Song, Y.; Qiao, X.; Iwamoto, Y.; Chen, Y.-w. Automatic Cephalometric Landmark Detection on X-ray Images Using a Deep-Learning Method. *Appl. Sci.* **2020,** *10*, 2547. https://doi.org/10.3390/app10072547

Spriggs, W. H. Essentials of Polysomnography; Jones & Bartlett Learning: Burlington, MA, USA, 2014.

Stevens, L. A.; Levey, A. S. Measured GFR as a Confirmatory Test for Estimated GFR. *J. Am. Soc. Nephrol.* **2009,** *20*, 2305–2313

Zorana, S.; Rutesic, P. An Improved Canny Edge Detection Algorithm for Detecting Brain Tumors in MRI Images. *Int. J. Signal Process.* **2018,** *3,* 11–15.

Stranges, S.; Tigbe, W.; Gómez-Olivé, F. X.; Thorogood, M.; Kandala, N.-B. Sleep Problems: An Emerging Global Epidemic? Findings from the INDEPTH WHO-SAGE Study Among More Than 40,000 Older Adults from 8 Countries Across Africa and Asia. *Sleep* **2012,** *35,* 1173–1181.

Tan, C.; Knight, Z. Regulation of Body Temperature by the Nervous System. *Neuron* **2018,** *98,* 31–48.

Tian, Z.; Shen, C.; Chen, H.; He, T. In *FCOS: Fully Convolutional One-Stage Object Detection*, Proceedings of the IEEE International Conference on Computer Vision, Seoul, Korea, 27 October–2 November 2019; pp. 9627–9636.

Togacar, M.; Cömert, Z.; Ergen, B.; Budak, Ü. In *Brain Hemorrhage Detection based on Heat Maps, Autoencoder and CNN Architecture*, Proceedings of the 2019 1st International Informatics and Software Engineering Conference (UBMYK), Ankara, Turkey, Nov 6–7 2019; pp. 1–5.

Tominari, S.; Morita, A.; Ishibashi, T.; Yamazaki, T.; Takao, H.; Murayama, Y.; Sonobe, M.; Yonekura, M.; Saito, N.; Shiokawa, Y.; et al. Prediction Model for 3-year Rupture Risk of Unruptured Cerebral Aneurysms in Japanese Patients. *Ann. Neurol.* **2015,** *77,* 1050–1059.

Tompson, J. J.; Jain, A.; LeCun, Y.; Bregler, C. In *Joint Training of a Convolutional Network and a Graphical Model for Human Pose Estimation*, Proceedings of the Advances in Neural Information Processing systems (NIPS), Montreal, QC, Canada, Dec 8–13 2014; pp. 1799–1807.

Kumar, U. Applications of Machine Learning in Disease Pre-screening. *Pre-Screening Systems for Early Disease Prediction, Detection, and Prevention*; Edoh, T., Ed.; et al., IGI Global, **2019**; pp 278–320. https://doi.org/10.4018/978-1-5225-7131-5.ch010

Vander, A. J.; Sherman, J. H.; Luciano, D. S. Human Physiology. The Mechanism of Body Function, 8th ed.; McGraw-Hill Publishing: New York City, NY, USA, 2001

Wu, Y.; Kirillov, A.; Massa, F.; Lo, W. Y.; Girshick, R. Detectron2, 2019. Available online: https://github.com/ facebookresearch/detectron2 (accessed on 1 May 2020).

Xiang, C.; Tian, S.; Zou, W.; Xu, C. SAIS: Single-stage Anchor-free Instance Segmentation. arXiv 2019, arXiv:cs.CV/1912.01176.

Yao, J.; Yu, Z.; Yu, J.; Tao, D. Single Pixel Reconstruction for One-stage Instance Segmentation. arXiv 2019, arXiv:1904.07426.

Yasugi, M.; Hossain, B.; Nii, M.; Kobashi, S. Relationship Between Cerebral Aneurysm Development and Cerebral Artery Shape. *J. Adv. Comput. Intell. Intell. Inform.* **2018,** *22,* 249–255.

Ye, H.; Gao, F.; Yin, Y.; Guo, D.; Zhao, P.; Lu, Y.; Wang, X.; Bai, J.; Cao, K.; Song, Q.; et al. Precise Diagnosis of Intracranial Hemorrhage and Subtypes Using a Three-dimensional Joint Convolutional and Recurrent Neural Network. *Eur. Radiol.* **2019,** *29,* 6191–6201.

Ying, H.; Huang, Z.; Liu, S.; Shao, T.; Zhou, K. EmbedMask: Embedding Coupling for One-stage Instance Segmentation. arXiv 2019, arXiv:cs.CV/1912.01954.

CHAPTER 3

Study of ML-Enabled Big Data Analysis for IoT-Based Smart Medico Systems

PRATIBHA PANDEY and UPENDRA KUMAR

Department of Computer Science and Engineering, Institute of Engineering and Technology, Lucknow, Uttar Pradesh, India

ABSTRACT

The spread of terrible diseases such as COVID-19 has led to a fresh call to provide immediate healthcare infrastructure to citizens around the world. The current epidemic reflects deficiencies in the traditional healthcare system, which means that only clinics and hospitals are able to recover from the problem. The key technique which helps in healthcare solutions is smart connected wearable devices. Advances in the Internet of Things have given access to these wearable devices to store information at a miraculous rate. These wearables collect relevant data about our behavioral, physical, and mental health. Big data management created by IoT's wearable and other smart healthcare devices is a difficult task that can adversely affect the assessment process in decision data centers. The application of Big Data Analytics in mining information, acquiring knowledge, and making estimates has gathered recent attention. Machine learning (ML) is a domain of research that has applications in addressing various network challenges such as resource allocation, routing, security, and traffic engineering. Recently, it has been observed that there is an increase in the application of machine learning-based technologies and algorithms for the enhancement of IoT devices and applications. Even though big data analytics and machine learning are widely worked on and well-researched, there is a scarcity of studies focusing specifically on the development of ML-based technologies for big data

Smart Electronic Devices and Systems for Biomedical and Healthcare Applications. Upendra Kumar, Rajesh K. Kesharwani, & Raj K. Keservani (Ed.)

DOI: 10.1201/9781003560852-3

analytics in the IoT smart healthcare sector. In this study, we provide a broad overview of machine learning techniques and their applications for big data analysis in the healthcare sector. In addition, the strengths and challenges of existing methods, along with various research weaknesses, were highlighted. Our study provides insight into how healthcare professionals and government agencies can better equip themselves with the current research in Machine Learning-based Big Data Analytics management for Smart Healthcare.

3.1 INTRODUCTION

In the past few years, wireless sensor networks have seen tremendous growth. Technological advances such as interfacing, scalability, applications, data computing, and interoperability, along with technological advancements in radio frequency identification (RFID) and wireless cellular networks, have provided a strong base for the Internet of Things (IoT). Kevin Ashton used the term Internet of Things (IoT) in the context of supply chain management (Evtodieva et al., 2020). It is a smart world of objects over a network integrated with the Internet (Abdollahzadeh et al., 2016). These objects are known as entities that have identities and are controlled. Due to the massive growth in smart device development, IoT has contributed in almost every aspect of our regular activities and continues to do so with a huge variety of new and intelligent products (Piccialli et al., 2017; Qin et al., 2014). These applications include Smart Infrastructure, Smart Healthcare, Smart Farming, Smart Cities, Smart Agriculture, Urban Computing, Crowd Sensing, Smart Grid, and Crowdsourcing. The applications of IOT are shown in Figure 3.1.

These advances with innovation in applications are very motivating and, on the one hand, show the bright future of IoT, but on the other hand, many challenges exist. Some of these problems include security, interoperability, energy management, and quality of service (QoS) (Dehkordi et al., 2020). Big data is important because of the interrelationship between the objects of IoT and the data bits they generate. A wide variety of IoT devices and applications generate vast amounts of information. Various large data analytics are used to obtain such data and to improve decision-making (Rodriguez et al., 2016). The most used of these is the 5V model. This model divides big data into five categories. These features include data volume (volume), real-time data collection (speed), data collection (diversity) from different sources, unpredictable data (truth), and the use of such data in different fields, industries, and educational institutions (value). We have observed extraordinary work in big data research and its application in wide fields. The integration of IoT with big data creates opportunities to improve work such as the healthcare infrastructure. A large number of large data technologies have been used in the Smart Healthcare domain to analyze large amounts of data from multiple sources. Machine learning (ML) is a key technology for creative problem solving on large datasets. The economic impact is estimated to be $ 5.2 trillion to $6.7 trillion per year by 2025 (Athey et al., 2018). It shows big data and the importance of ML in the IoT.

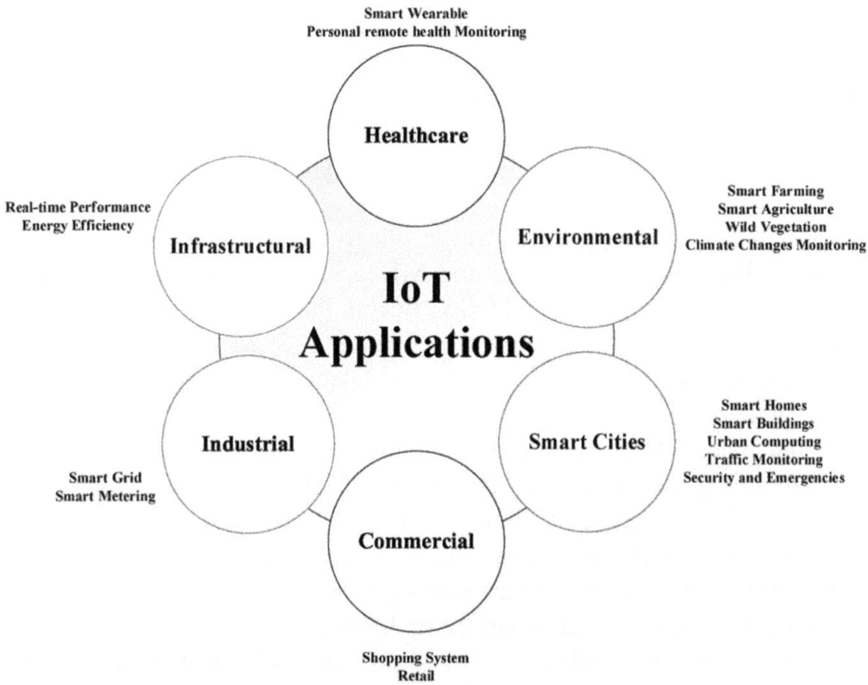

FIGURE 3.1 Application of the internet of things.

Source: Adapted from Hassan and Qamar, 2020

There are several literature reviews that identify research work in ML, big data, and IoT. Mohommadi et al. (2018) discussed the properties of huge data from different sizes, such as volume, speed, variability, accuracy, and value. In addition, they discussed recent and growing in-depth learning structures designed specifically for large data analysis across various IoT domains along with discussing in-depth practice methods for analyzing large data across multiple domains. Cui et al. (2018) studied the latest machine learning techniques for IoT traffic routing, security, edge-enabled computing infrastructure, device identification, and big data analysis used for managing networks. However, this survey is limited to applying large data analysis and ML technologies to a wide range of applications within a specific domain. Similarly, technologies based on big data in various fields, such as smart transportation, smart health and logistics, and smart agriculture, are discussed in Gem et al. (2018). The study allows readers to select the most relevant technology from the wide range of techniques available for data analysis across different domains. In addition, it partially discusses technologies from each domain. Some surveys target only one domain. For example, Mahadevinjad et al. (2018) demonstrated a classification of machine learning-based technologies for the smart cities. However, the security of data is not considered. It is clear from the literature that there is a shortage of big data analytics and research work specifically on machine learning in the Internet of Things healthcare sector. Some surveys mentioned above devote only one section to this topic, however, due to the lack of a comprehensive survey on these big technologies and ML techniques identifying the most suitable one for their application in IoT health. Furthermore, surveys on Big Data Analytics and Healthcare are still a matter of research. Similarly, there is no study that determines the importance of data aggregation and its role in this particular domain.

To understand these access gaps, we reviewed various papers, considering the challenging aspects of big data in IoT Healthcare. We also highlight the strengths and weaknesses of the future. It allows readers to choose the most appropriate technology from the big data analytics tools available for healthcare and explore more in the coming period.

It generally discusses the relationship between big data and IoT and Big Data research in IoT Healthcare. It will provide all interested parties with the most important features for further exploration in the years to come and explore the basic concepts of big data.

The remaining of the paper is organized as follows. In Section 3.3, an introduction to IoT and a specialization in the recent development and transformation of the traditional healthcare sector are discussed. Section 3.4 discusses the concept of big data challenges. Next, we provide a detailed discussion on the role of ML technologies for big data analysis in Section 3.5. Comprehensive and updated literature on various machine learning techniques for big data analysis in IoT eHealth is provided in the review Section 3.6. Research challenges in the field are presented in Section 3.7. Finally, the paper concludes with Section 3.8 outlining the limitations for further exploration and future work.

3.2 INTERNET OF THINGS (IoT)

IoT is a self-configurable content that communicates using the network. It is a cyber-physical system. The "IoT" by IEEE is defined as "each network of objects is integrated with sensors connected to the Internet" (Alfuqaha et al., 2015). Communication between objects is provided by the use of cheap sensors installed across objects that support universal and extensive applications (Lee et al., 2015). Technologies that have inspired the development of IoT are microelectro-mechanical systems (MEMS), wireless technologies, and the Internet. About 25 billion sensor-enabled devices will be installed by 2020 (Dohler et al., 2017). Furthermore, the size of the market of such devices is estimated to be approximately 210 billion (Agarwal et al., 2019) by 2025. This indicates that billions of devices or sensor-integrated objects are attached to the Internet. Many types of objects produce vast, different, and complex data in real-time. It is therefore imperative to extract useful samples in an efficient manner from these raw data and analyze the raw data collected from the physical environment and dig for novel, useful information. It is especially important with the development of intelligent IoT devices that can communicate and share information by making wise decisions, resulting in Big Data Analytics as an emerging new area of research (Ghosa et al., 2014). IoT has the potential to rebuild the landscape of the healthcare sector with promising economic, technological, and social opportunities. The health sector alone accounts for about 41%, followed by industrial automation with a huge share of 33%, and the energy IoT market for 7% (Shah et al., 2018).

3.3 IoT IN HEALTHCARE SYSTEM

Emerging technology related to eHealth and mHealth showed the role of growing technologies in the healthcare domain. Billions of sensors have been added that constantly monitor health using physical, environmental, and behavioral parameters in patients. Smart Healthcare is a key technology for monitoring the Wireless Body Sensor Network (WBSN) in IoT, i.e., eHealth and mHealth, data analysis, knowledge addition, data storage, management, and analysis. WBSN includes sensors placed on humans (Alkhyat et al., 2019). The WBSN consists of the communication layer, the sensing layer, the storage layer, the mining layer, the learning layer, and the processing layer (Nguyen et al., 2017). Each layer consists of different parts. The sensing layer contains multiple sensing devices such as in-body sensors and wearable sensors. Medical super sensors (MSS) are introduced, which have extra space with better communication capabilities. These are wearable or embedded inside the skin and can communicate remotely. They collect important data like blood pressure, temperature, respiratory rate, ECG, heart rate, and blood sugar (Abdullah et al., 2015). Recently, actuators have been used to modify environmental parameters as required. It is vital to take big data along with other issues into account while developing devices in the sensing layer. Some issues are the price, power consumption, size, processing, memory, extension, power, and organization of the various devices. The communication layer is responsible for objects sharing data on WBSN using specified communication protocols. Communication protocols specified at this level are provided for communication in the WBSN. These protocols also help in the exchange of data with the given infrastructure and are being used for in-communication via WBSNs. Each of these criteria has its own advantages and disadvantages and can be used depending on the requirements of the specific application (Shah et al., 2020). Congestion, delay, power efficiency, network management, security, and privacy are the various issues. Data aggregation and big data analysis methods save the power of the resource-hungry by significant reduction. The next layer is the processing layer, which analyzes the collected data, takes decisions, and invokes alarms. The basic components are as follows: (i) the processing unit; (ii) the h/w platform; and (iii) the OS. The challenge is the low processing h/w capacity. The analyzed data is passed to the storage layer. A huge number of devices may be attached to the body that provide complex information. It is the main function of the storage layer to efficiently store data for further use. To remove the issue, cloud-based systems are present, like Google Cloud, Thingworks, Nimbits, Amazon, OpenIoT, and JNI platforms. These platforms improve storage and data management. At last, the learning layer will provide knowledge extraction and huge data analysis. ML methods have been applied for huge data analysis in the healthcare IoT. Machine learning-based technologies may efficiently handle large datasets, understand from data, and enhance the learning. They mine large amounts of information to gather useful, interesting information. Clustering, time series analysis, association analysis, classification, and outlier analysis are major components.

3.4 BIG DATA ISSUES IN IoT-BASED SMART HEALTHCARE SYSTEM

Despite the hype surrounding eHealth and mHealth's smart applications in the IoT, it is still a challenging issue for big data. Various medical devices integrated with patients generate enormous amounts of diverse data (Jagdeshwari et al., 2018). These vast datasets include highly interrelated and redundant models. Digging into this data is essential in order to provide sustainable, effective, and sustainable healthcare facilities. However, processing and transmitting such information over the network can be challenging. These problems will lead to high power consumption, cause congestion, and shorten the power and lifespan of the network (Elhaymty et al., 2018). Therefore, it is vital to integrate data using data analytics by increasing the network to make an accurate decision. Furthermore, intelligent and efficient processing of data in the network becomes a major concern for all shareholders. The main purpose of Big Data Analytics (Wang et al., 2018) is to remove unnecessary and inaccurate data, identify and capture information, and gain large amounts of raw data. These technologies improve performance and save energy by using innovative power management methods (Vasaskis et al., 2018).

3.5 ML-BASED BIG DATA ANALYTICS FOR IoT

We have observed the application of machine learning for big data analytics, which will be discussed here. Machine learning comes from model recognition and computational learning (Huang et al., 2018). It is a kind of AI that allows the learning of machines and complex decision-making (Khattak et al., 2018). It has been applied in other fields like computer graphics, computer vision, speech recognition, natural language processing (NLP), intelligent control, and computer networks (Kremer et al.). Recently, we have observed its major role in Big Data Analytics and IoT as it has grown exponentially with a wide variety of applications, resulting in highly interconnected generation of information from these IoT devices. Therefore, the management of data in these systems can become very tedious and challenging for the researchers (Chen et al.). It is necessary for data to be managed from these large resources with better speed and scalable systems by developing new large data analysis techniques. Current technologies are not effective due to the low accuracy in results and high consumption of power that are not compatible with this wide range of devices. These technologies need to be improved to suit different devices. Machine learning technology plays a vital role in the IoT smart eHealth systems (Farhani et al., 2017). It strengthens to get in-depth analysis from the huge pool of information available. Furthermore, it supports us to develop effective and intelligent IoT applications. Wearable devices capture human health information. Different Machine Learning methods are applied for further analysis. It is evident from the literature survey that Machine Learning methods are applied to a huge pool of data for analysis in various domains of IoT, such as Smart Agriculture, Smart Traffic, Smart Human Activity Control, Smart Environment, Smart Cities, and Healthcare(Stupar et al.). A huge pool of data has been examined across a wide range of IoT infrastructure and devices. However, it is evident that IoT Healthcare lacks a literature review that specifically examines large data analyses. Although some of the surveys mentioned above assign only one section to this domain, there are no studies examining the importance of Machine Learning techniques for largescale analysis of data in IoT applications in healthcare. So, we present the comprehensive literature analysis of the latest Machine Learning techniques for analysis on IoT smart healthcare systems for big data. In addition, strengths, weaknesses, and challenges are discussed. It provides readers with insights that will encourage them to explore further.

3.6 A CLASSIFICATION OF ML ALGORITHMS FOR BIG DATA ANALYTICS IN IoT SMART HEALTHCARE

IoT improves the quality of life lived by people by automating tasks that humans can perform in manual mode. Supervision and decision-making are now being carried out by machines. Sensors are connected to a health monitoring section used by patients, as in IoT-based assisted living applications. The data collected through these devices is spread and made available to people. This helps in treating patients in a timely manner and improves the response and accuracy (Dantu et al.). The patient's medications are monitored, and the result of a new medication may be assessed if it occurs due to any allergic reaction (Farhani et al.). As a result, time is saved, and financial value is also maintained. In this section, we will only present a few ML algorithms/techniques for big data analysis in IoT smart eHealth. In addition, the major concepts are presented with their strengths and weaknesses, differences, and similarities. Refer to Table 3.1 for summary.

3.6.1 MACHINE LEARNING-BASED RECOMMENDATION SYSTEM

Asthana et al. (2017) presented a recommendation system to design the most viable wearable IoT devices based on an individual's needs. The system collects data related to the health of a patient, e.g., retrieves data stored from past history, population information, and patient-attached sensors. Multiple Machine Learning-based classification algorithms, such as decision trees, SVM, and logistic regression, assess the probability. Yao et al. (2019) have presented a diagnostic system using real-time electrocardiograph (ECG) analysis. Any signs of disease are assessed on the ECG and reported in real-time to the cloud via the IoT network, which is intended to be governed by the National Health Service in the UK. Results show that the accuracy of the scheme can go up to 99.4%. However, accuracy and performance should be assessed using time complexity, energy efficiency, etc.

In the Mossavi et al. research article, an IoT architecture consisting of five interconnected layers is discussed previously. The authors presented Pro-Trip, a recommendation system that performs activities before or during a trip (Subhramanyaswamy et al.). Pro-Trip collects the data from the patients, which is then used for recommendations to provide the necessary results. Pro-trip results are based on food datasets and weather collected in real-time. In the recommendation system, they assess performance taking into account energy efficiency, account latency, and safety. The novel recommendation system based on the obscure oncology datasets, specifically designated for health systems (Ali et al., 2020) helps in overcoming the difficulties encountered while capturing the risk factors in patient data. This helps in overcoming the difficulties encountered when capturing the risk factors in patient data. Therefore, the technique makes the patient observant and then recommends a routine of food intake and medication in isolated amounts. This procedure assesses the patient's risk, assesses the patient's health status by using wearable devices integrated with sensors, and then further prescribes medications and food prescriptions.

3.6.2 MACHINE LEARNING-BASED PREDICTION SYSTEM

Khan et al. have presented the framework of IoT to assess whether the observation of a person was stressful by monitoring their heart rate. Subsequently, the information collected over different time periods is integrated, and the estimate of stress is evaluated by applying various ML methods such as logistic regression and SVM. Results show that the accuracy of the framework can go up to 68%. However, its accuracy may be further increased by using an appropriately selected classification model. According to Borthakur et al., a smart telehealth monitoring system is proposed that uses speech recognition algorithms. Its goal is to detect the Parkinson's disease using the classification K-mean algorithm. The system is independent, and a variety of wearable devices can be used. Edge computing is used, which helps when resources are limited (Verma and Sood, 2019). A cloud-IoT framework has been presented to monitor multiple diseases. It assesses diseases in students, i.e., from high to low. The author proposed a gateway for the construction of a fog-assisted system. Gateway may handle data mining, data storage, and real-time data processing. The presentation of the proposed model demonstrates high-level features like the early warning score (Kumar and Gandhi, 2018). A three-layer structure stores large amounts of sensory data to predict cardiovascular disease. In the proposed structure, the first layer collects data. The second layer deals with storing large amounts of sensitive data in the cloud. The third and final layer is an assessment model for cardiovascular disease. At this stage, the receiver operating characteristic curve (ROC) identifies potential symptoms before the onset of heart disease is performed. Author Gelogo et al. introduced a new ML-based prototype based on simulations. It was concluded that the proposed framework would greatly enhance performance and detect diseases with good accuracy. The presented system consists of three distant but interconnected layers of fog computing: the sensor device layer, the edge computing device layer, and the first layer of the cloud computing layer, i.e., the sensor device layer, responsible for understanding and monitoring the phenomenon of interest. Edge computing device layer positioning is responsible for decision-making as well as system management. Finally, heavy training processes take place in the cloud layer. Results show that the presented system is better than traditional systems in terms of memory usage, response time, and bandwidth usage. However, accuracy is low. According to Kirtana et al., the system could detect a wide range of life-threatening diseases, including cardiovascular disease, hypertension, diabetes mellitus, and various chronic degenerative diseases. Patient data is captured using a zigzag pulse sensor. Captured data is transferred to the server using a specially designed IoT protocol, Message Query Telemetry (MQTT). On the application server, the HRV data is further analyzed and visualized, showing any abnormalities that may lead to timely actions. Similarly, the Neuro-Fuzzy Temporal Intelligent Medical Diagnosis System is proposed. It uses unambiguous rules that can classify and effectively diagnose malignant diseases.

3.6.3 MACHINE LEARNING-BASED DATA AGGREGATION

Alsheik et al. discussed a data compression algorithm called Adaptive Learner Vector Quantization. A unique feature of Adaptive Learner Vector

Quantization is that it works without any prior knowledge of the topology underlying it. Data are collected by wearables to ensure that only related data is passed to the cluster head (CH). It lowered the cost of processing on CH and the cost of communication over the channel. The technology proposed will not have any mechanism for aggregation at the Cluster Head level. Furthermore, the application of this method should be valued with a level of accuracy for difficult applications. Qie et al. proposed a data aggregation framework which is self-organizing for healthcare needs. An algorithm is implemented to classify healthcare data. The scheme would diminish the data transmitted over the channel and lower the high-dimensional space, which would increase the life of the network. Khan et al. discussed large data techniques using highly correlated data. The Hadoop framework was used to get information from data taken by sensors from patients. Once the redundant data is deleted, the data is sent to physicians for decision-making and timely action. It destroys the data transmitted over the network by improving response, network lifetime power conservation, QoS, and accuracy. The framework uses methods to extract useful samples from the raw data collected (Fang et al., 2020). Key features of the proposed framework are data capture, storage, analysis, and data discovery. The proposed framework eliminates interrelated data and transmits only the most sophisticated and useful features. These features allow the framework to make decisions with the help of a decision support system using various ML techniques.

3.6.4 MACHINE LEARNING-BASED LIVING ASSISTANCE

IoT solutions help the elderly people in the form of collaborative, preventive, and personalized care. Pinto et al. have introduced IoT life support for the elderly population. The proposed system monitors and stores important patient information using a cloud-connected wristband. A raised alarm signals patients by notifying health-concerned people to make the right decision in difficult situations. Similarly, Li et al. proposed a framework for monitoring patients' narcotics. Key features track drug intake from patients' histories, including missed doses. An alarm is designed to alert patients as well as medical personnel in the event of medication differences, such as a missing dose or an overdose. In addition, Prajapati et al. proposed a monitoring system for ill patients in the ICU. It will notify and assist all shareholders whenever there is a sudden change in predetermined conditions. Kim et al. proposed a new monitoring system for the movement of patients. Jan et al. clearly detect human presence without the use of motion detectors or cameras. Subsequently, the human presence was detected. The system was first trained and tested using datasets collected over a period of 3 days from 900 users. Simulation results show that the accuracy of the proposed approach can vary from 50 to 99% with different hierarchical classification algorithms. However, to study its behavior, it is necessary to test it in real-world scenarios in different settings. Vipallapalli et al. proposed an affordable healthcare monitoring system for patients. The model relies on lightweight sensor-enabled wearable devices to understand, analyze, and share real-time healthcare data from patients. An Arduino-based wearable device with a body sensor network is used for data collection. In addition, it is integrated with the system to facilitate remote monitoring of patients going home. The proposed system overcomes many of the shortcomings of the manual system.

3.6.5 MACHINE LEARNING-BASED SECURED ANALYSIS

Ensuring security and confidentiality is essential due to the sensitive nature of healthcare data. In this regard, Khan et al. (2020) demonstrated an online health monitoring system. This system stores and analyzes data related to health from patients using medical equipment and sensors that reject the conditions of death. Various technologies, such as watermarking, diagnostic error counting, and signal enhancement, will enhance safety and performance in the proposed scheme. Khan et al. (2020) proposed a model specifically for IoT-based environments. The goal is to reduce risks for a diverse range of IoT-integrated healthcare solutions. ML methods are used for the safe classification of patient data. Similarly, Gope et al. provided a WBSN-based IoT healthcare solution. The proposed approach monitors the patient using a wireless body network with small, light sensor nodes. Essa et al. proposed a framework that provides the availability and interactivity of healthcare data, which is shared among all interested parties. Various Machine Learning and Deep Learning methods have been used to classify and test the collected data. The confidentiality of healthcare data is discussed.

3.7 RESEARCH ISSUES AND OPEN CHALLENGES

In this section, we provide insights into the various challenges of ML-enabled big data analytics in the IoT smart healthcare sector.

TABLE 3.1 ML-based Techniques for Smart Healthcare.

S. No	Description	Features	Type	Strength	Weakness
1	Based on the needs of an individual.	1. Previous history of the patients 2. Demographic Data 3. Cardiac abnormalities in real-time	ML-Based Recommendation System	Efficient	Lack of numerical analysis
2	Prediction system of disease to perform real-time electrocardiograph	1. Sensing layer 2. Sending layer 3. Processing layer 4. Storage layer	ML-Based Recommendation System	Precise	real-time Monitoring, Portability
3	Layered-based IoT architecture	The mining layer helps the users to support the action regarding the trip.	ML-Based Recommendation System	Precise	complex system
4	Recommender system (Pro-Trip)	It recommends the food intake in portions and provides medical assistance. It evaluates the risk faced by the patient.	ML-Based Recommendation System	Precise	Food datasets evaluate pro-trip results.
5	Fuzzy ontology-aided RS, designed healthcare system	It observes the health of the patient and then recommends a diet with a discrete amount of food and medicines, evaluates the risks faced by the patient, and deduces the health state of the patient.	ML-Based Recommendation System	Precise	No support for sentiment analysis
6	Cardiovascular activities monitoring the remote system	It detects fatal diseases	ML-Based Prediction System	Cost-efficient, secure, adaptable, and usable	Integration with the network, low space
7	Inexpensive cardiac monitoring system	Cardio stress prediction using a support vector machine	ML-Based Prediction System	Precise	Efficiency, Security
8	Speech recognition-based smart telehealth system	Parkinson disease	ML-Based Prediction System	Energy-efficient, Lightweight	Privacy, Effectiveness

TABLE 3.1 *(Continued)*

S. No	Description	Features	Type	Strength	Weakness
9	Heart disease prediction model based on ROC	Cardiovascular disease	ML-Based Prediction System	Scalability, high throughput, and availability	Energy-efficiency, Precise
10	Disease detection using MAPE-K.	Arrhythmia disease	ML-Based Prediction System	Response time, memory utilization, and bandwidth	Precise
11	Medical diagnosis system	Issues related to the nervous system.	ML-Based Prediction System	Efficient	Precise,
12	Adaptive-learner vector quantization for real-time data compression	It does not have prior knowledge of topology.	ML-Based Data Aggregation	Efficient, compression, and signal reconstruction	No support of multidimensional data, Missing data
13	Self-organizing approach to transform high-dimensional space into low-dimensional space.	Quality of the data is enhanced	ML-Based Data Aggregation	Reliability, Communication cost, and Efficiency	Fault tolerant,
14	Hadoop-based framework	Improves response time, QoS, and energy efficiency	ML-Based Data Aggregation	Energy conservation	Flexible
15	Health informatics framework	Data storing, analysis, and data search.	ML-Based Data Aggregation	Precise	Low Optimization
16	Cloud-connected monitoring system	Old patient assistance	ML-Based Assisted living	Life time and performance	Connectivity, Limited functionalities
17	Patient monitoring system	Medicine schedule	ML-Based Assisted living	Energy efficiency	Qos
18	Real-time monitoring system	ICU Healthcare monitoring system	ML-Based Assisted living	Precise, Efficient	Load balancing, Availability

TABLE 3.1 *(Continued)*

S. No	Description	Features	Type	Strength	Weakness
19	Intelligent monitoring system for patients based on their movement.	Motion Sensor	ML-Based Assisted living	Energy-efficient	Error, Noise,
20	ML-based patient monitoring system	Human presence detection w/o cameras and motion sensors	ML-Based Assisted living	Low cost, Feasible	Performance, Interoperability, Precision
21	Arduino real-time monitoring system	Patient health monitoring system	ML-Based Assisted living	Simple, Inexpensive	Single subject, Low accuracy, Real time
22	IoT-enabled real-time monitoring system using watermarking and signal enhancements	Fault tolerance for clinical error detection	ML-Based Secured Analysis	Secure, Precise, and QoS	Implementing and testing on real-world patients
23	IoT-enabled healthcare	Secure healthcare data	ML-Based Secured Analysis	Secured, Lightweight	Fault tolerant,
24	WBSN-based intelligent monitoring system	Security of the healthcare system	ML-Based Secured Analysis	Precise, Secure	Energy- efficiency Performance
25	Mobile Cloud Computing Architecture	Big data analytics for the integration of health data	ML-Based Secured Analysis	Interoperable, Availability	Precise, Performance

Source: Asthana et al., 2017; Yao et al., 2019; Moosavi et al., 2015; Subramaniyaswamy et al., 2019; Ali et al., 2018; Khan et al., 2019; Borthakur et al., 2017; Verma and Sood, 2019; Rahmani et al., 2018; Kumar and Gandhi, 2018; Gelogo et al., 2015; Azimi et al., 2017; Kirtana and Lokeswari, 2017; Ganapathy et al., 2014; Alsheikh et al., 2016; Qiu et al., 2016; Khan et al., 2019; Fang et al., 2016; Pinto et al., 2017; Li et al., 2020; Prajapati et al., 2017; Kim and Chung, 2015; Jan et al., 2019; Vippalapalli and Ananthula, 2016; Khan et al., 2020; Khan et al., 2018; Gope and Hwang, 2015; Essa et al., 2018.

3.7.1 RESOURCE DEARTH

Sensors, gateways, smartphones, RFIDs, and microcontroller actuators have restricted processing and computing power in the IoT (Khan et al., 2018). Furthermore, the data that came from this compression is used by resource-hungry tools with redundant designs. Transferring such data results in higher power consumption, lower Quality of Service, and low output (Zhou et al., 2017). Integrating IoT with the cloud computing model will solve the resource constraint problem to some extent. However, it does add cost and complexity. Furthermore, due to the unique feature of the IoT connection, other challenges related to the management of resources, such as modeling, resource discovery, monitoring, scheduling, provisioning, and evaluation, are still a matter of concern (Ali et al., 2018).

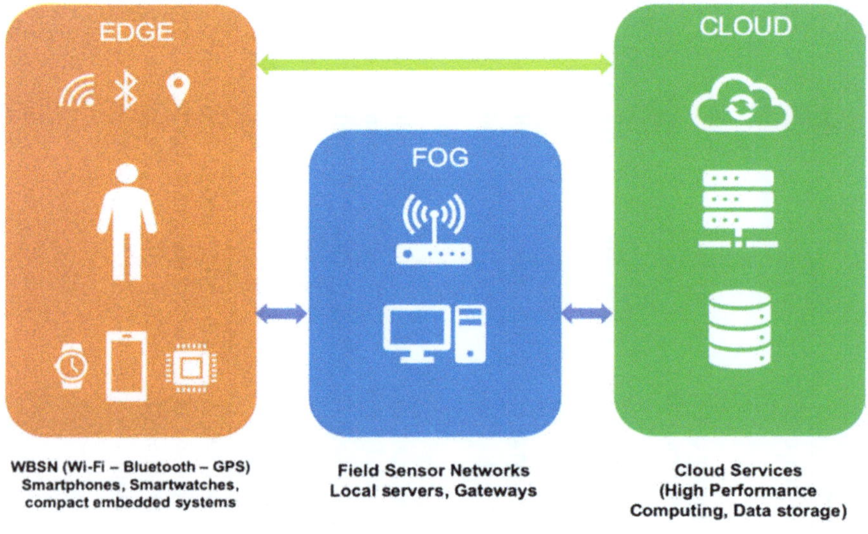

FIGURE 3.2 Possible solutions to internet of things-based healthcare system challenges (kh).

Furthermore, optimization in resource allocation methods is an area that needs to be explored further in this context. It is imperative to develop innovative, lightweight, and energy-efficient data aggregation technologies based on ML as most existing technologies are not energy-efficient. In addition, new schemes must be created to distribute tasks across different IoT components that not only match the resource limits of these networks but also provide acceptable accuracy (Khan et al., 2020).

3.7.2 SECURITY

IoT's implementation in the smart healthcare sector offers in person facilities, i.e., instant access to and customization of healthcare. Applications of both healthcare and technology equipment together provide a wide range of services. With a largescale market of US$ 1368 million by last year, it is estimated that about forty percent of IoT-concerned technology will soon be health-associated (Marr et al., 2018). However, such developments in this area are revolutionary and should be approached with caution because of the security challenges posed by smart healthcare data.

It is estimated that 72% of harmful elements target healthcare data (Jan et al., 2019). Therefore, it is important to protect such data from hackers (Bhattacharya et al., 2020) by providing both physical and virtual security. The remaining challenges include incorrectly configured devices and the setting of the network. Furthermore, the data on these different categories of machines is very diverse in nature and is generally controlled by third parties, and the security of such data can be difficult. The current or existing security technology is not viable. Creating an energy-efficient lightweight data aggregation method is not only secure but also ensures privacy, which is an interesting domain for further research.

3.7.3 INFORMATION INTEGRATION AND SHARING

We have observed an increase in development in both h/w and s/w, but the actual difficulty is the compromise in global standards that are appreciated by people around the world. Therefore, healthcare IoT devices face interoperability issues. Designers should not only focus on developing but also strive for interaction between all aspects of IoT smart health and advanced integrated healthcare to promote a healthy lifestyle. The advantages associated with these devices are increased output, low interruptions, and low maintenance costs.

3.7.4 ENERGY OPTIMIZATION

Wearable objects and sensors are equipped with a limited power supply. Replacing the cell in these sensors or devices frequently can be cumbersome, and sometimes it is impossible. Replacing batteries whenever power exceeds certain limits requires the additional cost of constant care by healthcare professionals about these devices and sensors. It is caused by fatigue and management errors as the environment is dynamic. Energy efficiency is a vital part of determining the performance of the underlying applications. In order to do energy conservation, it is necessary to prepare sensors that carry low power and do not need frequent battery changes while having a reliable power input. Furthermore, energy optimization algorithms with smart energy management techniques have received very little attention and therefore require serious scrutiny from researchers in the IoT healthcare sector.

3.7.5 BIG DATA ANALYTICS

The next challenge for IoT smart healthcare is Big Data Analytics, which works with huge structured data. Significant improvements have been observed in a wide domain of hardware and software for IoT applications. In addition, future IoT growth forecasts are even more extended with the large amount of connected data resources that include global platforms.

A large amount of unnecessary data is transferred over the network for decision-making and analysis. Transferring such huge amounts of data can have a negative impact on network bandwidth. This may cause many difficult issues that need to be addressed very carefully (Yang et al., 2020). In this context, it would be interesting to look at how to make better decision-making using different ML and DL-enabled techniques and get insight into the vast information available for optimized operations (Gill et al., 2019). It is vital to develop new tools for big data analytics and technologies that analyze and collect relevant information. Innovative techniques for noise removal are required to improve the data signal, upgrade the quality of the collected data, and preserve the power of the network (Wan et al., 2019). It will be interesting to see in the future the novel ML technologies that implement real-time analysis by monitoring current situations and responding accordingly as healthcare devices monitor real-time devices. New aggregation methods should be developed with external constraints for better QoS, security, and lower computational complexity. In this case, clustering is more efficient on stable connections, where the network configuration remains the same for a long time. However, they should be studied in diverse environments. Further research should be done to find the right location for these devices so that IoT can receive the wide variety of healthcare applications that will be developed in the coming years.

3.8 CHALLENGES AND FUTURE WORK

We have provided a detailed study of Big Data Analytics in the IoT Healthcare sector. We have done a comprehensive study of the literature and picked up the most relevant and current surveys to fill the research gaps. Additionally, we have provided literature on Machine Learning-based technologies for Big Data Analytics-enabled Internet of Things Smart Healthcare. An overview of their findings and challenges is also gathered. It gave readers a perception of this area and enabled them to begin their study by choosing their topic from the pool of technologies. Various current research challenges are the prime focus of this study, which motivate enthusiastic researchers to dig further. Furthermore, the various issues raised for IoT emerging structures, namely, Web of Things (WoT) and Nano-Things (IoNT), are fully integrated to make the whole IoT scenario a reality.

KEYWORDS

- **smart healthcare**
- **big data**
- **machine learning**
- **IoT**

REFERENCES

Abdollahzadeh, S.; Navimipour, N. J. Deployment Strategies in the Wireless Sensor Network: A Comprehensive Review. *Comput. Commun.* **2016,** *91,* 1–16.

Abdullah, A.; Ismael, A.; Rashid, A.; Abou-Elnour, A.; Tarique, M. Real Time Wireless Health Monitoring Application Using Mobile Devices. *Int. J. Comput. Networks Commun.* **2015,** *7* (3), 13–30.

Aggarwal, M.; Saxena, N.; Roy, A. Towards Connected Living: 5g Enabled Internet of Things (IoT). *IETE Tech. Rev.* **2019,** *36* (2), 190–202.

Al-Fuqaha, A.; Guizani, M.; Mohammadi, M.; Aledhari, M.; Ayyash, M. Internet of Things: A Survey on Enabling Technologies, Protocols, and Applications. *IEEE Commun. Surv. Tutorials* **2015,** *17* (4), 2347–2376.

Al-Garadi, M. A.; Mohamed, A.; Al-Ali, A.; Du, X.; Ali, I.; Guizani, M. A Survey of Machine and Deep Learning Methods for Internet of Things (IoT) Security. *IEEE Commun. Surv. Tutor.* **2020.**

Ali, F.; Islam, S. M. R.; Kwak, D.; Khan, P.; Ullah, N.; Yoo, S.-J.; Kwak, K. S. Type-2 Fuzzy Ontology–Aided Recommendation Systems for IoT–Based Healthcare. *Comput. Commun.* **2018,** *119,* 138–155.

Ali, S. A.; Ansari, M.; Alam, M. Resource Management Techniques For Cloud-Based IoT Environment. In *Internet of Things (IoT)*; Springer: Cham, 2020; pp 63–87.

Alkhayyat, A.; Thabit, A. A.; Al-Mayali, F. A.; Abbasi, Q. H. WBSN in IoT Health-Based Application: Toward Delay and Energy 3–13 Consumption Minimization. *J. Sens.* **2019**.

Almolhis, N.; Alashjaee, A. M.; Duraibi, S.; Alqahtani, F.; Moussa, A. N. In *The Security Issues in IoT-Cloud: A Review*, 2020 16th IEEE International Colloquium on Signal Processing & Its Applications (CSPA). IEEE, 2020; pp 191–196.

Alsharif, M. H.; Kelechi, A. H.; Yahya, K.; Chaudhry, S. A. Machine Learning Algorithms for Smart Data Analysis in Internet of Things Environment: Taxonomies and Research Trends. *Symmetry* **2020,** *12* (1), 88.

Alsheikh, M. A.; Lin, S.; Niyato, D.; Tan, H.-P. Rate-Distortion Balanced Data Compression for Wireless Sensor Networks. *IEEE Sens. J.* **2016,** *16* (12), 5072–5083.

Aman, W.; Khan, F. In *Ontology-based Dynamic and Context Aware Security Assessment Automation for Critical Applications*, The IEEE 8th Global Conference on Consumer Electronics (GCCE). IEEE, Japan, 2019; pp 644–647.

Ansari, S.; Aslam, T.; Poncela, J.; Otero, P.; Ansari, A. In *Internet of Things-based Healthcare Applications*, IoT Architectures, Models, and Platforms for Smart City Applications; IGI Global, 2020; pp 1–28.

Asthana, S.; Megahed, A.; Strong, R. In *A Recommendation Sysrem for Proactive Health Monitoring Using IoT and Wearable Technologies*, 2017 IEEE International Conference on AI & Mobile Services (AIMS). *IEEE,* 2017; pp 14–21.

Athey, S. The Impact of Machine Learning on Economics. In *The Economics of Artificial Intelligence: An Agenda*; University of Chicago Press, 2018; pp 507–547.

Ayoubi, S.; Limam, N.; Salahuddin, M. A.; Shahriar, N.; Boutaba, R.; Estrada-Solano, F.; Caicedo, O. M. Machine Learning for Cognitive Network Management. *IEEE Commun. Mag.* **2018,** *56* (1), 158–165.

Azimi, I.; Anzanpour, A.; Rahmani, A. M.; Pahikkala, T.; Levorato, M.; Liljeberg, P.; Dutt, N. Hich: Hierarchical Fog Assisted Computing Architecture for Healthcare IoT. *ACM Transac. Embed. Comput. Syst.* **2017,** *16* (5), 1–20.

Babar, M.; Khan, F.; Iqbal, W.; Yahya, A.; Arif, F.; Tan, Z.; Chuma, J. M. A Secured Data Management Scheme for Smart Societies in Industrial Internet of Things Environment. *IEEE Access.* **2018,** *6,* 43088–43099.

Babu, G. C.; Shantharajah, S. P. Survey on Data Analytics Techniques in Healthcare Using IoT Platform. *Int. J. Reas. Based Intell. Syst.* **2018,** *10* (3–4), 183–196.

Baker, S. B.; Xiang, W.; Atkinson I. Internet of Things for Smart Healthcare: Technologies, Challenges, and Opportunities. *IEEE Access.* **2017,** *5,* 26521–26544.

Balakrishna, S.; Thirumaran, M.; Solanki, V. K. IoT Sensor Data Integration in Healthcare Using Semantics and Machine Learning Approaches. In *A Handbook of Internet of Things in Biomedical and Cyber Physical System*; Springer; Cham, 2020; pp 275–300.

Bansal, S.; Kumar, D. IoT Ecosystem: A Survey on Devices, Gateways, Operating Systems, Middleware and Communication. *Int. J. Wirel. Inform. Netw.* **2020,** 1–25.

Bhattacharjya, A.; Zhong, X.; Wang, J.; Li, X. Present Scenarios of IoT Projects with Security Aspects Focused. In *Digital Twin Technologies and Smart Cities*; Springer: Cham, 2020; pp 95–122.

Borthakur, D.; Dubey, H.; Constant, N.; Mahler, L.; Mankodiya, K. In *Smart Fog: Fog Computing Framework for Unsupervised Clustering Analytics in Wearable Internet of Things*, 2017 IEEE Global Conference On Signal and Information Processing (Globalsip). IEEE, 2017; pp 472–476.

Boubiche, S.; Boubiche, D. E.; Bilami, A.; Toral-Cruz H Big Data Challenges and Data Aggregation Strategies in Wireless Sensor Networks. *IEEE Access* **2018,** *6,* 20558–20571.

Bowya, M.; Karthikeyan, V. A Novel Secure IoT Based Optimizing Sensor Network for Automatic Medicine Composition Prescribe System. In *Inventive Communication and Computational Technologies*; Springer; Singapore, 2020; pp 1109–1118.

Chen, H.; Chiang, R. H. L.; Storey, V. C. Business Intelligence and Analytics: From Big Data to Big Impact. *MIS Quarterly* **2012,** 1165–1188.

Choo, J.; Liu, S. Visual Analytics for Explainable Deep Learning. *IEEE Comput. Graphics Appl.* **2018,** *38* (4), 84–92.

Crane-Droesch, A. Machine Learning Methods for Crop Yield Prediction and Climate Change Impact Assessment in Agriculture. *Environ. Res. Lett.* **2018,** *13* (11), 114003.

Cui, L.; Yang, S.; Chen, F.; Ming, Z.; Lu, N.; Qin, J. A Survey on Application of Machine Learning for Internet of Things. *Int. J. Mach. Learn. Cybernet.* **2018,** *9* (8), 1399–1417.

Dantu, R.; Dissanayake, I.; Nerur, S. Exploratory Analysis of Internet of Things (IoT) in Healthcare: A Topic Modelling & Co-citation Approaches. In *Information Systems Management*; Taylor & Francis, 2020; pp 1–17.

Dehkordi, S. A.; Farajzadeh, K.; Rezazadeh, J.; Farahbakhsh, R.; Sandrasegaran, K.; Dehkordi, M. A. A Survey on Data Aggregation Techniques in IoT Sensor Networks. *Wireless Netw.* **2020,** *26* (2), 1243–1263.

Diène, B.; Rodrigues, J. J. P. C.; Diallo, O.; Ndoye, E. L. H. M.; Korotaev, V. V. Data Management Techniques for Internet of Things. *Mech. Syst. Sig. Process.* **2020,** *138,* 106564.

Elhayatmy, G.; Dey, N.; Ashour, A. S. Internet of Things Based Wireless Body Area Network in Healthcare. In *Internet of Things and Big Data Analytics Toward Next-Generation Intelligence*; Springer: Cham, 2018; pp 3–20.

Enginkaya, E.; AkgüL, A. K. The Consumers'life Simlifiers: Innovative Developments and Transformations. *Bus. Stud.* **2018,** *83.*

Essa, Y. M.; Attiya, G.; El-Sayed, A.; Elmahalawy, A. Data Processing Platforms for Electronic Health Records. *Health Technol.* **2018,** *8* (4), 271–280.

Evtodieva, T. E.; Chernova, D. V.; Ivanova, N. V.; Wirth, J. The Internet of Things: Possibilities of Application in Intelligent Supply Chain Management. In: *Digital Transformation of the Economy: Challenges, Trends and New Opportunities*; Springer: Cham, 2020; pp 395–403.

Fang, R.; Pouyanfar, S.; Yang, Y.; Chen, S.-C.; Iyengar, S. S. Computational Health Informatics in The Big Data Age: A Survey. *ACM Computing Surveys* (*CSUR*) **2016,** *49* (1), 1–36.

Farahani, B.; Firouzi, F.; Chakrabarty, K. Healthcare IoT. In *Intelligent Internet of Things*; Springer: Cham, 2020; pp 515–545.

Firouzi, F.; Farahani, B.; Weinberger, M.; Depace, G.; Aliee, F. S. IoT Fundamentals: Definitions, Architectures, Challenges, and Promises. In *Intelligent Internet of Things*; Springer: Cham, 2020; pp 3–50.

Firouzi, F.; Rahmani, A. M.; Mankodiya, K.; Badaroglu, M.; Merrett, G. V.; Wong, P.; Farahani, B. Internet-of-Things and Big Data for Smarter Healthcare: From Device To Architecture, Applications and Analytics. *Future Gener. Comput. Syst.* **2018,** *78,* 583–586.

Flynn, T.; Grispos, G.; Glisson, W.; Mahoney, W. In *Knock! Knock! Who Is There? Investigating Data Leakage from a Medical Internet of Things Hijacking Attack*, Proceedings of the 53rd Hawaii International Conference on System Sciences, 2020.

Ganapathy, K.; Sethukkarasi, R.; Yogesh, P.; Vijayakumar, P.; Kannan, A. An Intelligent Temporal Pattern Classification System Using Fuzzy Temporal Rules and Particle Swarm Optimization. *Sadhana* **2014,** *39* (2), 283–302.

Ge, M.; Bangui, H.; Buhnova, B. Big Data for Internet of Things: A Survey. *Fut. Gen. Comput. Syst.* **2018,** *87,* 601–614.

Gelogo, Y. E.; Oh, J.-W.; Park, J. W.; Kim, H.-K. In *Internet of Things (IoT) Driven U-Healthcare System Architecture*, 8th International Conference On Bio-Science and Bio-Technology (BSBT); 2015; pp 24–26.

Ghate, V. V.; Vijayakumar, V. Machine Learning for Data Aggregation in WSN: A Survey. *Int. J. Pure Appl. Math.* **2018,** *118* (24), 1–12.

Ghose, A.; Pal, A.; Choudhury, A. D.; Chattopadhyay, T.; Bhowmick, P. K.; Chattopadhyay, D. Internet of Things (IoT) Application Development. *U.S. Patent Application 14/286,068*, Filed November 27, 2014.

Gill, S. S.; Buyya, R. Bio-Inspired Algorithms for Big Data Analytics: A Survey, Taxonomy, and Open Challenges. In *Big Data Analytics For Intelligent Healthcare Management*; Academic Press, 2019; pp 1–17.

Gope, P.; Hwang, T. Bsn-Care: A Secure IoT-Based Modern Healthcare System Using Body Sensor Network. *IEEE Sens. J.* **2015,** *16* (5), 1368–1376.

Hashem, I. A. T.; Nor Badrul Anuar, Y. I.; Mokhtar, S.; Gani, A.; Khan, S. U. The Rise of "Bi Data" on Cloud Computing: Review and Open Research Issues. *Inform. Syst.* **2015,** *47,* 98–115.

Hassan R, Qamar F, Hasan MK, Aman AHM, Ahmed AS. Internet of Things and its Applications: A Comprehensive survey. *Symmetry.* **2020;** *12*(10):1674. doi:10.3390/sym12101674

Huang, X.-L.; Ma, X.; Hu, F. Machine Learning and Intelligent Communications. *Mob. Netw. Appl.* **2018,** *23* (1), 68–70.

Hussain, F.; Hassan, S. A.; Hussain, R.; Hossain, E. Machine Learning For Resource Management in Cellular and IoT Networks: Potentials, Current Solutions, and Open Challenges. *IEEE Commun. Surv. Tutor.* **2020,** *22* (2), 1251–1275.

Ishtiaq, M.; Rehman, A. U.; Khan, F.; Salam, A. In *Performance Investigation of SR-HARQ Transmission Scheme in Realistic Cognitive Radio System*, The IEEE 9th Annual Computing and Communication Workshop and Conference (CCWC). IEEE, 2019; pp 0258–0263.

Jagadeeswari, V.; Subramaniyaswamy, V.; Logesh, R.; Vijayakumar, V. J. H. I. S. A Study on Medical Internet of Things and Big Data in Personalized Healthcare System. *Health Inform. Sci. Syst.* **2018,** *6* (1), 14.

Jan, M. A.; Khan, F.; Alam, M.; Usman, M. A Payload-based Mutual Authentication Scheme for Internet of Things. *Future Gener. Comput. Syst.* **2019,** *92,* 1028–1039.

Jan, M. A.; Zhang, W.; Usman, M.; Tan, Z.; Khan, F.; Luo, E. Smartedge: An End-To-End Encryption Framework for an Edge-Enabled Smart City Application. *J. Netw. Comput. Appl.* **2019,** *137,* 1–10.

Jayaraman, P. P.; Perera, C.; Georgakopoulos, D.; Dustdar, S.; Thakker, D.; Ranjan, R. Analyticsas-A-Service in A Multi-Cloud Environment Through Semantically-Enabled Hierarchical Data Processing. *Software: Practice Exp. 47* (8), 1139–1156, **2017**

Karim, A.; Siddiqa, A.; Safdar, Z.; Razzaq, M.; Gillani, S. A.; Tahir, H.; Kiran, S.; Ahmed, E.; Imran, M. Big Data Management in Participatory Sensing: Issues, Trends and Future Directions. *Future Generation Computer Systems* 107:942–955, **2020**

Kaur, H.; Atif, M.; Chauhan, R. An Internet of Healthcare Things (Ioht)-Based Healthcare Monitoring System. *In: Advances in Intelligent Computing and Communication. Springer, Singapore*, pp 475–482, **2020**

Ker, J.; Wang, L.; Rao, J.; Lim, T. Deep Learning Applications in Medical Image Analysis. *Ieee Access* 6:9375–9389, **2017**

Khan, F. Fairness and Throughput Improvement in Multihop Wireless Ad Hoc Networks. In: The *IEEE 27th Canadian Conference On Electrical and Computer Engineering (CCECE). IEEE*, pp 1–6, **2020**

Khan, F.; Jan, M. A.; Rehman A U, Mastorakis, S.; Alazab, M.; Watters, P. "A Secured and Intelligent Communication Scheme For Iiot-Enabled Pervasive Edge Computing", in *IEEE Transaction On Industrial Informatics*. Early Access, **2020**

Khan, F.; Rehman A U, Usman, M.; Tan, Z.; Puthal, D. Performance of Cognitive Radio Sensor Networks Using Hybrid Au- Tomatic Repeat Request: Stop-And-Wait. Mobile Networks and Applications **2018,** *23* (3), 479–488,

Khan, F.; Rehman, A. U.; Jan, M. A.; Rahman, I. U. Efficient Re- Source Allocation For Real Time Traffic in Cognitive Radio Internet of Things. In: in The *International Conference On Internet of Things (Ithings). IEEE, pp* 1143–1147, **2019**

Khan, F.; Rehman, A. U.; Zheng, J.; Jan, M. A.; Alam, M. Mobile Crowdsensing: A Survey On Privacy-Preservation, Task Manage- Ment, Assignment Models, and Incentives Mechanisms. *Future Generation Computer Systems* 100:456–472, **2019**

Khan, F.; Yahya, A.; Jan, M. A.; Chuma, J.; Tan, Z.; Hussain, K. A Quality of Service-Aware Secured Communication Scheme Forinternet of Things-Based Networks. *MDPI Sensors* **2019,** *19* (19), 4321,

Khan, I. H.; Khan, M. I.; Khan, S. Challenges of IoT Implementation in Smart City Development. In *Smart Cities—Opportunities and Challenges*; Springer: Singapore, 2020; pp 475–486.

Khattak, M. I.; Edwards, R. M.; Shafi, M.; Ahmed, S.; Shaikh, R.; Khan, F. Wet Environmental Conditions Affecting Narrow Band on Body Communication Channel For Wbans. **2018,** *40,* 297–312.

Kim, S.-H.; Chung, K. Emergency Situation Monitoring Service Using Context Motion Tracking of Chronic Disease Patients. *Cluster Comp.* **2015,** *18* (2), 747–759.

Kirtana, R. N.; Lokeswari, Y. V. An IoT Based Remote HRV Monitoring System for Hypertensive Patients. In *2017 International Conference On Computer, Communication and Signal Processing (ICCCSP). IEEE*; 2017; pp 1–6.

Kremer, J.; Stensbo-Smidt, K.; Gieseke, F.; Pedersen, K. S.; Igel, C. Big Universe, Big Data: Machine Learning and Image Analysis for Astronomy. *IEEE Intell. Syst.* **2017,** *32* (2), 16–22.

Kumar, P. M.; Gandhi, U. D. A Novel Three Tier Internet of Things Architecture with Machine Learning Algorithm for Early Detection of Heart Diseases. *Comput. Electr. Eng.* **2018,** *65,* 222–235.

Lashkari, B.; Rezazadeh, J.; Farahbakhsh, R.; Sandrasegaran, K. Crowdsourcing and Sensing For Indoor Localization in IoT: A Review. *IEEE Sensors J.* **2018,** *19* (7), 2408–2434.

Latif, S.; Afzaal, H.; Zafar, N. A. Intelligent Traffic Monitoring and Guidance System for Smart City. In *International Conference On Computing, Mathematics and Engineering Technologies (Icomet). IEEE*; 2018; pp 1–6.

Lee, I.; Lee, K. The Internet of Things (IoT), Applications, Investments, and Challenges for Enterprises. *Bus. Horiz.* **2015,** *58* (4), 431–440.

Lee, S.; Kim, Y.; Kahng, H.; Lee, S.-K.; Chung, S.; Cheong, T.; Shin, K.; Park, J.; Kim, S. B. Intelligent Traffic Control for Autonomous Vehicle Systems Based on Machine Learning. *Expert Syst. Appl.* **2020,** *144,* 113074.

Li, J.; Cai, J.; Khan, F.; Rehman, A. U.; Balasubramaniam, V.; Sun, J.; Venu, P. A Secured Framework for SDN-based Edge Computing in IoT-Enabled Healthcare System. *IEEE Access* **2020,** *8,* 135479–135490.

Liu, J.; Shen, H.; Narman, H. S.; Chung, W.; Lin, Z. A Survey of Mobile Crowd Sensing Techniques: A Critical Component for the Internet of Things. *ACM Transac. Cyber Phys. Syst.* **2018,** *2* (3), 1–26.

Mahdavinejad, M. S.; Rezvan, M.; Barekatain, M.; Adibi, P.; Barnaghi, P.; Sheth, A. P. Machine Learning for Internet of Things Data Analysis: A Survey. *Digit. Commun. Netw.* **2018,** *4* (3), 161–175.

Malakis, S.; Psaros, P.; Kontogiannis, T.; Malaki, C. Classification of Air Traffic Control Scenarios Using Decision Trees: Insights from a Field Study in Terminal Approach Radar Environment. Cognition. *Technol. Work* **2020,** *22* (1), 159–179.

Marr, B. Why the Internet of Medical Things (IOMT) Will Start to Transform Healthcare, **2018**.

Mehta, N.; Pandit, A.; Kulkarni, M. Elements of Healthcare Big Data Analytics. In *Big Data Analytics in Healthcare*; Springer, 2020; p 23.

Mittal, M.; Tanwar, S.; Agarwal, B.; Goyal, L. M.; Eds. *Energy Conservation for IoT Devices: Concepts, Paradigms and Solutions*; Springer; vol. 2062019.

Mohammadi, M.; Al-Fuqaha, A.; Sorour, S.; Guizani, M. Deep Learning for IoT Big Data and Streaming Analytics: A Survey. *IEEE Commun. Surv. Tutor*. **2018,** *20* (4), 2923–2960.

Moosavi, S. R.; Gia, T. N.; Rahmani, A.-M.; Nigussie, E.; Virtanen, S.; Isoaho, J.; Tenhunen, H. SEA: A Secure and Efficient Authentication and Authorization Architecture for IoT-Based Healthcare Using Smart Gateways. *Procedia Comput. Sci.* **2015,** *52,* 452–459.

Naha, R. K.; Garg, S.; Chan, A.; Battula, S. K. Deadline-Based Dynamic Resource Allocation and Provisioning Algorithms in Fog Cloud Environment. *Future Gener. Comput. Syst.* **2020,** *104,* 131–141.

Nassif, A. B.; Shahin, I.; Attili, I.; Azzeh, M.; Shaalan, K. Speech Recognition Using Deep Neural Networks: A Systematic Review. *IEEE Access.* **2019,** *7,* 19143–19165.

Nguyen, H. H.; Mirza, F.; Naeem, M. A.; Nguyen, M. In *A Review on IoT Healthcare Monitoring Applications and a Vision for Transforming Sensor Data Into Real-Time Clinical Feedback*, IEEE 21st International Conference On Computer Supported Cooperative Work in Design (CSCWD). IEEE; 2017; pp 257–262.

Obermeyer, Z.; Emanuel, E. J. Predicting the Future-Big Data, Machine Learning, and Clinical Medicine. *New Eng. J. Med.* **2016,** *375* (13), 1216–1219.

Olaković, A. Č.; Hadžlalić, M. Internet of Things (IoT), a Review of Enabling Technologies, Challenges, and Open Research Issues. *Comput. Netw.* **2018,** *144,* 17–39.

Park, J.; Bhat, G.; Nk, A.; Geyik, C. S.; Ogras, U. Y.; Lee, H. G. Energy Per Operation Optimization For Energy-Harvesting Wearable IoT Devices, Multidisciplinary Digital Publishing Institute. *Sensors (Basel)* **2020,** *20* (3), 764.

Pflanzner, T.; Kertész, A. A Survey of IoT Cloud Providers. *Croatian Society for Information and Communication Technology Electronics*; 2016; pp 730–735.

Piccialli, F.; Jung, J. E. Understanding Customer Experience Diffusion on Social Networking Services by Big Data Analytics. *Mob. Netw. Appl.* **2017,** *22* (4), 605–612.

Pinto, S.; Cabral, J.; Gomes, T. In *We-Care: An IoT-Based Health Care System For Elderly People*, 2017 IEEE International Conference On Industrial Technology (ICIT). IEEE, 2017; pp 1378–1383.

Pourghebleh, B.; Navimipour, N. J. Data Aggregation Mechanisms in The Internet of Things: A Systematic Review of The Literature and Recommendations for Future Research. *J. Netw. Comput. Appl.* **2017,** *97*, 23–34.

Pouryazdan, M.; Fiandrino, C.; Kantarci, B.; Soyata, T.; Kliazovich, D.; Bouvry P. Intelligent Gaming for Mobile Crowd-Sensing Participants to Acquire Trustworthy Big Data in the Internet of Things. *IEEE Access.* **2017,** *5,* 22209–22223.

Prajapati, B.; Parikh, S.; Patel, J. In *An Intelligent Real Time IoT Based System (IRTBS) for Monitoring ICU Patient*, International Conference On Information and Communication Technology For Intelligent Systems; Springer: Cham, 2017; pp 390–396.

Qadri, Y. A.; Nauman, A.; Zikria, Y. B.; Vasilakos, A. V.; Kim, S. W. The Future of Healthcare Internet of Things: A Survey of Emerging Technologies. *IEEE Commun. Surv. Tutor. 22* (2), 1121–1167.

Qian, B.; Jie, S.; Wen, Z.; Jha, D. N.; Li, Y.; Guan, Y.; Puthal, D.; et al Orchestrating the Development Lifecycle of Machine Learning-based IoT Applications: A Taxonomy and Survey. *ACM Comput. Surv.* **2020,** *53* (4), 1–47.

Qin, J. S. Process Data Analytics in the Era of Big Data. *Alche. J.* **2014,** *60* (9), 3092–3100.

Qiu, T.; Liu, X.; Lin, F.; Yu, Z.; Zheng, K. An Efficient Tree-based Self-Organizing Protocol for Internet of Things. *IEEE Access* **2016,** *4,* 3535–3546.

Rahmani, A. M.; Gia, T. N.; Negash, B.; Anzanpour, A.; Azimi, I.; Jiang, M.; Liljeberg, P. Exploiting Smart e-Health Gateways at the Edge of Healthcare Internet-of-Things: A Fogcomputing Approach. *Future Gener. Comput. Syst.* **2018,** *78,* 641–658.

Ramathulasi, T.; Rajasekhara Babu, M. Comprehensive Survey of IoT Communication Technologies. In *Emerging Research in Data Engineering Systems and Computer Communications*; Springer: Singapore, 2020; pp 303–311.

Ray, P. P. A Survey of IoT Cloud Platforms. *Future Comput. Inform. J.* **2016,** *1* (1–2), 35–46.

Rodríguez-Mazahua, L.; Rodríguez-Enríquez C-A José Luis Sánchez-Cervantes, Jair Cervantes, Jorge Luis García Alcaraz, and Giner Alor-Hernández. A General Perspective of Big Data: Applications, Tools, Challenges and Trends. *J. Supercomput.* **2016,** *72* (8), 3073–3113,

Hassan, R.; Qamar, F. Internet of Things and Application: a Comprehensive Survey. *Symmetry,* **2020.**

Selvaraj, S.; Sundaravaradhan, S. Challenges and Opportunities in IoT Healthcare Systems: A Systematic Review. *SN Appl. Sci>* **2020,** *2* (1), 139.

Shah, J. L.; Bhat, H. F. Cloudiot for Smart Healthcare: Architecture, Issues, and Challenges. In *Internet of Things Use Cases For The Healthcare Industry*; Springer: Cham, 2020; pp 87–126.

Shah, S.; Ververi, A. In *Evaluation of Internet of Things (IoT) and Its Impacts on Global Supply Chains*, 2018 IEEE International Conference on Technology Management, Operations and Decisions (ICTMOD). IEEE, 2018; pp 160–165.

Sharma, D.; Tripathi, R. C. Performance of Internet of Things Based Healthcare Secure Services and Its Importance: Issue and Challenges. *Technical Report*, Easychair, 2020.

Sheikhnejad, Y.; GonçAlves, D.; Oliveira, M.; Martins, N. Can Buildings be More Intelligent than Users?—The Role of Intelligent Supervision Concept Integrated into Building Predictive Control. *Energy Rep.* **2020,** *6,* 409–416.

Shirvanimoghaddam, M.; Dohler, M.; Johnson, S. J. Massive Non-Orthogonal Multiple Access for Cellular IoT: Potentials and Limitations. *IEEE Commun. Mag.* **2017,** *55* (9), 55–61.

Stupar, S.; Ćar, M. B.; Kurtović, E.; Vico, G. In *Theoretical and Practical Aspects of Internet of Things (IoT) Technology*, International Conference "New Technologies, Development An Applications"; Springer: Cham, 2020; pp 422–431.

Subramaniyaswamy, V.; Manogaran, G.; Logesh, R.; Vijayakumar, V.; Chilamkurti, N.; Malathi, D.; Senthilselvan, N. An Ontology-Driven Personalized Food Recommendation in IoT-Based Healthcare System. *J. Supercomput.* **2019,** *75* (6), 3184–3216.

Tsai, C.-W.; Lai, C.-F.; Chao, H.-C.; Vasilakos, A. Big Data An Alytics: A Survey. *J. Big Data* **2015,** *2* (1), 21.

Vassakis, K.; Petrakis, E.; Kopanakis, I. Big Data Analytics: Applications, Prospects and Challenges. In *Mobile Big Data*; Springer: Cham, 2018; pp 3–20.

Verma, P.; Sood, S. K. A Comprehensive Framework for Student Stress Monitoring in Fog-Cloud IoT Environment: M-Health Perspective. *Med. Biol. Eng. Comput.* **2019,** *57* (1), 231–244.

Vippalapalli, V.; Ananthula, S. In *Internet of Things (IoT) Based Smart Health Care System*, 2016 International Conference on Signal Processing, Communication, Power and Embedded System (SCOPES). IEEE, 2016; pp 1229–1233.

Wan, R.; Xiong, N.; Hu, Q.; Wang, H.; Shang, J. Similarity Aware Data Aggregation Using Fuzzy C-Means Approach for Wireless Sensor Networks. EURASIP *J. Wireless Commun. Netw.* **2019,** *2019* (1), 59.

Wang, Y.; Kung, L. A.; Byrd, T. A. Big Data Analytics: Understanding Its Capabilities and Potential Benefits for Healthcare Organizations. *Technol. Forecast. Soc. Change* **2018,** *126,* 3–13.

Williams, P. A. H.; Mccauley, V. In *Always Connected: The Security Challenges of the Healthcare Internet of Things*, 2016 IEEE 3rd World Forum On Internet of Things (WF-IoT). IEEE; 2020; pp 30–35.

Yang, C.; Shen, W.; Wang, X. In *Applications of Internet of Things in Manufacturing*, IEEE 20th International Conference on Computer Supported Cooperative Work in Design (CSCWD). IEEE 2016; pp 670–675.

Yang, K.; Shi, Y.; Zhou, Y.; Yang, Z.; Fu, L.; Chen, W. Federated Machine Learning for Intelligent IoT via Reconfigurable Intelligent Surface. *Arxiv Preprint Arxiv*: 2004.05843, 2019.

Yao, W.; Yahya, A.; Khan, F.; Tan, Z.; Rehman, A. U.; Chuma, J. M.; Jan, M. A.; Babar, M. A Secured and Efficient Communication Scheme for Decentralized Cognitive Radio-based Internet of Vehicles. *IEEE Access* **2019,** *7,* 160889–160900.

Young, T.; Hazarika, D.; Poria, S.; Cambria, E. Recent Trends in Deep Learning Based Natural Language Processing. *IEEE Comput. Intell. Mag.* **2018,** *13* (3), 55–75.

Yuehong, Y. I. N.; Zeng, Y.; Chen, X.; Fan, Y. The Internet of Things in Healthcare: An Overview. *J. Indus. Inform. Integr.* **2016,** *1,* 3–13.

Zhou, J.; Cao, Z.; Dong, X.; Vasilakos, A. V. Security and Privacy for Cloud-Based IoT: Challenges. *IEEE Commun. Mag.* **2017,** *55* (1), 26–33.

CHAPTER 4

Predictive Modeling of Vector-Borne Diseases and Their Pathogens Using Artificial Intelligence

ISHITA RAI, HIMANGI SHARMA, HARSH ARORA, ARIHANT KULSHRESTHA, and ASHUTOSH SINGH

Department of Computer Science, Keshav Mahavidyalaya, University of Delhi, India

ABSTRACT

Health system research is concerned with the health system, and its findings give managers, policymakers, and community members the foundation they need to make decisions based on the best available evidence. The definition of a health system is subject to debate. When used in its narrowest sense, the term "health system" refers to the various levels of healthcare services, including those provided at the federal, state, provincial, local, and regional/zonal/district levels. In a broad sense, the health system encompasses a variety of societal factors that influence social epidemiology, including socioeconomic position, culture, religion, education, politics, and the public and private sectors. In addition, it contributes to the creation of solid, ethically acceptable, and socially relevant recommendations for more efficient, effective, and long-lasting health policies and systems. This information is used to enhance how societies organize themselves to attain health goals. This chapter aims to suggest a perceptive healthcare system that uses machine learning models to identify, monitor, and inform dengue virus, short for DeV, affected patients, and other associates at the current time to manage the DeV infection epidemic. Due to the rise in fatal viral illnesses,

Smart Electronic Devices and Systems for Biomedical and Healthcare Applications. Upendra Kumar, Rajesh K. Kesharwani, & Raj K. Keservani (Ed.)

DOI: 10.1201/9781003560852-4

the requirement for an in-depth investigation into their causes, treatment, and prevention strategies has become essential. These illnesses can sometimes be fatal, and some of them are life-threatening. Two of such diseases causing serious public health concerns in various locations throughout the world are dengue and chikungunya. The pandemic viral virus called dengue is spread to people by female Aedes albopictus or Aedes aegypti mosquitoes. It offers a serious risk to human morbidity and death and gradually worsens the health of affected people. On the other hand, a viral disease that is also mosquito-borne is termed Chikungunya. The virus that causes Chikungunya is spread by mosquitoes, and it was first identified in 1952 during an epidemic in southern Tanzania. Apart from the RNA virus being spread by mosquitoes, it is also a virus that is a member of the Togaviridae family's alphavirus genus. This paper presents numerous machine learning regression algorithms that analyze a dataset and predict the population affected by dengue. In order to do this, we used and compared various machine learning regression techniques, and we measured performance using the Mean Absolute Error (MAE). Consequently, they attained the best ratios.

4.1 INTRODUCTION

Research is the methodical gathering, examination, and interpretation of data to provide an answer to a question or resolve a problem. It is incorporated into the theoretical framework and epistemological positions. It demands a precise description of the issue, as well as specific goals and a strategy. To address the original research objectives, researchers gather data and conduct methodical analysis. Most researchers use qualitative or quantitative approaches, and occasionally they combine inductive and deductive reasoning. Nowadays, researchers in the social sciences and healthcare systems prefer hybridized methods to simple ones. A subset of health research is health system research. It applies its findings for the improvement of the healthcare system and closely links the available scientific evidence. It was first referred to as "health service research," but by the 1980s, the term had changed to "health system research." The research on health services concentrated on issues that occurred in public healthcare facilities. Health system research examines the health system as a whole while taking into account that health is greatly dependent on factors outside of the realm of healthcare.

For the purpose of guaranteeing quality and containing costs, the majority of healthcare research has been done in developed nations with well-established health facilities. However, there is a pressing need for health system research in less developed nations to improve the complete health system, including healthcare issues. Health System Research is defined as "the multidisciplinary field of scientific investigation that studies how social factors, financing systems, organizational structures and processes, health technologies, and personal behaviors affect access to healthcare, the quality and cost of healthcare, and ultimately our health and well-being." Its research areas include populations, families, organizations, and institutions. Health system research is the creation of knowledge to enhance how societies structure themselves to attain health goals, including how they plan, manage, and enhance efforts to improve health, as well as the roles, viewpoints, and interests of many actors in this endeavor.

The inflating viruses and diseases that threaten human life have become a matter of serious concern these days and have hence become another crucial topic of 'healthcare' research. To control this widespread phenomenon, research is an important step, and thereby, research in biomedical science becomes important. Two of such life-threatening mosquito-borne diseases are as follows:

4.1.1 DENGUE VIRUS

Among distinct vector-borne diseases, one such disease in the world is dengue fever. It is brought on by the bites of contaminated mosquitoes leading to the dengue virus. The main vectors of DeV transfer into humans are Aedes albopictus mosquitoes or female Aedes aegypti (McGregor and Connelly, 2021). Now the question arises as follows: how are mosquitoes affected by viruses? It happens when they accidentally bite a person already affected by that disease, i.e., having a virus in their blood. Now, whenever the affected mosquito bites any healthy individual, it spreads the respective disease to his body too. Do you know? Female Aedes aegypti mosquitoes can carry the DeV for 4–10 days after an infected individual has symptoms. In the event that the necessary care is not given, the DeV incidence may result in human mortality and morbidity. Not only the tropical but also the subtropical regions are the worst affected by the DeV infection (Omodior et al., 2018), particularly the progressing and undeveloped nations where subpar housing, insufficient garbage disposal, ineffective vector control, fatigued water management, and failing disease prevention efforts become its main carriers. DeV infection symptoms include a headache, body rashes, high fever, eye pain, bleeding from the gums or nostrils, and anorexia (Tzadok et al., 2015).

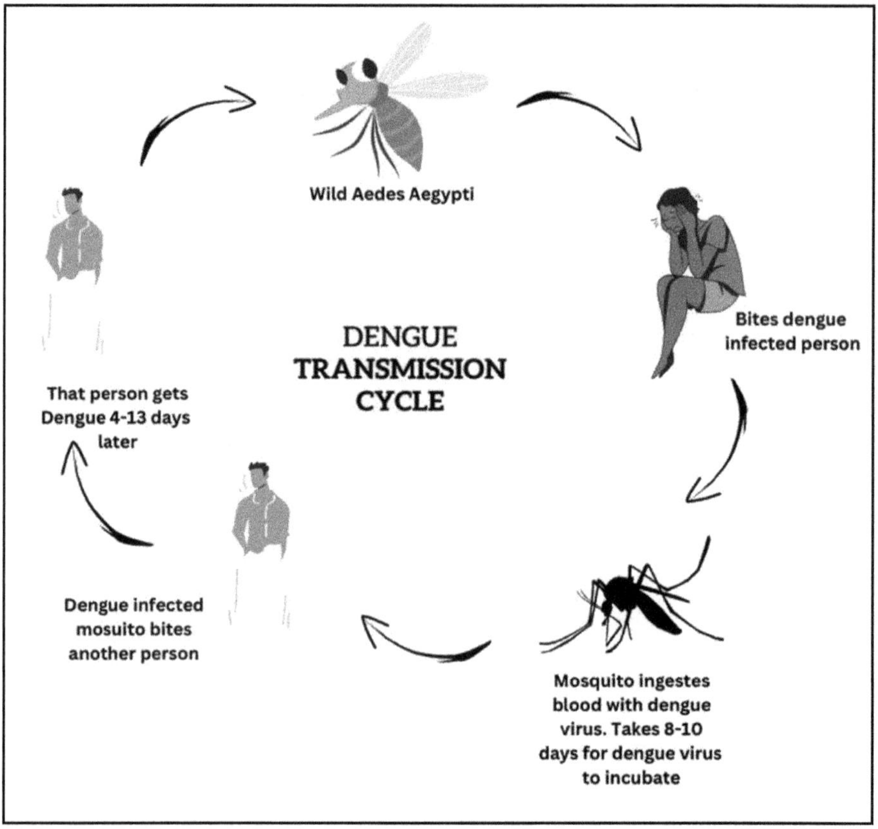

FIGURE 4.1 Dengue transmission cycle.

The respective infection can occasionally result in numerous serious illnesses such as dengue shock syndrome, hemorrhagic fever, and a number of long-term health consequences. As per the World Health Organization's data, around 40% of the world's population, which is approximately 2.5 billion people, resides in locations where the threat of DeV transmission is high (Guzman et al., 2009). According to the report, the prevalence of DeV infection has multiplied thirty-fold over the past fifty years. According to an estimate, more than 50 million DeV infections occur annually throughout the world, and 2.5% of those hospitalized for severe dengue cases die. The sad reality, as stated by one of the prestigious reports, is that out of the 390 million dengue-infected patients, only 96 million are recorded and given appropriate treatment. The research also claims that there is a significant danger of DeV infection for the 3.9 billion individuals living in 128 nations

(Waggoner, 2016). Therefore, it is becoming increasingly important for healthcare organizations, particularly in poor and undeveloped nations, to diagnose and control it in real-time.

4.1.2 CHIKUNGUNYA VIRUS

An alphavirus that is from the Togaviridae family and spreads by mosquito bites is termed the Chikungunya virus, short for CHIKV. The Chikungunya sickness is spread in this way only. A dengue-like sickness that had been spreading in Tanzania during 1952 and 1953 led to the virus's initial isolation from a patient there. Earlier years have seen frequent accounts of similar sickness in Mozambique and other distinct parts of Africa. The phrase "bent over in pain", which describes the forward-bent body of various Chikungunya patients due to arthralgias as well as terrible joint pains, both hallmark symptoms of the disease, is taken from the local dialect of the Makonde region of Tanzania, where the virus originated (Paolucci et al., 2019). Between 3 and 28% of CHIKV infections during outbreaks have been observed to be asymptomatic. Patients often experience severe joint pain along with a prolonged fever as well as febrile exanthema, which are signs of infection with the dengue virus and the Zika virus. This makes identification of CHIKV challenging, especially given the numerous reports of CHIKV, ZIKV, and DENV cocirculating in various places with other infected patients (Li et al., 2021).

A positive-strand RNA virus with an envelope, CHIKV, typically has a genome of less than 12,000 nucleotides. The genome is made by the combination of structural and nonstructural proteins. The structural proteins are five in number, which are termed E1, E2, E3, C (capsid protein), and 6K, while the nonstructural proteins are four in number, which are termed nsP1, nsP2, nsP3, and nsP4. As per the virus' evolutionary history, the West African descent of the virus began in Africa and later traveled to Asia, where it changed to become a unique Asian lineage. An Indian Ocean sublineage within the East/Central/South African (ECSA) lineage was formed by the latest strains formed in the Indian Ocean according to the respective virus isolate in 2005 on Reunion Island (Phadungsombat and Imad, 2021). According to reports, the Chikungunya epidemic, which spread in the last quarter of 2013, became worse for Brazil and was brought on by two independent imports of the Asian and ECSA CHIKV lineages.

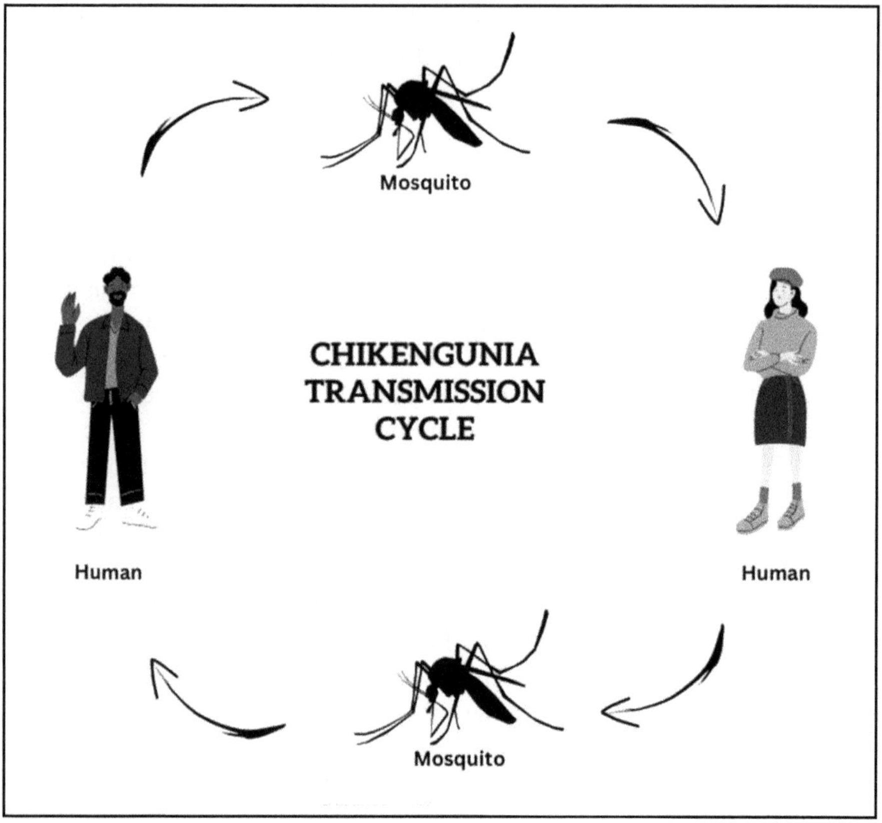

FIGURE 4.2 Chikungunya transmission cycle.

4.2 RELATED WORK

Based on the major streams, this section discusses important associated works. The pre-existing theories pertaining to dengue-based healthcare are covered in the initial subsection, while the work pertaining to Chikungunya healthcare is covered in the later subsection.

4.2.1 DENGUE-BASED HEALTHCARE

In this subsection, we have discussed the research on all the factors, whether they are climatic, socioeconomic, or sociogeographical. The temperature, humidity, etc. are all covered by climatic factors. These works have examined the clinical symptoms and laboratory characteristics of dengue fever as well as the impact of these characteristics on the spread of the following infection. Huang et al.'s identification of clinical signs and laboratory findings for separating individuals with influenza from those with DeV or other hectic diseases was helpful in this approach (Mehdi El-Radhi, 2012). The contribution of climate conditions to the DeV outbreak has been discussed by Sirisena and Noordeen. Both social and climatic conditions were considered by Machado-Machado in determining DeV infection. The study's findings show that factors linked to climate are more important as predictors of DeV infection than factors connected to socioeconomics. Numerous studies have revealed that human movement and sociogeographical elements have an influence in the spread of viruses and their outbreaks, in addition to symptoms and environmental factors. In one such study, Tao et al. (Adams and Kapan, 2009) provided a model that shows how the movement of infected people contributes to DeV infection outbreaks and recommended that restricting travel in DeV-infested areas can reduce the probability of DeV infection spread.

In certain fields, a number of research papers have developed computer-aided DeV systems which not only detect but also prevent based on the aforementioned characteristics and numerous improvements in data analysis-based methodologies. The notable scientists Rao and Kumar developed a computer-based diagnostic approach to diagnose dengue fever, which detected it on the basis of laboratory and clinical symptoms (Yanase and Triantaphyllou, 2019). The system makes use of wrapper-based genetic searches, feature selection methods, and decision tree methods to extract and categorize symptoms in order to determine the prevalence of dengue fever. The eminent personalities Vinarti and Hederman, in order to predict the probability of contracting dengue infection using a Bayesian network, put forward a DeV infection risk anticipation system that analyzes individual personal characteristics such as gender, age, and visiting locations. Apart from these, it also depends on environmental attributes, including specific location features, i.e., Aedes aegypti mosquito dwell, climatic conditions, i.e., favorable mosquito breeding surroundings, and visiting locations (Dey et al., 2022).

In order to identify dengue outbreaks, Zhang et al. presented a DeV infection epidemic surveillance system that examines information from online news sources about dengue. The system used a text mining technique to extract articles about dengue and do statistical studies to calculate the count of dengue cases. The sole goal of this research was to investigate additional sources of information about dengue other than clinical patient examinations in order to identify the dengue outbreak. In this regard, another study by

Missier et al. (2016) showed the abstraction of dengue-related information from Twitter, a social networking site used for online information sharing, for assessing the dengue infection outbreak in specific places. A user-friendly framework for delineating mosquito-borne illnesses and breeding grounds was created by Sousa et al., which supports the government's efforts to prevent and manage mosquito-borne illnesses. However, the platform has also investigated other sources of knowledge through the mining of geolocated content from the internet community in an effort to reduce dependence on user-provided information, which is dependent on user willingness (Hannay and Baatard, 2011).

These motivations look into information sources outside of clinical investigations, which require the remote perception of dengue infection. It has also prompted the incorporation of distributed computing paradigms, including cloud computing and the Internet of Things (IoT) wireless sensor network prototype, in order to create a variety of settings for agile healthcare systems.

4.2.2 CHIKUNGUNYA-BASED HEALTHCARE

Machine learning techniques were previously also used by researchers to categorize the Chikungunya virus. Hossain et al.'s accurate diagnosis of Chikungunya disease made use of the various patient symptoms. Their suggested framework gathered information from patient interviews. When attempting to forecast the severity of Chikungunya, they used a belief-based rule system. Their model achieved a 92% accuracy rate (Mukti et al., 2021). Yang created a system for making Chikungunya disease diagnoses. In order to classify the data, the neural network took into account the ambiguity of the disease's symptoms. However, this method does not take the ambiguity of some symptoms into account. Three different models for the diagnosis of the Chikungunya disease were presented by Ganesan et al. (Johnson et al., 2016). However, in order to assess this disease, these models need human input. A machine learning-based classifier for separating Chikungunya patients from dengue patients and vice versa was proposed by Caicedo-Torres et al. 447 patients underwent testing with their classifier. The model for logistic regression performed better than the other models. 87% accuracy was found through logistic regression.

The backpropagation method for forecasting the epidemic disease was presented by Ibrahim et al. (Abu Alasal and Watzlaf, 2021). For forecasting, they used epidemic disease factors. The clustering technique was then used with these factors. Their approach can use feature classification to find the epidemic disease. Coelho et al. employed a transfer learning model to forecast diseases spread by mosquitoes. They considered Brazil for taking the data and took time-series data of the two cities. The random quantile forest model and the long short-term memory neural network models were considered for approach, and both performed predictions with the same accuracy. Machine learning techniques were used by Caicedo-Torres et al. to project the foreboding of Chikungunya in Colombia (Mascarenhas et al., 2018). The forecasting of the Chikungunya cases was done using kernel ridge regression. We employed mean absolute error and cross-validation.

On the Machala dataset, Sippy et al. created two prediction models. The severity index for suspected arbovirus (SISA) model made use of demographic information. The severity index for suspected arbovirus (SISAL) with a laboratory model made use of the lab data (Sippy et al., 2020). SISA and SISAL yielded accuracy rates of 91% and 95%, respectively. Both models can consider hospitalization due to an arbovirus.

A backpropagation algorithm was used by Shimpi et al. to forecast the Chikungunya disease. The backpropagation algorithm for classification used the preprocessed features. 95% accuracy was achieved using this model. A small dataset, though, was utilized for legitimization. Machine learning methods were used by Eng et al. to forecast the binding empathy of Chikungunya T-cell epitopes (Sánchez-Burgos and Montalvo-Marin, 2021). To distinguish between binders and nonbinders, they created prediction models. The development of vaccines will benefit from using this model.

According to the existing literature, the majority of models are plagued by issues with overfitting, data insensitivity, and hyperparameter tuning. Therefore, it is necessary to create an effective model for classifying the extremities of the Chikungunya virus (Schratz et al., 2019).

4.3 MATERIALS AND METHODS

This section comprises the dataset used in this experiment, how we have visualized the data, the preprocessing methods we applied, and the machine learning techniques used.

4.3.1 DATA EXPLORATION

The dataset is in the CSV file format and is separated by semicolons. We perform some basic data exploration and reach the following conclusions:

- The dataset has 27 columns.
- It contains 17,172 records.
- The respective dataset contained no null values.

Following are the columns in the dataset shown in Table 4.1.

TABLE 4.1 Features for Disease Prediction.

Attribute	Description
CS_GESTANT	Gestational Age of the Patient (Quarter) in Case CS_SEXO=Female
CS_RACA	Patient Race
NU_IDADE_N	Patient age
CS_SEXO	Patient sex
CS_ZONA	Residence area
FEBRE	Fever (Symptom)
MIALGIA	Myalgia (Symptom)
CEFALEIA	Headache (Symptom)
EXANTEMA	Rash (Symptom)
VOMITO	Vomiting (Symptom)
NAUSEA	Nausea (Symptom)
DOR_COSTAS	Back Pain (Symptom)
CONJUNTVIT	Conjunctivitis (Symptom)
ARTRITE	Arthritis (Symptom)
ARTRALGIA	Arthralgia (Symptom)
PETEQUIA_N	Petechiae (Symptom)
LACO	Tourniquet test (Symptom)
DOR_RETRO	Retro-orbital pain (Symptom)
DIABETES	Diabetes (Pre-existing disease)
HEMATOLOG	Hematological diseases (Pre-existing disease)
HEPATOPAT	Liver diseases (Pre-existing disease)
RENAL	Kidney disease (Pre-existing disease)
HIPERTENSA	Hypertension (Pre-existing disease)
ACIDO_PEPT	Peptic acid disease (Pre-existing disease)
AUTO_IMUNE	Autoimmune disease (Pre-existing disease)
DIAS	Number of days the patient is experiencing symptoms
CLASSI_FIN	Final patient classification

4.3.2 PREPROCESSING OF DATASET

After carrying out the appropriate data exploratory measures, the data has now been preprocessed to be ready to be fed into the Machine Learning

Models for proper training (Sarker, 2021). First of all, the names of the columns were changed from the present one to the one given in attributes. csv. Moving onto the first column, namely, "Patient Age", the format of the age was in the following manner.

The attribute's order of magnitude is defined according to the first digit

- 1. Hour
- 2. Day
- 3. Month
- 4. Year

Ex. 3009 -> 9 months | 4021 -> 21 years old

So, all the records were converted to years, and after that, since there were float values too, the continuous values were converted into categorized values. Five categories were devised as follows:

- **Infant:** Patients having age less than one year
- **Children:** Patients having age varying from 1 to 15 years
- **Young Adults:** Patients having age varying from 15 to 30 years
- **Middle-Aged Adults:** Patients having age varying from 30 to 45 years
- **Old Adults:** Patients having age greater than 45 years

Next column is patient's sex, which consists of binary values, originally 1 for Male and 2 for Female, which is later changed to 0 for Female and 1 for Male.

Next is the Gestational Period, in which the values are shown in Table 4.2.

TABLE 4.2 Gestation Period.

Continuous values	Categorized values
1st Quarter	1
2nd Quarter	2
3rd Quarter	3
Gestational age ignored	4
No	5
Not Applicable	6
Ignored	9

This column is too complex for a feature that does not highly impact the label we are trying to predict, so the categories were changed from

Gestational Period to Pregnancy, which is a Binary Field with a value of 0 if the patient is not pregnant and 1 if the patient is pregnant.

Next column is Patient Race, having values as shown in Table 4.3.

TABLE 4.3 Gestation Period According to the Patient's Race.

Continuous values	Categorized values
White	1
Black	2
Yellow	3
Brown	4
Indigenous	5
Ignored	9

However, since the race does not make a difference in the Viral Fever Trend, this column will not be included in the data used for training the models.

Up next: Residence area of the patient. This is a really crucial factor in determining the chances of a person suffering from said diseases. Values vary as shown in Table 4.4.

TABLE 4.4 Gestation Period According to Residence Area.

Continuous values	Categorized values
Urban	1
Rural	2
Peri-Urban	3
Ignored	9

The ignored values were changed from 9 to 0, and the rest of the values were kept untouched.

The next 13 columns are shown in Table 4.5.

All of them mean whether or not the patient was experiencing the said symptom.

- 1 if Yes
- 2 if No

The categories were changed to 0 for No and 1 for Yes for all of the "Symptom" columns.

TABLE 4.5 Gestation Period According to Residence Area.

Symptoms			
1	Fever	8	Conjunctivitis
2	Myalgia	9	Arthritis
3	Headache	10	Arthralgia
4	Rash	11	Petechiae
5	Vomiting	12	Tourniquet test
6	Nausea	13	Retro-orbital pain
7	Back Pain		

The next seven columns focus on whether or not the patient had a history of a certain disease, as shown in Table 4.6.

TABLE 4.6 Pre-existing Diseases.

Pre-existing diseases			
1	Diabetes	5	Hypertension
2	Hematological diseases	6	Peptic acid disease
3	Liver diseases	7	Autoimmune disease
4	Kidney disease		

The same procedure that was carried out in the Symptoms column is carried out in these columns too.

The next column is the Number of Days the patient has been experiencing the symptoms. The values are in integers.

The last and final column is Final Patient Classification, and this column has three categorized values:

- Chikungunya
- Dengue
- Discarded/Inconclusive

Here, the first two categories mean that the patient was diagnosed with that disease, and the last category means that the patient was classified as "inconclusive" or "negative" for both Dengue and Chikungunya. These were changed to 1 for Chikungunya, 2 for Dengue, and 0 for Inconclusive/Discarded (Kaur et al., 2018).

A few other measures taken to make the data suitable for model fitting were as follows:

- Converting all the values' data types from float to integer (int32)
- One-Hot Encoding of the multicategorical features, i.e., Age and Residence Area
- Normalizing (scaling) the continuous feature, i.e., the Number of Days the patient has been experiencing the symptoms

4.3.3 DATA VISUALIZATION

Before moving on to Machine Learning Techniques, we have made use of plots and figures to show the trend of diseases with respect to various factors for a better understanding of our dataset.

Data visualization refers to a pictorial representation of the data, making the data and the results easy to understand. Graphical representation of the data and its features gives us an idea of how the data is affecting the result as well as helps us spot the patterns and relationships that previously we were not aware of and can make different relationships between the data by making slight changes (Sadiku et al., 2016). It is easier to anticipate enormous quantities of perplexing data in the form of graphs or charts than to read numerous papers and files because of the way in which a person processes information. Being a swift and simple technique, it communicates ideas in a global style. It lets you try out the distinct conditions just by making minor changes.

The data visualization packages used here are Matplotlib and Seaborn from Python. With the aid of additional libraries like NumPy and Pandas, Matplotlib is a Python library that is used to plot graphs. It is an effective Python tool for data visualization (Oberoi and Chauhan, 2019).

In Figure 4.1, the graphical representation shows the relationship between different age groups, namely, infants, children, young adults, middle-aged adults, and old adults of ages 0–2 years, 3–16 years, 17–30 years, 31–45 years, and above 45 years, respectively, and the occurrence of diseases in them, which are classified into Chikungunya, dengue, and discarded/inconclusive, meaning those people in whom either of the diseases could not be detected. In Figure 4.1, it can be clearly observed that people in the age group of old adults are the ones in whom the occurrence of both the diseases is high compared to the other age groups. Therefore, it can be substantiated that age plays a vital role in the occurrence of Chikungunya or dengue.

In Figure 4.2, it shows how the gender of a patient affects the occurrence of diseases in them. In this figure, it can be observed that the occurrence of both the diseases, namely, Chikungunya and dengue, is more commonly observed in females than in males. Hence, it shows that females are more vulnerable to the diseases in comparison to males.

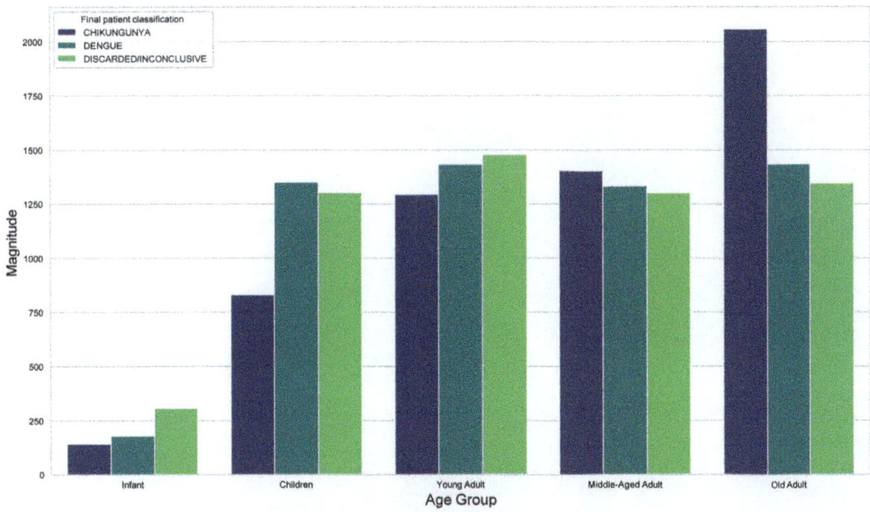

FIGURE 4.3 Age-wise trend of diseases.

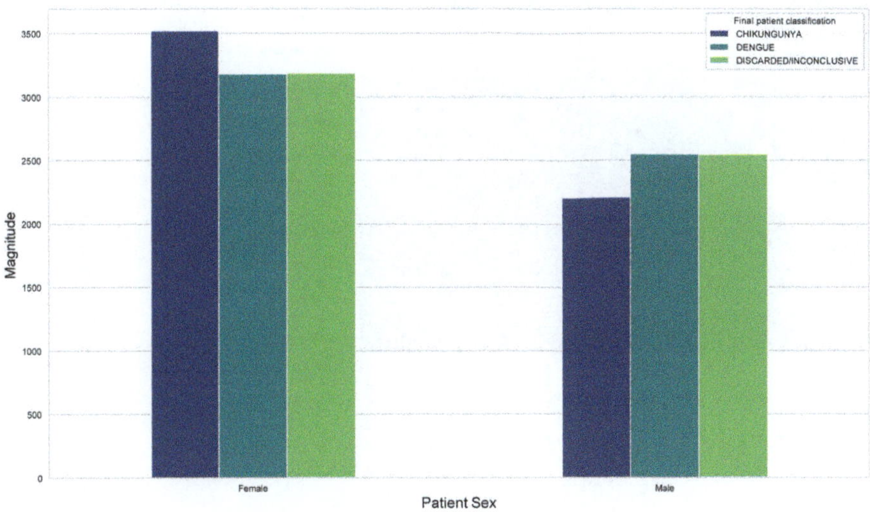

FIGURE 4.4 Gender-wise trend of diseases.

In Figure 4.3, the given plot is known as the heat map, which shows the correlation of the columns of the datasets, such as patient sex, patient race,

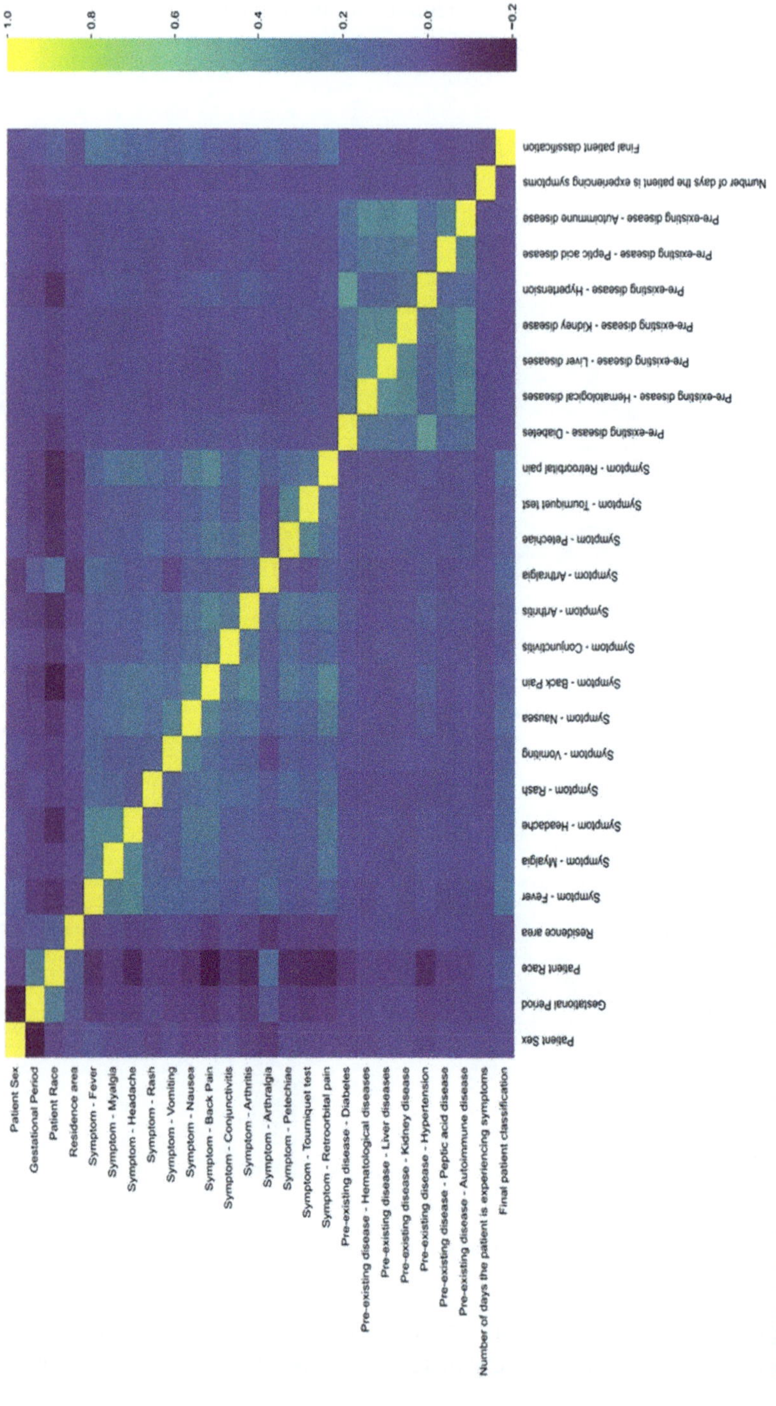

FIGURE 4.5 Heat map of the correlation of columns.

symptoms, etc., with each other. Hence, the yellow part of the heat map is in a straight line because the correlation with itself is always 1, but the color variation shows how one factor depends on the other factor, which helps us in understanding the dependency in a better manner.

4.3.4 MACHINE LEARNING TECHNIQUES

4.3.4.1 K-FOLD CROSS VALIDATION

A resampling technique or machine learning model that is used on a small dataset is known as K-Fold Cross Validation. For determining the creation of groups from a given data sample, a single parameter, k, is used. Due to the use of the k variable, it is also known as k-fold cross validation (Darapureddy et al., 2019). To exemplify, a value chosen for k can be used as its replacement in the model's reference, such as when the value of k is equal to 10, and then it will be referred to as cross-validation by a 10-fold factor.

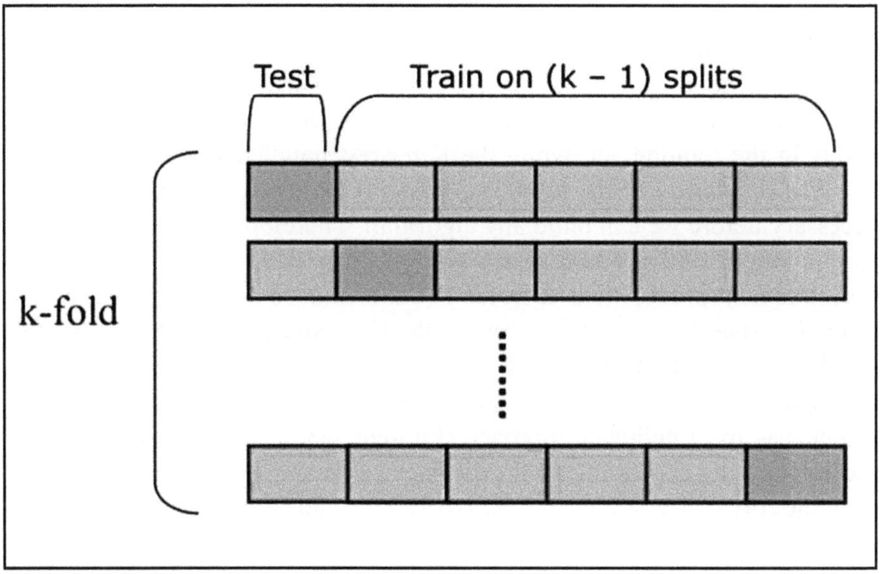

FIGURE 4.6 K-fold cross-validation model.

The motive of K-Fold Cross Validation is to determine the performance of a machine learning model on untrained data. What it basically does is take a small sample of data and apply machine learning models on it to see if it generates appropriate predictions. Not only is it easy to understand but also a less biased technique than other techniques available; it is one of the most prominent techniques used nowadays. The peculiar group allotted to each observation in the data sample remains there throughout the process. Thus, the number of times each sample can be used to train the model is k-1, while the number of times in the hold-out set is 1. A k-fold cross-validation run is formed by adding the mean of all the model skill scores of the dataset. Do not forget to consider a measure of the skill scores' variance, whether it is a standard error or standard deviation (Yadav and Shukla, 2016).

4.3.4.2 KNN (K-NEAREST NEIGHBORS)

The simplest machine learning algorithm based on a supervised machine learning model that deduces the similarity between the new data and available cases and puts the new case into the category that is most similar to the available categories is known as K-nearest neighbors, short for KNN. This algorithm uses entire data for the training of the model and does not have a dedicated training phase; thus, it is also known as a lazy learning algorithm. It can also be referred to as a nonparametric learning algorithm because it makes no hypothesis about the underlying data. The feature similarity states that the new data points are allotted values based on their match with the points in the training set, while the K-nearest neighbors (KNN) approach assumes the values of the new data points (Guo et al., 2004). A dataset is necessary before we can build any algorithm. Therefore, it is crucial to load the training as well as test data in the initial phase of KNN.

In the next phase, the user needs to either choose the closest data points or the K value. Further, choose any of the techniques among the Euclidean, Manhattan, or Hamming and calculate the distance between each row of test data and the training data. The most commonly used technique to calculate distance is the Euclidean approach. Further, we need to arrange them in ascending order depending on the distance value and pick up the top K rows from the sorted array (Taunk et al., 2019). From this point onward, select the most relevant class of these rows and allot it the test point.

4.3.4.3 RFC (RANDOM FOREST CLASSIFIER)

An ensemble method which is composed of numerous little decision trees is known as a Random Forest Classifier. For making the technique reliable, classification and regression are used. Do you know what the little decision trees called RFCs are made of? They are known as estimators which generate a distinct set of predictions and are combined by the respective model to give a more apt prediction (Sarica et al., 2017).

The proneness of overfitting to the dataset was one of the reasons why the traditional decision tree classifiers failed. The Random Forest Classifier overcomes this issue by deriving it well to hidden as well as missing data. Additionally, the technique not only excels at managing enormous datasets with larger dimensionality but also with diverse feature types such as the dataset having a mixture of categorical and numerical columns. There might be some circumstances where the technique fails for regression issues, while it is always proficient for classification difficulties (Ali et al., 2012). Unlike linear regression, a random forest regressor always generates predictions within the bounds of training data.

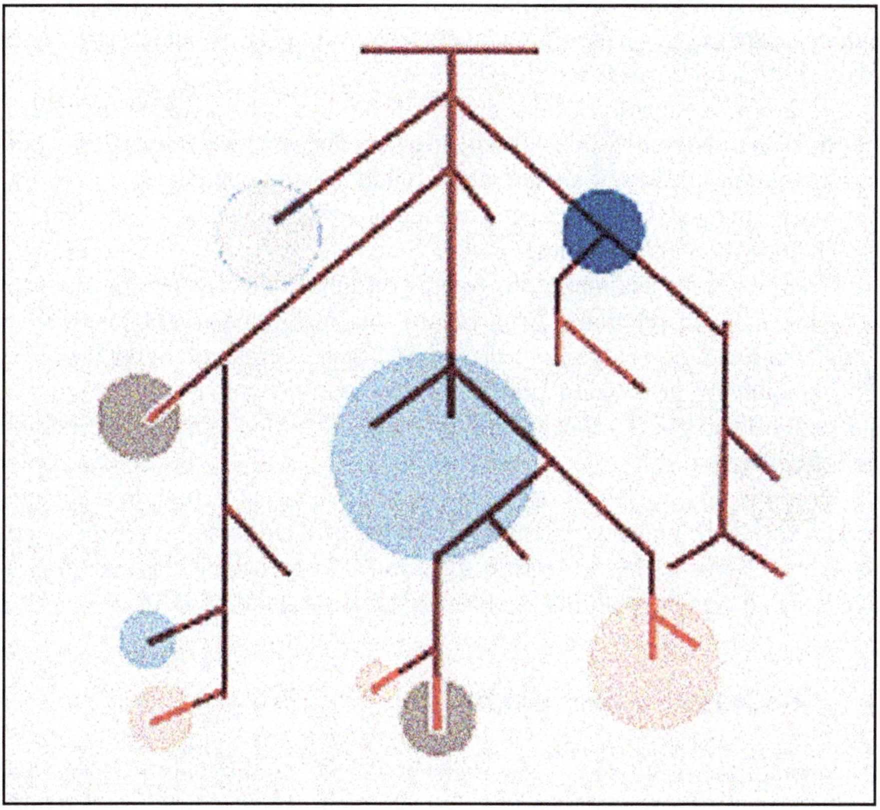

FIGURE 4.7 Random forest model.

However, despite having various advantages, it is still referred to as a black box due to its complexity in understanding the logic behind its choices, producing enormous file sizes, training, and running slowly. Being a simple and sturdy technique, it is the first choice for data scientists to try while developing a new machine learning system. In contrast to other machine learning algorithms, this algorithm has fewer hyperparameters and works extremely well with heterogeneous data types (Biau and Yu, 2012). Nonetheless, the model data scientist finalizes applying it to his dataset, but he still applies this model once as it determines the accuracy of the model in one go.

4.3.4.4 DEEP NETWORK

When the trends in the data are not well-defined, deep networks can be useful. They consist of an input layer, many completely connected, concealed layers, and an output layer. These output layers frequently perform a regression or a classification (Alzubaidi et al., 2021).

Groups of comparable nodes that are completely connected to the nodes in the layers above comprise a hidden layer. These nodes convert an input into an output by way of an activation function and a weight. The network is trained, and the dataset's trends are modeled by adjusting the weights to match known inputs to outputs.

Deep networks use numerous rounds of these hidden layers to produce complex data abstractions. As a result, this technique is challenging to understand and expensive in terms of processing demands (Sarker, 2021). The amount of processing power needed to train and test the network substantially grows as the network's complexity and number of hidden layers both grow.

Deep models, such as those used for computer vision, natural language processing, and image augmentation, can be utilized to solve complicated data processing problems where the ratio of processing time to model accuracy and interpretability is optimized (Vargas et al., 2017).

4.4 RESULTS AND DISCUSSION

Underneath this heading, we have discussed the results that the discussed methods yielded when our data was fed into them. The data was divided into 50 splits and fed iteratively to classifiers.

4.4.1 RANDOM FOREST CLASSIFIER

As discussed, Random Forest Classifiers are excellent at classification problems; they yielded a maximum of 91.66% accuracy rate on our data in one of the K-fold splits. The following is the confusion matrix produced when comparing the prediction of the model with our true test data.

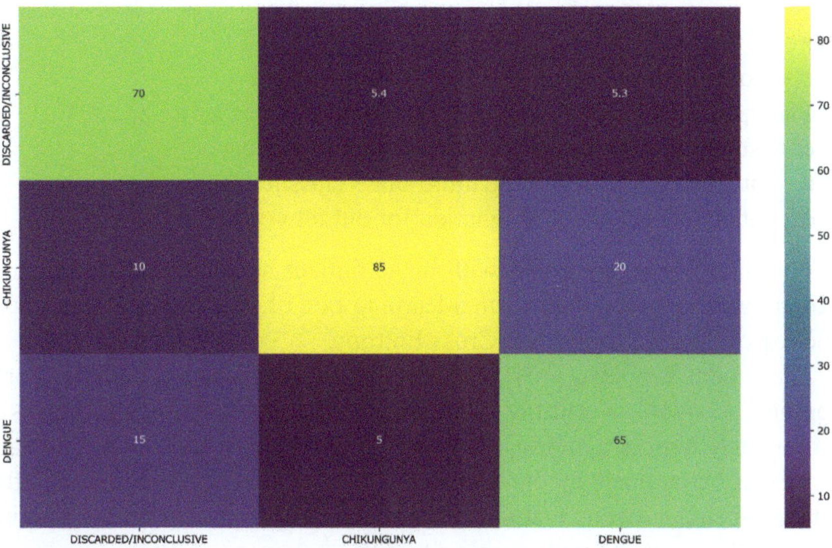

FIGURE 4.8 Confusion matrix for the prediction model.

4.4.2 K-NEAREST NEIGHBORS

We iterated a loop to find the most suitable n_neighbors parameter for the model, but all being done, the model did not yield much promising results on our data. It scored a maximum of 79.1% accuracy.

4.4.3 DEEP NETWORK MODEL

A model was constructed using TensorFlow and the Keras API consisting of seven layers and having a maximum depth of 150 neurons. The activation function of every layer except the last one was set to Rectified Linear Unit Function, short for ReLU, whereas the activation function of the last layer was kept to Softmax Function.

ReLU function

$$f(x) = max\ (0, x),$$

where x is input vector or signal

Softmax function

$$\sigma(z)_i = \frac{e^{z_i}}{\sum_{j=1}^{K} e^{z_j}}\ for\ i = 1,\ldots,K\ and\ z = (z_1,\ldots,z_K) \in \mathbb{R}^K$$

where

σ = Softmax function having z as input argument
z = input vector (consisting of K real numbers, i.e., z_1, z_2, z_3,...z_K)
e^{zi} = standard exponential function for input vector
K = number of classes in the multi-class classifier
e^{zj} = standard exponential function for output vector

The model was compiled with an optimizer set to Adaptive Moment Estimation, short for Adam, with a learning rate of 3e-4 (0.0003) and a loss function of Sparse Categorical Cross Entropy.

The model was also provided with a callback function, namely, Early Stopping, to avoid overfitting of the data. What Early Stopping does is it continuously notices a metric, "val_loss" in our case, which shows the validation loss being made by the data, and if the val_loss starts increasing after certain epochs, it means that our model is starting to overfit on the training data, which is quite an undesirable situation.

The model was fit on our data and then we tried to predict the results on our test data, but unfortunately, the model did not yield excellent results. It scored 74% recall and 74% precision in the classification report. The following is the history of the model while it was being trained.

4.4.4 *FINAL RESULT*

The best accuracy rate among all the Machine Learning models used was given by the Random Forest Classifier, which was 91.66%.

4.5 APPLICATIONS AND FUTURE SCOPE

The invention of machine learning dates back to the 1950s when Alan Turing proposed the first artificially intelligent system that could learn. Since its inception, machine learning has been utilized for a wide range of purposes, including face detection for security purposes, improving safety and reducing risk in public transportation, and more recently, different elements of healthcare and biotechnology. Similar changes are expected to occur in healthcare and medicine. Artificial intelligence and machine learning have significantly changed business processes and daily lives. Recent developments in this field have demonstrated remarkable progress and opportunities to relieve physicians' workloads and enhance precision, predictability, and treatment quality. The current advances in machine learning in healthcare have mainly supported the ability of a doctor or analyst to carry out their duties, recognize healthcare trends, and create disease prediction models.

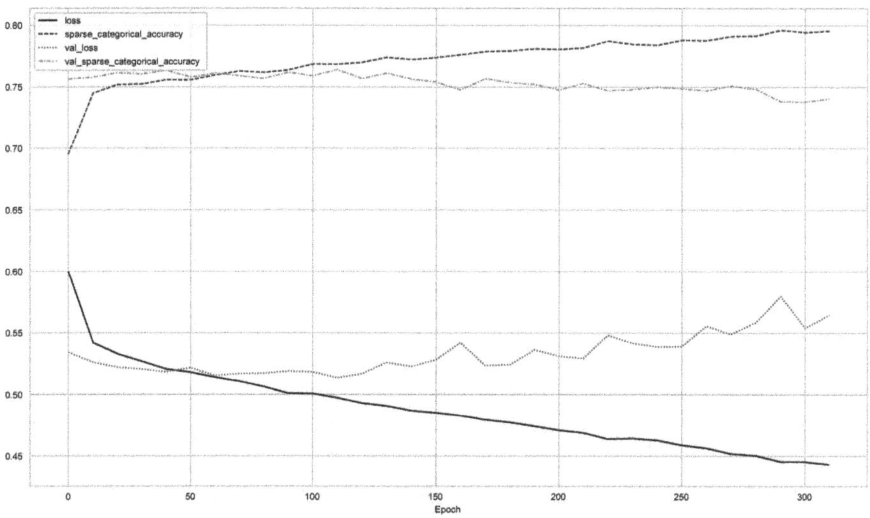

FIGURE 4.9 Accuracy and loss plot for the prediction model.

In the respective book chapter, we have predicted that the person has been affected by Dengue or Chikungunya, on the basis of symptoms like Headache, Rash, Fever, Conjunctivitis, Myalgia, Vomiting, Nausea, Back Pain, Arthritis, etc. Further, in the future, if more of the symptoms are observed in Dengue or Chikungunya-affected people, then we can make the model more precise by adding those columns and then applying the respective machine learning algorithms. The same applies for the pre-existing diseases such as Diabetes, Hematological diseases, Liver diseases, etc. If a person is affected by any of these diseases, he has a greater chance of getting affected by Dengue or Chikungunya.

The conversion of continuous values to categorical values was done in the preprocessing of the datasets on the columns such as age, gender, etc. Here, we categorized age into categories like Old Adults, Middle-Aged Adults, Young Adults, Children, and Infants. We have not categorized the category embryo as the disease has not been observed there till now, but in the future, if it is observed in embryos transferred through mothers, then we have to devise one more category, i.e., embryo. The same applies for the gender too; we have categorized the data as male and female and not transgender due to the unavailability of the transgender data. Thus, if we get the transgender data too, then we can make the model more precise.

We tried and implemented the several machine learning models, such as K-nearest neighbors, K-Fold Cross Validation, Random Forest Classifiers, Deep Networks, etc., but in our case, the best accuracy was given by the Random Forest Classifier, which was 91.66%. Hereafter, with the addition of more data, we can try implementing the various models again for accuracy. Thus, whichever model will give more accuracy with the addition of more data, we can proceed with that model.

4.6 CONCLUSIONS

The disease known as dengue fever is regarded as one that primarily affects tropical and semitropical regions and spreads to numerous nations. Dengue fever has dramatically increased in frequency over the past 50 years. Dengue fever is a serious illness that has an impact on both human life and national economies. The virus known as CHIKV, which spreads through mosquito bites, is the cause of the virus-related illness Chikungunya. When patient blood was isolated, this virus was discovered for the first time in Tanzania. Fever, swelling in joints, pain in joints and muscles, and headaches are among the signs and symptoms of Chikungunya that are frequently experienced.

The doctor will typically make a preliminary diagnosis of these diseases based on an examination of these signs and symptoms. However, the doctor is unable to measure them precisely. The majority of the time, the preliminary diagnosis will therefore be inaccurate, which will result in the wrong course of treatment. In order to predict the people who will contract Dengue or Chikungunya, we have used a bulk of machine learning regression models in this work, and the best accuracy was given by the Random Forest Classifier, which was 91.66%.

KEYWORDS

- **dengue fever**
- **Chikungunya fever**
- **machine learning**
- **deep learning**

REFERENCES

Abu Alasal, S. I.; Watzlaf, J. M. A Simulation Study of Coronavirus as an Epidemic Disease Using Agent-Based Modeling. *Perspect. Health Inf. Manag.* **2021,** *18* (Winter).

Adams, B.; Kapan, D. D. Man Bites Mosquito: Understanding the Contribution of Human Movement to Vector-Borne Disease Dynamics. *PLoS One* **2009,** *4* (8), e6763.

Ali, J.; Khan, R.; Ahmad, N.; Maqsood, I. Random Forests and Decision Trees. *IJCSI Int. J. Comput. Sci. Issues.* **2012,** *9.*

Alzubaidi, L.; Zhang, J.; Humaidi, A. J.; et al. Review of Deep Learning: Concepts, CNN Architectures, Challenges, Applications, Future Directions. *J. Big. Data.* **2021,** *8,* 53.

McGregor, B. L.; Connelly, C. R. A Review of the Control of Aedes aegypti (Diptera: Culicidae) in the Continental United States. *J. Med. Entomol.* **2021,** *58* (1), 10–25.

Dey, S. K.; Rahman, M. M.; Howlader, A.; Siddiqi, U. R.; Mohi Uddin, K. M.; Borhan, R.; Rahman, E. U. Prediction of Dengue Incidents Using Hospitalized Patients, Metrological and Socio-economic Data in Bangladesh: A Machine Learning Approach. *PLoS One* **2022,** *17* (7).

Biau, G.; Yu, B. Analysis of a Random Forests Model. *J. Mach. Learn Res.* **2012,** *13,* 1063–1095.

Guo, G.; Wang, H.; Bell, D.; Bi, Y. KNN Model-Based Approach in Classification 2004.

Guzman, M. G.; Vázquez, S.; Kouri, G. Dengue: Where are We Today? *Malaysian J. Med. Sci.* **2009,** *16* (3), 4–11.

Hannay, P.; Baatard, G. *GeoIntelligence: Data Mining Locational Social Media Content for Profiling and Information Gathering.*

Johnson, B. W.; Russell, B. J.; Goodman, C. H. Laboratory Diagnosis of Chikengunia Virus Infections and Commercial Sources for Diagnostic Assays. *J. Infect. Dis.* **2016,** *214* (Suppl 5), S471.

Phadungsombat, J.; Imad, H.; Rahman, M.; Nakayama, E E.; Kludkleeb, S.; Ponam, T.; Rahim, R.; Hasan, A.; Poltep, K.; Yamanaka, A.; Matsee, W.; Piyaphanee, W.; Phumratanaprapin, W.; Shioda, T. A Novel Sub-Lineage of Chikengunia Virus East/Central/South African Genotype Indian Ocean Lineage Caused Sequential Outbreaks in Bangladesh and Thailand. *Viruses* **2020,** *12* (11), 1319.

Kaur, M.; Singh, K.; Sidhu, S. K.; Devi, P.; Kaur, M.; Soneja, S.; Singh, N. Coinfection of Chikengunia and Dengue Viruses: A Serological Study from North Western Region of Punjab, India. *J. Lab. Physicians.* **2018,** *10* (4), 443–447.

Li, Z.; Wang, J.; Cheng, X.; Hu, H.; Guo, C.; Huang, J.; et al. The Worldwide Seroprevalence of DENV, CHIKV and ZIKV Infection: A Systematic Review and Meta-Analysis. *PLoS Negl. Trop. Dis.* **2021,** *15* (4), e0009337.

Mascarenhas, M.; Garasia, S.; Berthiaume, P.; Corrin, T.; Greig, J.; Ng, V.; Young, I.; Waddell, L. A Scoping Review of Published Literature on Chikengunia virus. *PLoS One* **2018,** *13* (11).

Mehdi El-Radhi, A. S. Fever management: Evidence vs Current Practice. *World J. Clin. Pediatr.* **2012,** *1* (4), 29–33.

Missier, P.; et al. Tracking Dengue Epidemics Using Twitter Content Classification and Topic Modelling. In *Current Trends in Web Engineering. ICWE 2016. Lecture Notes in Computer Science*; Casteleyn, S., Dolog, P., Pautasso, C., Eds.; Springer: Cham, 2016; Vol. 9881.

Mukti, Z.; Nahar, L.; Basnin, N.; Hossain, M. Inference and Learning Methodology of Belief Rule Based Expert System to Assess Chikengunia. 2021. 10.1007/978-3-030-82269-9_1.

Darapureddy, N.; Karatapu, N.; Battula, T. K. Research of Machine Learning Algorithms Using K-Fold Cross Validation. *Int. J. Eng. Adv. Technol.* **2019,** *8* (6S).

Oberoi, A.; Chauhan, R. Visualizing Data Using Matplotlib and Seaborn Libraries in Python for Data Science. *Int. J. Sci. Res. Public.* **2019,** *9,* 8733. 10.29322/IJSRP.9.03.2019.p8733.

Omodior, O.; Luetke, M. C.; Nelson, E. J. Mosquito-borne Infectious Disease, Risk-Perceptions, and Personal Protective Behavior Among U.S. International Travelers. *Prev. Med. Rep.* 12, 336–342.

Paolucci, T.; Attanasi, C.; Cecchini, W.; Marazzi, A.; Capobianco, S. V.; Santilli, V. Chronic Low Back Pain and Postural Rehabilitation Exercise: A Literature Review. *J. Pain Res.* **2019,** *12,* 95–107.

Schratz, P.; Muenchow, J.; Iturritxa, E.; Richter, J.; Brenning, A. Hyperparameter Tuning and Performance Assessment of Statistical and Machine-learning Algorithms Using Spatial Data. *Ecol. Model.* **2019,** *406,* 109–120,

Sippy, R.; Farrell, D. F.; Lichtenstein, D. A.; Nightingale, R.; Harris, M. A.; Toth, J.; Hantztidiamantis, P.; Usher, N.; Aponte, C. C.; Aguilar, J. B.; Puthumana, A.; Lupone, C. D.; Endy, T.; Ryan, S. J.; Ibarra, A. M. S. Severity Index for Suspected Arbovirus (SISA): Machine Learning for Accurate Prediction of Hospitalization in Subjects Suspected of Arboviral Infection. *PLoS Negl. Trop. Dis.* **2020,** 14.

Yadav, S.; Shukla, S. In *Analysis of k-Fold Cross-Validation over Hold-Out Validation on Colossal Datasets for Quality Classification*, 2016 IEEE 6th International Conference on Advanced Computing (IACC), 2016, pp 78–83, doi: 10.1109/IACC.2016.25.

Sadiku, M.; Shadare, A.; Musa, S.; Akujuobi, C.; Perry, R. Data Visualization. *Int. J. Eng. Res. Adv. Technol.* **2016,** *12,* 2454–6135.

Sánchez-Burgos, G. G.; Montalvo-Marin, N. M.; Díaz-Rosado, E. R.; Pérez-Rueda, E. In Silico Identification of Chikengunia Virus B- and T-Cell Epitopes with High Antigenic Potential for Vaccine Development. *Viruses* **2021,** *13* (12).

Sarica, A.; Cerasa, A.; Quattrone, A. Random Forest Algorithm for the Classification of Neuroimaging Data in Alzheimer's Disease: A Systematic Review. *Front. Aging Neurosci.* **2017**.

Sarker, I. H. Deep Learning: A Comprehensive Overview on Techniques, Taxonomy, Applications and Research Directions. *SN Comput. Sci.* **2021,** *2,* 420.

Sarker, I. H. Machine Learning: Algorithms, Real-World Applications and Research Directions. *SN Comput. Sci.* **2021,** *2,* 160.

Taunk, K.; De, S.; Verma, S.; Swetapadma, A. A Brief Review of Nearest Neighbor Algorithm for Learning and Classification, **2019,** 1255–1260. 10.1109/ICCS45141.2019.9065747.

Tzadok, T.; Toledano, R.; Fuchs, L.; Bartal, C.; Novack, V.; Ifergane, G. Headache in the Presentation of Noncephalic Acute Illness. *J. Neurosci. Rural Prac.* **2015,** *6* (4), 494–498.

Vargas, R.; Mosavi, A.; Ruiz, R. Deep Learning: A Review. *Adv. Intell. Syst. Comput.* **2017**.

Waggoner, J. J.; et al. Viremia and Clinical Presentation in Nicaraguan Patients Infected With Zika Virus, Chikengunia Virus, and Dengue Virus. *Clin. Infect. Dis.* **2016,** *63* (12): 1584–1590.

Yanase, J.; Triantaphyllou, E. A Systematic Survey of Computer-Aided Diagnosis in Medicine: Past and Present Developments. *Expert Syst. Appl.* **2019,** *138,* 112821. 10.1016/j.eswa.2019.112821.

CHAPTER 5

Machine Learning with Cloud–IoT in Health Prediction System

PRATIBHA PANDEY and UPENDRA KUMAR

Department of Computer Science and Engineering, Institute of Engineering and Technology, Lucknow, Uttar Pradesh, India

ABSTRACT

The Internet of Things (IoT) is a very powerful tool like machine learning (ML). ML is used to improve decision-making processes in different fields such as health care. This hybrid technology works smarter to improve the decision-making process in various categories such as security, education, the healthcare sector, and business. ML empowers IoT to verify hidden patterns in a collection of data for good forecasting systems and recommendations. Health professionals have utilized IoT and ML to automate machines that may create medical records, predict disease diagnosis, and most importantly, perform patient monitoring in real time. Each ML algorithm acts differently on different datasets. Due to various predictive effects, this can affect the overall results. Variation of predictive outcomes is very close to the treatment decision-making process. Therefore, it is necessary to acknowledge the various ML algorithms that manage IoT data in the healthcare domain. This chapter highlights the ML algorithms known for classification and prediction and shows how they are utilized in the healthcare domain. The aim of this chapter is to focus on an overview of ML methods and their application to cloud–IoT medical data. In a comprehensive analysis, we see that various ML algorithms have different concerns. We need to select the appropriate method for predicting sensitive health data depending on different IoT.

Smart Electronic Devices and Systems for Biomedical and Healthcare Applications. Upendra Kumar, Rajesh K. Kesharwani, & Raj K. Keservani (Ed.)

DOI: 10.1201/9781003560852-5

This chapter gives a glimpse of cloud–IoT and ML to forecast future healthcare system trends.

5.1 INTRODUCTION

Health forecasting systems help hospitals to deliver patients faster as they can go to small clinics. They increase the number of people receiving real medical care. The health forecasting system addresses the common problem of sudden changes in the flow of patients to hospitals. The need for healthcare facilities in many healthcare systems is driven by emergencies such as the arrival of ambulances during natural disasters and car accidents, as well as the general demand for patients (Mtonga et al., 2019). Hospitals that lose real-time data on a patient's flow often have difficulty meeting needs, while nearby facilities may have fewer people. The Internet of Things (IoT) enhances connections between virtual computers and virtual reality to facilitate communication and enables faster data collection using new microprocessor chips.

It is noteworthy that health care is improved and maintained by the diagnosis and prevention of diseases. The anomalies or cracks that occur under the skin can be treated with devices such as positron emission tomography (PET), single-photon emission computed tomography (SPECT), computed tomography (CT), and MRI. Similarly, certain unpleasant situations such as heart attack and epilepsy can be considered (Mosenia et al., 2017). Population growth and the unfavorable distribution of chronic conditions have disrupted modern health facilities. The overall need for medical services, such as doctors, nurses, and beds in hospitals is quite high (Iqbal et al., 2021), resulting in the need to reduce the pressure on healthcare systems while maintaining the quality and standards of health facilities (Wu et al., 2017). The IoT outlines existing ways to reduce the burden on healthcare systems. For example, radio frequency identification (RFID) is used in medical services to reduce medical costs and to improve health care. Significantly, patients' heartbeats are easily monitored by physicians through healthcare monitoring schemes, thus assisting physicians in providing an accurate diagnosis (Birje and Hanji, 2020). To provide consistent wireless data transmission, a variety of wearable devices have been developed. Both information technology (IT) professionals and medical professionals are concerned about data security for additional benefit (Shahbazi and Byun, 2020). As a result, many studies have examined the integration of IoT and machine learning (ML) management of patients with medical problems as a measure to protect data integrity.

IoT has started a new trend in the healthcare industry that enables professionals to communicate with patients more effectively. IoT with ML assesses emergency care needs to develop a strategy to deal with this situation. Many inefficient sections face the issue of overloading in their hospital waiting rooms (Kumar and Gandhi, 2018). Patients visiting hospitals experience a variety of conditions, some of which require immediate medical attention. The situation is exacerbated when patients with emergency services have long waiting queues. The problem is exacerbated in developing countries with few staff hospitals. Many patients often return home without medical attention because of overcrowding.

5.2 LITERATURE SURVEY

Yuvaraj and SriPreethaa (2017) have developed a wearable medical nerve (WMS) platform with a variety of applications and resources. The authors have thoroughly analyzed the use of WMS and their development. The authors present the benefits that come with the use of these resources in observing health conditions such as heart and Alzheimer's disease in patients. A monitoring system based on the wireless cable network (WSN) and the wireless logic network has been proposed by Miotto et al. (2018). Specifically, researchers are using microelectromechanical systems (MEMS) established with WSN to form a body sensor network (BSN) that often monitors abnormal changes in patients' lives. Significantly, the authors developed a system for measuring clinical data using devices such as pulse meters, microcontrollers, and temperature sensors (Shahbazi and Byun, 2021). In addition, the proposed system was integrated with the basic station equipment to remotely control heart rate and temperature of patients and to transfer patient information to a physician. Significantly, the system can send short messaging service (SMS) to both patients' relatives and medical professionals in emergencies (Iqbal et al., 2021). Therefore, patients can get a prescription from a distance from the medical staff who use the program.

In addition, the IoT application has enabled hospitals to monitor critical symptoms of patients with chronic conditions (Shoaib et al., 2017; Gao et al., 2022). The system uses that information to predict the patient's health status. IoT nerves are placed in the patient's body to detect and detect their activity and to predict possible health conditions. For example, the IoT sensor system monitors diabetic patients to predict incidence and any abnormalities in patients. Patients can get suggestions from other hospitals where they can seek treatment with a health forecasting system. Those unable to get an appointment at other hospitals may choose to stay in the area but face the possibility of standing in long lines or returning home without treatment. Rajkumar et al. proposed ZigBee Technology and body sensor network (BSN) healthcare monitoring platforms to monitor patients remotely. In particular, they used standards such as the ZigBee Institute of Electrical and Electronics Engineers (IEEE) protocol 802.15.4, spirometer data, temperature signals, electrocardiogram (ECG), and heart rate to assess patients' health status (Jamil et al., 2020). The information obtained is transmitted through radio waves and shown on visual objects such as mobile devices or desktops. Therefore, the presented scenario can monitor patients' qualifications including heat, glucose, respiration, ECG, electroencephalogram (EEG), blood pressure (BP), EEG, and transmit them to the database via Wi-Fi or GPRS. When sensory information is provided to ZigBee, it is transmitted to a separate network, which allows them to be seen through objects such as emergency equipment and cell phones of doctors and relatives (Shahbazi and Byun, 2021). Similarly, the integration of IoT and machine learning facilitates patient health management by increasing communication between patients and physicians.

IoT offers patient monitoring and monitoring systems through sensory networks built into software and hardware. The latter include such things as BP sensors, the Raspberry Pi board, heart rate sensors, and temperature sensors. The software process includes sensor data recording, cloud storage, and cloud-based data analysis to assess health instability (Jadhav et al., 2019). However, abnormalities often develop when there are unknown functions in unknown body parts. For example, the heart rate often rises when a rush occurs in the brain (Ahmad et al., 2018). For this reason, ML techniques are used to combine heartbeat sensors with Raspberry Pi boards to show unusual results with an LCD or serial monitor. Due to the large amount of data, cloud computing is used to store data and improve data analysis (Panch et al., 2018). Various open cloud platform platforms are compatible with Raspberry Pi boards and Raspbian Jessi (Jamil et al., 2019). These devices use ML capabilities to scan the database to see if something is wrong (Wiens et al., 2019). Thus, the use of machine learning in IoT contributes to erroneous assumptions arising from unknown functions in different parts of the body.

The main purpose of ML is to learn the happenings and patterns. Big data is included in a standard algorithm and analysis is performed using available data (Hung et al., 2017) instead of classical methods of simply generating code. Big data allows IoT and ML systems to easily train the system using simple data for predicting medical malpractice. The accuracy of the predictions is exactly the same as the amount of big data trained (Ngain et al., 2019).

Therefore, big data improves the predictive ability of ML strategies used in health forecasting policy.

Patient crowd prediction ML models share patient load information quickly between hospitals. At the hospital, historical data is captured and used to predict future patient uploads to ensure adequate correction. IoT-embedded ML Models are used to train the differentiation that may result in detection of certain health condition in elderly patients. Interaction ML algorithms can identify patterns of abnormal behavior in patients effectively and send alarms to health care providers. Similarly, a patient's daily activity is monitored by modeling daily routines with IoT microchips. The data are used to identify irregularities among adults.

This chapter focuses to understand and discuss various ML techniques for the prediction and classification of IoT devices and data in the healthcare domain. We analyzed their performance and compared them based on various parameters. This chapter further compares the literature, focus on their features and their shortcomings, and discusses the possible gaps in each method to select the appropriate techniques for building an effective predictive model. In this study, we found that K-nearest neighbor (KNN) could be the most popular algorithm for performing classification and predictive work. However, it may take longer to predict outcomes in real world devices. Hence, some authors say that combining long short-term memory (LSTM) and recurrent neural networks (RNNs) may improve predictive performance. In this study we focus on: how can we use IoT data with machine-based algorithms to develop a better healthcare forecasting system.

5.3 ML ALGORITHMS AND THEIR METHODS OF CLASSIFICATION

ML is an integral part of the field of artificial intelligence (AI) field. ML equips the model with an understanding to analyze the integration of neural inputs (Dike et al., 2018). There are two critical stages to building an effective ML model. The first one is training and the other is the testing phase. The training phase involves the provision of labeled or non-labeled inputs to the system. The program then saves the reading to the feature space for reference in future predictions. Finally, in the test phase, the system is provided with unwritten inputs that must guess the accurate output.

ML utilizes well-known data in its input space to forecast unlabeled data results. Therefore, an effective ML model can refer to its experience and understanding to predict results. The correctness of such a model is based on the accuracy of its output and effective training of the model.

In more recent times, ML models have been broadly used in health service programs. Such ML algorithms are used in many treatment decision support systems to establish high-quality learning models to improve health service programs. Support vector machines (SVMs) and artificial neural networks (ANNs) are examples of ML integration into health service applications. These types are used in various cancer planning programs to get a correct diagnosis of the kind of cancer. These algorithms work by analyzing data obtained from sensory devices and various data repositories. These algorithms analyze patient behavior patterns and clinical conditions. For example, these algorithms point to patient improvement, habits, daily variations, changes in sleep patterns, diet, digestion, and drinking. Patterns identified through these ML algorithms can then be utilized by healthcare applications and treatment decision support programs to promote change in patient lifestyle and practices and to recommend specialized treatment options and patient healthcare programs. This enables clinicians to develop a plan to make sure that patients are delivered with the recommended changes in their lives.

ML technology has three types of models. Each type of ML has many common algorithms. This section presents the most common ML methods used to predict and classify. These routes are neighbors of KNN, decision trees, Naïve Bayes, neural networks, SVM, random forest, and gradient boosted regression tree in learning surveillance. These methods will be presented in Section 5.4. The next subsection will present the concept of data labels and data points in machine learning.

5.3.1 DATA IN ML

Data and databases should be collected from multiple sources for modelling analysis using ML technology. The obtained datasets are stored in a standard central cloud storage. Data points or observations are basic units of data. These data points represent the unit of the system. This unit of the system is being tested to create training datasets.

Recently an increase in access to data points from health facilities was noticed. Such data points can be written or not written. Label data has a separate feature assigned to it, which can also be called output. It is also called the flexibility that depends on him in relation to ancient mathematical texts. The label can be categorized or ordinal, where the separated one does not have a predetermined order of predefined values.

While most ML models can use both data types, labeled and nonlabeled, in some cases, labeled data is used in supervised learning. Conversely, it has no label. Recently, an increase in data points from data are monitored for unregulated learning, while learning can use both health facilities. Such data points can be written or not written. Details with the label are available in labeled data without a label. In the next section, we will try to go further separate feature assigned to them (called a label), which can be reposted for each type of data and their use in all ML types. Labels are given input sets and input or output examples. The input objectives, the input vector, and the required output value are called the administrative signal. These models are used to train ML algorithms to detect segmentation functions after analyzing training datasets. The purpose of algorithm training in supervised learning is to estimate the significance of one or more outcomes with different input features.

Despite the obvious performance of supervised reading, there is the problem that it requires more labeled data to develop a larger labeled database (Osiaswan et al., 2017). Supervised learning models are widely used in classifying and reversing objectives. This paper, however, will discuss just how to differentiate.

Separation or prediction is the main purpose of the use of ML algorithm methods. They classify and predict class labels using a set of examples of examples. Separation samples may belong to a particular category or not at all. They are not part of a class. Missing values adversely affect planning and speculation.

There are two types of partitions: binary and multiclass. In binary separation, the input details are organized into these two class sets. For example, to make yes/no predictions or split emails into two categories—spam or no spam. This division categories are translated as 0 and 1. On the contrary, multiclass classification deals with more than three predictable categories. An example would be identifying the stage of cancer, such as 0, 1, 2, etc.

To solve a problem using the ML method of supervised reading, there are usually certain steps that need to be taken. First, the type of training model should be determined. Next, a set of training needs to be collected, either from human experience or rating. It also needs to represent the real-world use of the work. After that, the representation of the input element should be determined. The introduction should contain enough facts to correctly guess the outcome. This is followed by selecting a learning algorithm. After completing the construction, we implement the learning algorithm in an integrated training set. At this stage, determining certain control parameters in other algorithms in supervised learning is required for the user, mostly for determining predictive problems (Praveena and Jaiganesh, 2017). These parameters are called the verification set, which can be adjusted by enhancing the performance of the training set subset, or by cross-validation (Golub et al., 1979). Troubleshooting is of two types: complete and noncomplete. For complete cross-validation, methods learn and test all methods to divide real data into training sets and validation sets. Leave-p-out cross-validation and leave-one-out cross-validation are fully-fledged authentication methods. On the contrary, cross-validation methods do not apply to all methods of separating the original sample. Those are how permission is granted to secure leave. Examples of such methods are *k*-fold cross-validation, 2-fold cross-validation, and repeat subsampling validation. Finally, the study activity should be assessed the accuracy. After adjustment and parameter learning, the user uses a separate set of tests in the training set for measuring the performance.

One of the tasks of unseen machine learning is to identify hidden structures and properties within the unmarked data. Several successful applications have it within unmarked data. Many successful programs have been used; however, these applications are not easy to measure. It is because of the second step is to configure all cluster centers as the purpose of the data points assigned to them. One major problem with unsupervised learning is to evaluate its effectiveness. The success of unregulated learning tells us whether the algorithm has learned accurately. Labels or one major problem with unsupervised learning is to evaluate its effectiveness. Success for results is not provided for unsupported reading; therefore, the exact release is unknown. Unsupervised learning tells us whether the algorithm has learned useful things or not. Therefore, it is difficult to determine the effectiveness of algorithms. That's why results are not provided for unsupported reading; therefore, the exact release is unknown. Unsupervised learning is used only in the form of assessment, for example, better understanding. Therefore, it is difficult to determine the effectiveness of algorithms. That is why another critical feature of uncontrolled algorithms is the step-by-step unsupervised learning used only in the form of assessment, for example, a better understanding of supervised algorithms. Getting a new data representation form can improve data. Another critical aspect of algorithms that can be controlled is the preparation step for the accuracy of the monitored algorithms. In real-life situations, small-scale data should be used with a large amount of unlabeled data to obtain better reading accuracy. The marking process is time consuming, which can prevent the creation of fully labeled training and incur high costs. Because of this, in some cases, mass-read learning may be the best possible solution.

When limited labeled samples are there, randomized reading is widely used to improve model performance. Currently, there are many samples without labels. These unlabeled samples can be used to maximize model performance. Poor performance of the model is more obvious than improvement and is due to the implementation of unlabeled sample data in unattended reading (Van Engelen and Hoos, 2019).

5.4 COMMON MACHINE LEARNING TECHNIQUES

To differentiate and predict, researchers have developed and adopted learning models of several machines. Some of these are described in the following sections.

5.4.1 K-NEAR NEIGHBOR (KNN)

The basic model of learning equipment known for classification and retrieval functions is KNN. The main focus of this model determines the distance between the existing training and a new unlabeled data point database stored in the feature space. It is important in section predictions (Hussain and Sohaib, 2019). In this process, nearby data points will be according to the k value of the new recognition. The k-value is the hyperparameter of the model that can also be used to filter data points near *k*. The KNN separator votes and assigns a limited category to a sample of new unlabeled data depending on the class label volume in *k*-neighborhoods. In a well-designed KNN algorithm, the neighbor has only a good relationship with the applicant. Such algorithms are widely used in many types of research (Hussain et al., 2016; 2020).

In the said approach, it is important to assess the severity of the influence of the immediate neighbor. A close neighbor can touch more. The distance function is used by the KNN algorithm to determine the severity of the influence of nearby neighbors. Manhattan distance, Euclidean distance, Pearson correlation, and Spearman correlation are standard distance functions used for continuous flexibility. The Hamming distance function may often be used to assess the amount of variability in features such as two-point data features. The KNN separator predicts a new class point data label using Euclidean distance. The *k*-value is six data points, so the generated sample rate class is the most repetitive class of these six data points, meaning that a sample of the new data point may be the green category here.

There are pros and cons to using the KNN algorithm. The nonparametric method is this in each new comparison process for labeled views. This has one of the advantages of using K-NN isolation, meaning that there is no hypothesis of low performance when used in indivisible data sets. Additionally, KNN is the basic distribution of data. For example, model design is developed depends heavily on its hyperparameter *k*. It will generate results with the most accuracy in the database. Another advantage is that it is easy to understand and easy to know the training is set using only one neighbor in stages. It can update its label-set to modification because KNN does not exist as the model is very complex and there are many potentials for computer complications. Therefore, it is important to accurately select the number of neighbors for this purpose. One of the cons is that it requires a lot of time to perform in stages; so, the KNN model is not too complicated or simplified again due to each new comparative process for label viewing. This can be easily done. It is also necessary to install the corresponding features results in lower performance when used in indivisible datasets. In addition, KNN model has various features. When the model uses the KNN algorithm, the middle distance NN relies heavily on its hyperparameter k. It will produce the most accurate results samples with various characteristics can be easily affected. This is because of the size of the training set only one neighbor is separated, but the problem is that and changes in certain factors which cause loss of information from other factors. The model is very complex and has high probability of calculation, for example, there are three elements, ×1, ×2, and ×3 where ×1 difficulty and ×2 are standard variables with values ranging between 0 and 1. Therefore, it is important to accurately select the number of neighbors on purpose. The ×3 is lying at a distance of 1–100 so that the KNN model is not too complicated or simplified Euclidean distance work, will be limited to the distance between and can be easily made. It is also necessary to use the same features as two data points. There are various factors when a model uses the KNN algorithm. On the contrary, the ×3 will be more than the rated range values. The price between samples with various characteristics can be easily affected. This should ×3 is stored within the range 0 and 1, which also highlights the tested range size and changes in some aspects, causing loss of information from others prices. Similarly, it is possible to customize and use the distance function, that is, parts. Consider the following examples. There are three elements, ×1, ×2, and ×3, which do not depend on the various factors. In summary, it is simple and minimal ×1 and ×2 standard variables with values ranging from 0 to 1.

By comparison, it takes time to upgrade the KNN model, but the predictions set can go very slowly in video recognition, image recognition, and pattern recognition.

5.4.2 NAÏVE BAYES CLASSIFIER (NBC)

Bayes classifier is a separator used for segmentation; developed based on the concept of Bayes theorem. The model assumes that each factor is independent of the rest of the factors and is not relevant to the training set (Kaur and Oberoi, 2020). This assumption is used to predict the category of new views using statistics. The NB separator assesses the probability of a new sample without label B (B = b1, b2, b3, etc.). A model release is predicated on a probability basis with a large P (A | B). The value of P(B) does not affect the selection of the category with the largest P(A | B).

$$P(A \mid B) = P(B \mid A) \times P(A)\ (1)\ P(B)$$

Supervised learning is used for NB classification to improve the efficiency of learning classification activities. This high efficiency is due to the fact that NBC can be trained to understand the limits by extracting the basic statistics of each class into features by simply looking at each of these features. The NB model uses the parameter authorization tool using the most advanced method for a variety of applications. The performance of the NB model can therefore remain independent of the Bayes algorithm.

The NB separator has some serious problems, although they are characterized by simple structure and thinking. One of these issues is that they measure the parameters of the required categories. In addition, there is no need to examine the entire matrix of covariance, but it is important to assess the diversity of each class.

5.4.3 DECISION TREE (DTC) CLASSIFIER

A simple dividing algorithm consisting of an internal node and a single node with a classified name is called decision tree (DT) (Hussain et al., 2020). The solution to the DT-partition problems is based on the constant fragmentation of the input space to form a tree with pure and straight points related to one phase. As the tree descends, the new point is separated by selecting one branch (side) of the tree in each area. The development of DTs depends on the type of variation intended for the model. The algorithm uses a variation of the variance method to create a drug model with continuous test scores. Conversely, the Gini contamination method is used for phase-specific variability.

The DT process starts in the root zone and cuts through the storage area, known as the terminal or tree node. Internal nodes exist between the roots and leaf areas. There is only one possible result for each internal node. For example, there is a new unwritten data point that starts with the root node. It reaches to the next node according to the root test result. The same strategy is applied and the next node is the internal node. Thereafter, a new unlabeled data point is assigned to the leaf node and the classification of new unlabeled data points is made based on the section label corresponding to the leaf node concerned.

There are several real-world applications of DTs in various fields. In the field of healthcare, DTs helps in diagnosis of mental retardation, which increases the effectiveness of positive case assessments; it also determines the risk factors for a variety of dementia.

The DT model has advantages and disadvantages, as do other ML models. The advantage is that it is easy to translate. It can also deal with continuous and isolated symptoms and requires minimal information correction. In addition, this approach is important when it comes to dividing and predicting activities. However, in the case of unequal data, DT can have a completely down performance. This model is also sensitive to noise and can also lead to overload when there is noise in the training database. In addition, any increase in database entries will lead to the complexity of the larger model.

There are several real-world applications of DTs in various fields. Large memory is important for large datasets in the case of healthcare. This challenge will be overcome by restrictions DTs are used for early diagnosis of mental retardation, which increases the effectiveness of the depth of the tree or adjust the value of the important sample in addition to positive case assessment; it also determines the main risk factors for the occurrence of separation. Low-energy DTs are important; however, it can lead to various forms of dementia. Sophia's robot built in Saudi Arabia is one of the DT model constraints, including low variance and high selection, while the opposite algorithm applications have been developed to interact with people. And it is a well-known algorithm it is true of trees that cut deep. Wrapping and lifting DTs is common in ML ways to reduce high model variability or high bias. The DT model has advantages and disadvantages, as do other ML models. The advantage of the DT algorithm is that it is easy to translate. It can also work with persistent and fragmented features and requires little in the processing of information. Random forest integration that includes multiple models as well. In addition, this approach is important when it comes to segregation and performance tasks. It shows compatibility with a wide range of dataset for planning or regression (Kaur et al., 2020) predictions. In the case of unequal data, DT can have it all Bootstrap integration or wrap are examples of such models. Betting can reduce the model's improper performance. This model is also sensitive to noise and can lead to overload diversity and as a result, improves overall performance to prevent overeating.

In addition, any increase in database entries will lead to the complexity of the larger model. Therefore, the DT model has advantages and disadvantages, as do other ML models. The advantage of the DT algorithm is that it is easy to translate. It can also deal with continuous and isolated symptoms and requires minimal information correction. In addition, this approach is important when it comes to dividing and predicting activities. However, in the case of unequal data, DT can have a completely negative performance. This model is also sensitive to noise and can also lead to overload when there is noise in the training database. In addition, any increase in database entries will lead to the complexity of the larger model.

There are several real-world applications of DTs in various fields. In the field of healthcare, large memory is important for large datasets. This challenge can be overcome by restrictions DTs are used for early diagnosis of mental retardation, which increases the effectiveness of the height or depth of the tree or adjust the value of the important sample in addition to positive case assessment; also determines the main risk factors for the occurrence of separation. Low-energy DTs are important; however, they can lead to certain forms of dementia. Sophia's robot built in Saudi Arabia is one of the DT model constraints, including low variance and high selection, while the opposite algorithm applications are developed to interact with people. And it is a well-known algorithm it is true of trees that cut deep. Wrapping and lifting DTs is common in ML. ways to reduce high model variability or high bias.

The DT model has advantages and disadvantages, as do other ML models. The advantages of the DT algorithm are that it is easy to translate. It can also work with persistent and fragmented features and requires little in the processing of information. One of the models is a random forest integration that includes multiple models as well In addition, this approach is important when it comes to segregation and performance tasks shows compatibility with a wide range of dataset for planning or regression (Oshiro et al., 2012) predictions. However, in the case of unequal data, DT can have it all Bootstrap integration or wrap are examples of such models. Betting can reduce the model's improper performance. This model is also sensitive to noise and can lead to overload diversity and as a result improves overall performance to prevent overeating. This model can work if there is noise in the training database, overcoming a major conflict even over the trees of decision. In addition, any increase in database entries will lead to the complexity of the larger model. Therefore, there are several deciduous trees (DTs) present in the random and individual forest model large memory is important for large datasets. This challenge can be overcome by restrictions as it shows little difference from others. For a specific data point, different results the height or depth of the tree or adjusting the value of the important sample in addition to the findings of each DT will be included. In the process of integration, the majority is separated. Low-energy DTs are important; however, it can lead to certainty the vote is found in the case of a split, and the normal number is found there model constraints, including low variance and high selection, while the opposite it comes to backup activities. The performance of integrated DTs is better than that it is true of trees that cut deep. Wrapping and lifting DTs are common in a single decision-making tree in terms of forecasting due to different training for all ways to reduce high model variability or high bias. DTs on random samples are taken from the training database.

5.4.4 SUPPORT VECTOR MACHINES (SVMs)

SVMs can be defined as a machine monitoring system that expands the boundary of a decision (or hyperplane) between multiple classes by performing various input analyses within the database; therefore, it is possible to guess labels based on vectors for one or more features. Its position is to ensure the highest distance from the data points closest to each class. The name vector support machine is based on the closest points, also known as supporting vectors.

The main purpose of this process is to classify binary options and forecasts. Many biological applications have successfully used this method (Nobel et al., 2006). SVMs are widely used in biomedical engineering to automatically differentiate microarray gene expression profiles, in addition to substances with high variability such as protein sequence and DNA. The kernel method is another important SVM application, which helps to model the most advanced and nonlinear models. In the case of offline predictions, divisions can be made using the kernel process based on SVMs, which help with mapping or case planning within a higher space. Counting in a space of high magnitude can be time consuming; however, the use of kernel function leads to faster calculation.

In the mapping process, SVM shows the boundary of the decision between the two classes by determining the point of each data point, in particular the supporting points of the vector at the boundary between the affected sections. SVM learns the value of all data points when mapping; in particular, the importance of points to support vector boundaries of classes is evident in SVMs to indicate the boundary of decision between those classes. The distance of the entire support vector is measured by the predicting new point. Knowledge-based support vector machines (KSVMs) make a distinction based on the velocity of the vector support and the value of the supporting vectors. It is important to note that significant improvements can be made to KSVM performance with data growth from 0 to 1.

In the case for partition function when using a simple SVM algorithm, we think there are many cases in the database. In addition, it is thought that one or more classes are related to each case. The internal cases of the given database are still distributed in several phases by SVMs by creating a model that chooses the most accurate hyperplane. In the context of SVMs, each case is marked with a point instead of a point that arranges the cases into two separate categories with a gap between them. In addition, the location of a new case in terms of space is used to organize these cases into space and facilitates their division into one or more categories.

The center line in the space represents the location of the SVM decision. Any data point above the decision area will be called section 1; otherwise, it will belong to another category. Sometimes, real datasets cannot be easily separated.

Finally, RNN is naturally repetitive because it performs the same process for all data entries, while the predicted result of the current input will be read from the previous computer. After receiving the result, it will be copied and returned to the duplicate network. Due to limitations, the LSTM neural network is integrated with the RNN to overcome long-term predictive instability due to explosion or disappearing gradient (Kim et al., 2019). Reviewing these methods in detail is beyond this review.

5.5 MACHINE LEARNING APPLICATIONS

5.5.1 CLINICAL RESEARCH

Clinical trials are conducted to assess the efficacy and safety of cattle interventions, surgery, and treatment. Since clinical trials always include human studies and are the final step in the research process, they should be done carefully to avoid harming people. ML improves the clinical trial process by enabling access to information on the effectiveness of interventions in publicly available clinical and biological database data, data obtained from health records, and valid evidence from senators (Zame et al., 2020).

5.5.2 SMART ELECTRONIC HEALTH RECORDS

Electronic health records provide timely access to patient information that enables care providers to provide quality care. In other words, instead of acting as a patient recorder, electronic health records can be improved with ML to incorporate intelligent functions (Wang et al., 2019).

5.5.3 PREDICTING HEART DISEASE

Heart disease is the cause of death in many parts of the world. As a result of lifestyle changes and other risk factors, heart disease is increasing worldwide. In 2016, cardiovascular diseases accounted for 17.6 million deaths worldwide, an increase of 14.5% compared to 2006 (Almustafa, 2020).

5.5.4 COVID-19 DIAGNOSTIC AND GUIDANCE MODELS

ML can also be used in the identification and prediction of COVID-19. The concept is to design an algorithm that takes predictive predictions and diagnoses and provides accurate results. Highly reported predictions include body temperature, age, lung thinking factors, and lymphocyte counts (Zoabi et al., 2021).

5.5.5 CUSTOMIZED CARE

Providing personalized services is at the heart of patient-centered care. Patients need care that fits their needs, beliefs and expectations. Personal care improves patient satisfaction and improves the use of official health services. ML algorithms can play a role in providing customized care by allowing healthcare professionals to evaluate individual patient data and develop personalized care programs (Ahmed et al., 2020). Machine learning systems combine health records and integrate different data sources to find specific human-specific patterns of disease progression (Vanani and Amirhosseini, 2021).

5.6 IoT AND MACHINE LEARNING APPLICATIONS IN HEALTHCARE PROGRAMS FOR THE FUTURE

As discussed earlier, IoT and AI learning devices have improved the health sector by having patients wear devices such as premium jackets and smart belts used to monitor their condition and send general reports to a database accessible to physicians and medical professionals (Wang et al., 2018). Devices can monitor important patient symptoms and organs and send progress reports to specific information. The system also collects and reports the presence of pathogens and expression (Shahzad et al., 2018). This is an important precaution to help the healthcare system deliver better alternatives.

The availability of smart pills, nerves, and wearable monitors in healthcare adds value to the industry. These tools help to identify and predict future signs and symptoms in disease patterns. The context of performing patient and disease monitoring activities saves time and steps when all physicians are busy—for example, in crisis (Mocnik et al., 2020). The use of smart technology in this field is essential for saving lives during epidemics such as COVID-19. Portable monitoring devices that capture and send information to a doctor's database can analyze and diagnose a patient and mail a prescription.

Patients can use smart tablets and smart bands (IoT) that monitor and collect certain database feed information during a pandemic. These devices help physicians and other machines to learn patterns and signs of disease, giving physicians the opportunity to understand the symptoms and analyze the symptoms to improve a faster and safer diagnosis (Lundeviod et al., 2019).

Cloud computing is also an effective part of the IoT industry. It helps to connect various AI learning devices to understand the data by analyzing and storing it. Another important aspect of cloud computing is that we can store a large amount of data and, therefore, maintain the requirements of the healthcare system. Because of its ability to share information, cloud computing works on different devices to access information. On the contrary, cloud computing is currently facing some challenges. These challenges help scientists and researchers who want to improve the use of ML and IoT in the healthcare industry. Medical records in the healthcare industry are very sensitive and require careful protection as they contain protected public health information (PHI). Therefore, the Health Insurance Portability and Accountability Act (HIPAA) (Xu et al., 2021), regulate the process of obtaining and analyzing this information. This poses a major challenge to data mining and ML technology, such as in-depth learning, which requires a large amount of training detail. Sharing this sensitive information to improve the quality of care can jeopardize a patient's privacy.

TABLE 5.1 Summary of the Reviewed Applications.

Application name	ML algorithm	Issues	Challenges	Work to be addressed in the future	Comparison with existing literature
Clinical trial research (Zame et al., 2020)	Deep learning algorithm	It addressed the issue of findings of clinical data by analyzing human capacity.	Utilization of deep learning model on a difficult dataset. Larger labeled datasets are required.	To improve the applicability of deep learning in clinical research trials.	Zame et al. (2020) compare the use of ML in trials of clinics in the COVID-19 scenario.
Smart electronic health records (Wang et al., 2019)	The deep learning algorithm Supervised ML and natural language processing	Present record issue clinical data but it does not support any decision making.	Data preparation before use in ML algorithms is still competitive. It is not convenient to add the factors in ML models.	The acceptance of smart electronic records for health to support different diseases.	Reviews are very few in the case of health records. Shinozaki (2020) reviewed the use of ML in electronic health records to help in drug development.
Heart disease prediction (Almustafa et al., 2020)	Deep learning and ANNs	It is required to predict accurate cardiovascular diseases as well as implementation of algorithm for good outcomes.	Missing ethical guidelines for adoption of heart disease prediction algorithms. ML algorithms is unable to determine highly abstract reasoning problems.	Extends the use of ML in decision making to involve patient-centric predictive analytics.	Reviews report the findings of suggested areas for further research.
Diagnostic and prognostic models for COVID-19 (Zoabi et al., 2021)	Deep learning models	It is required to review prediction models for the diagnosis and prognosis of COVID-19 to support decision making.	Poor training datasets were used hence the datasets are problematic.	Collection of high-volume quality datasets train COVID-19 models for prediction.	Since the knowledge of the disease is improved, better and more efficient models can emerge, hence there is a need for more reviews.

TABLE 5.1 *(Continued)*

Application name	ML algorithm	Issues	Challenges	Work to be addressed in the future	Comparison with existing literature
Personalized care (Vanani et al., 2021)	Deep neural networks, deep learning, supervised and unsupervised learning, and many others	The inability to provide personalized care despite the increasing accumulation of personal data. It can't provide personal care instead of carrying the multiple personal data.	Accumulation of high-quality training datasets is required.	Create systems that can be joined into electronic health records to promote medicine for each individual.	Fröhlich et al. (2018) reviewed and identified some of the challenges associated; as such, the proposed review explore how studies have shown these challenges.

Some devices can monitor BP, body temperature, and heart rate. They are useful for storing and collecting data about patients and can therefore help diagnose. IoT and machine learning can help keep healthcare professionals aware of changes, which are important for a healthy society. Retaining diagnostic data and symptoms of COVID-19 is key to ensuring that the disease is eradicated or that the vaccine is available and can be obtained by scientists and medical professionals to ask questions, analyze, and indeed -Share time results.

5.7 CONCLUSION

The healthcare sector is complex in terms of commitment and strict rules, making it an important and important sector of innovation. The IoT has opened up the ways in the field of healthcare and solves many problems. Applying medical IoT will bring excellent opportunities for telemedicine, remote patient monitoring, and much more. In this chapter, we have summarized the most powerful ML algorithms, listed other ML programs in the healthcare sector, and analyzed IoT and ML in the healthcare system to predict future trends.

KEYWORDS

- **Internet of Things**
- **machine learning**
- **positron emission tomography**
- **computed tomography**
- **radio frequency identification**
- **electrocardiogram**
- **support vector machines**

REFERENCES

Ahmad, M. A.; Eckert, C.; Teredesai, A. Interpretable Machine Learning in Healthcare. In *Proceedings of the 2018 ACM International Conference on Bioinformatics, Computational Biology, and Health Informatics*, Washington, DC, USA, August 24–29, 2018; pp 559–560.

Ahmed, Z.; Mohamed, K.; Zeeshan, S.; Dong, X. Artificial Intelligence with Multi-functional Machine Learning Platform Development for Better Healthcare and Precision Medicine. *Database J. Biol. Databases Curation* **2020**, *2020*.

Albawi, S.; Mohammed, T. A.; Al-Zawi, S. Understanding of a Convolutional Neural Network. In *Proceedings of the 2017 International Conference on Engineering and Technology (ICET)*, Antalya, Turkey, August 21–23, 2017; pp 1–6.

Aliberti, A.; Bagatin, A.; Acquaviva, A.; Macii, E.; Patti, E. Data Driven Patient-Specialized Neural Networks for Blood Glucose Prediction. In *Proceedings of the 2020 IEEE International Conference on Multimedia & Expo Workshops (ICMEW)*, London, UK, July 6–10, 2020; pp 1–6.

Alharbi, R.; Almagwashi, H. The Privacy requirements for Wearable IoT Devices in Healthcare Domain. In *Proceedings of the 2019 7th International Conference on Future Internet of Things and Cloud Workshops (FiCloudW)*, Istanbul, Turkey, August 26–28, 2019; IEEE: Piscataway, NJ, USA, 2019; pp 18–25.

Almustafa, K. M. Prediction of Heart Disease and Classifiers' Sensitivity Analysis. *BMC Bioinform.* 2020, *21*, 278.

Battineni, G.; Sagaro, G. G.; Chinatalapudi, N.; Amenta, F. Applications of Machine Learning Predictive Models in the Chronic Disease Diagnosis. *J. Pers. Med.* **2020,** *10*, 21.

Ben Ali, J.; Saidi, L.; Harrath, S.; Bechhoefer, E.; Benbouzid, M. Online Automatic Diagnosis of Wind Turbine Bearings Progressive Degradations under Real Experimental Conditions based on Unsupervised Machine Learning. *Appl. Acoust.* 2018, *132*, 167–181.

Birje, M. N.; Hanji, S. S. Internet of Things Based Distributed Healthcare Systems: A Review. *J. Data Inf. Manage.* **2020,** *2*, 149–165.

Bisong, E. *Building Machine Learning and Deep Learning Models on Google Cloud Platform*; Springer: Berlin/Heidelberg, Germany, 2019.

Cahyadi, A.; Razak, A.; Abdillah, H.; Junaedi, F.; Taligansing, S. Y. Machine Learning Based Behavioral Modification. *Int. J. Eng. Adv. Technol.* **2019,** *8*, 1134–1138.

Chae, S.; Kwon, S.; Lee, D. Predicting Infectious Disease Using Deep Learning and Big Data. *Int. J. Environ. Res. Public Health* **2018,** *15*, 1596.

Chauhan, R., Ghanshala, K. K., Joshi, R. C. Convolutional Neural Network (CNN) for Image Detection and Recognition. In *Proceedings of the 2018 First International Conference on Secure Cyber Computing and Communication (ICSCCC)*, Kolkata, India, November 22–23, 2018; pp 278–282.

Cortes, C.; Vapnik, V. Support-Vector Networks. *Mach. Learn.* **1995,** *20*, 273–297.

Datilo, P. M.; Ismail, Z.; Dare, J. A Review of Epidemic Forecasting Using Artificial Neural Networks. *Int. J. Epidemiol. Res.* **2019,** *6*, 132–143.

Desai, S. B.; Pareek, A.; Lungren, M. P. Deep Learning and its Role in COVID-19 Medical Imaging. *Intell. Med.* **2020,** *3*, 100013.

Dike, H. U.; Zhou, Y.; Deveerasetty, K. K.; Wu, Q. Unsupervised Learning Based on Artificial Neural Network: A Review. In *Proceedings of the 2018 IEEE International Conference on Cyborg and Bionic Systems (CBS)*, Shenzhen, China, October 25–27, 2018; pp 322–327.

Friedman, J.; Hastie, T.; Tibshirani, R. *The Elements of Statistical Learning (No. 10)*; Springer Series in Statistics: New York, NY, USA, 2001.

Gao, H.; Qin, X.; Barroso, R. J. D.; Hussain, W.; Xu, Y.; Yin, Y. Collaborative Learning-Based Industrial IoT API Recommendation for Software-Defined Devices: The Implicit Knowledge Discovery Perspective. *IEEE Trans. Emerg. Top. Comput. Intell.* **2022,** *6*(1), 66–76.

Golub, G. H.; Heath, M.; Wahba, G. Generalized Cross-Validation as a Method for Choosing a Good Ridge Parameter. *Technometrics* **1979,** *21*, 215.

Hung, C. Y.; Chen, W. C.; Lai, P. T.; Lin, C. H.; Lee, C. C. Comparing Deep Neural Network and Other Machine Learning Algorithms for Stroke Prediction in a Large-Scale Population-Based Electronic Medical Claims Database. In *Proceedings of the 2017 39th Annual International Conference of the IEEE Engineering in Medicine and Biology Society*, Jeju Island, Korea, July 11–15, 2017; pp 3110–3113.

Hussain, W.; Sohaib, O. Analysing Cloud QoS Prediction Approaches and Its Control Parameters: Considering Overall Accuracy and Freshness of a Dataset. *IEEE Access* **2019,** *7*, 82649–82671.

Hussain, W.; Hussain, F. K.; Hussain, O. K.; Chang, E. Provider-Based Optimized Personalized Viable SLA (OPV-SLA) Framework to Prevent SLA Violation. *Comput. J.* **2016,** *59*, 1760–1783.

Hussain, W.; Hussain, F. K.; Hussain, O.; Bagia, R.; Chang, E. Risk-Based Framework for SLA Violation Abatement from the Cloud Service Provider's Perspective. *Comput. J.* **2018,** *61*, 1306–1322.Hussain, W.; Sohaib, O.; Naderpour, M.; Gao, H. Cloud Marginal Resource Allocation: A Decision Support Model. *Mob. Netw. Appl.* **2020,** *25*, 1418–1433.

Ibrahim, N.; Akhir, N. S. M.; Hassan, F. H. Predictive Analysis Effectiveness in Determining the Epidemic Disease Infected Area. *AIP Conf. Proc.* **2017,** *1891*, 20064.

Iqbal, N.; Jamil, F.; Ahmad, S.; Kim, D. A Novel Blockchain-Based Integrity and Reliable Veterinary Clinic Information Management System Using Predictive Analytics for Provisioning of Quality Health Services. *IEEE Access* **2021,** *9*, 8069–8098.

Jadhav, S.; Kasar, R.; Lade, N.; Patil, M.; Kolte, S. Disease Prediction by Machine Learning from Healthcare Communities. *Int. J. Sci. Res. Sci. Technol.* **2019,** 29–35.

Jamil, F.; Hang, L.; Kim, K.; Kim, D. A Novel Medical Blockchain Model for Drug Supply Chain Integrity Management in a Smart Hospital. *Electronics* **2019,** *8*, 505.

Jamil, F.; Ahmad, S.; Iqbal, N.; Kim, D. H. Towards a Remote Monitoring of Patient Vital Signs based on IoT-Based Blockchain Integrity Management Platforms in Smart Hospitals. *Sensors* **2020,** *20*, 2195.

Kaur, G.; Oberoi, A. Novel Approach for Brain Tumor Detection Based on Naïve Bayes Classification. *Data Management, Analytics and Innovation*; Springer: Berlin/Heidelberg, Germany, 2020; pp 451–462.

Kim, M.; Yun, J.; Cho, Y.; Shin, K.; Jang, R.; Bae, H. J.; Kim, N. Deep Learning in Medical Imaging. *Neurospine* **2019,** *16*, 657–668.

Kumar, P. M.; Gandhi, U. D. A Novel Three-Tier Internet of Things Architecture with Machine Learning Algorithm for Early Detection of Heart Diseases. *Comput. Electr. Eng.* **2018,** *65*, 222–235.

Lin, W. C.; Chen, J. S.; Chiang, M. F.; Hribar, M. R. Applications of Artificial Intelligence to Electronic Health Record Data in Ophthalmology. *Transl. Vis. Sci. Technol.* **2020,** *9*, 13.

Lundervold, A. S.; Lundervold, A. An Overview of Deep Learning in Medical Imaging Focusing on MRI. *Z. Med. Phys.* **2019,** *29*, 102–127.

Michie, S.; Thomas, J.; John, S. T.; Mac Aonghusa, P.; Shawe-Taylor, J.; Kelly, M. P.; Deleris, L. A.; Finnerty, A. N.; Marques, M. M.; Norris, E.; et al. The Human Behaviour-Change Project: Harnessing the Power of Artificial Intelligence and Machine Learning for Evidence Synthesis and Interpretation. *Implement. Sci.* **2017,** *12*, 1–12.

Miotto, R.; Wang, F.; Wang, S.; Jiang, X.; Dudley, J. T. Deep Learning for Healthcare: Review, Opportunities and Challenges. *Brief. Bioinform.* **2018,** *19*, 1236–1246.

Mocnik, F. B.; Raposo, P.; Feringa, W.; Kraak, M. J.; Köbben, B. Epidemics and Pandemics in Maps—the Case of COVID-19. *J. Maps* **2020,** *16*, 144–152.

Mosenia, A.; Sur-Kolay, S.; Raghunathan, A.; Jha, N. K. Wearable Medical Sensor-Based System Design: A Survey. *IEEE Trans. Multi-Scale Comput. Syst.* **2017,** *3*, 124–138.

Mtonga, K.; Kumaran, S.; Mikeka, C.; Jayavel, K.; Nsenga, J. Machine Learning-Based Patient Load Prediction and IoT Integrated Intelligent Patient Transfer Systems. *Future Internet* **2019,** *11*, 236.

Ngiam, K. Y.; Khor, W. Big Data and Machine Learning Algorithms for Health-Care Delivery. *Lancet Oncol.* **2019,** *20*, e262–e273.

Noble, W. S. What Is a Support Vector Machine? *Nat. Biotechnol.* **2006,** *24*, 1565–1567.

Olmedo, M. T. C.; Paegelow, M.; Mas, J. F.; Escobar, F. In *Geometric Approaches for Modeling Land Change Scenarios*; Springer: Berlin/Heidelberg, Germany, 2018.

Oshiro, T. M.; Perez, P. S.; Baranauskas, J. A. How Many Trees in a Random Forest? *International Workshop on Machine Learning and Data Mining in Pattern Recognition*; Springer: Berlin/Heidelberg, Germany, 2012; pp 154–168.

Osisanwo, F.; Akinsola, J.; Awodele, O.; Hinmikaiye, J.; Olakanmi, O.; Akinjobi, J. Supervised Machine Learning Algorithms: Classification and Comparison. *Int. J. Comput. Trends Technol.* **2017,** *48*, 128–138.

Panch, T.; Szolovits, P.; Atun, R. Artificial Intelligence, Machine Learning and Health Systems. *J. Glob. Health* **2018**, *8*, 020303.

Praveena, M.; Jaiganesh, V. A Literature Review on Supervised Machine Learning Algorithms and Boosting Process. *Int. J. Comput. Appl.* **2017,** *169*, 32–35.

Rajkomar, A.; Hardt, M.; Howell, M. D.; Corrado, G.; Chin, M. H. Ensuring Fairness in Machine Learning to Advance Health Equity. *Ann. Intern. Med.* **2018,** *169*, 866–872.

Shah, P.; Kendall, F.; Khozin, S.; Goosen, R.; Hu, J.; Laramie, J.; Ringel, M.; Schork, N. Artificial Intelligence and Machine Learning in Clinical Development: A Translational Perspective. *NPJ Digit. Med.* **2019,** *2*, 1–5.

Shahbazi, Z.; Byun, Y. C. Towards a Secure Thermal-Energy Aware Routing Protocol in Wireless Body Area Network Based on Blockchain Technology. *Sensors* **2020,** *20*, 3604.

Shahbazi, Z.; Byun, Y. C. Improving Transactional Data System Based on an Edge Computing–Blockchain–Machine Learning Integrated Framework. *Processes* **2021,** *9*, 92.

Shahzad, A.; Lee, Y. S.; Lee, M.; Kim, Y. G.; Xiong, N. Real-Time Cloud-Based Health Tracking and Monitoring System in Designed Boundary for Cardiology Patients. *J. Sens.* **2018,** *2018*, 1–15.

Shinozaki, A. Electronic Medical Records and Machine Learning in Approaches to Drug Development. *Artificial Intelligence in Oncology Drug Discovery and Development*; IntechOpen: London, UK, 2020; p 51.

Singh, A.; Thakur, N.; Sharma, A. A Review of Supervised Machine Learning Algorithms. In *Proceedings of the 2016 3rd International Conference on Computing for Sustainable Global Development (INDIACom)*, New Delhi, India, March 16–18, 2016; pp 1310–1315.

Sohaib, O.; Lu, H.; Hussain, W. Internet of Things (IoT) in E-commerce: For People with Disabilities. In *Proceedings of the 2017 12th IEEE Conference on Industrial Electronics and Applications (ICIEA)*, Siem Reap, Cambodia, June 18–20, 2017; pp 419–423.

Solares, J. R. A.; Raimondi, F. E. D.; Zhu, Y.; Rahimian, F.; Canoy, D.; Tran, J.; Gomes, A. C. P.; Payberah, A. H.; Zottoli, M.; Nazarzadeh, M.; et al. Deep Learning for Electronic Health Records: A Comparative Review of Multiple Deep Neural Architectures. *J. Biomed. Inform.* **2020,** *101*, 103337.

Tambuskar, D.; Narkhede, B.; Mahapatra, S. S. A Flexible Clustering Approach for Virtual Cell Formation Considering Real-Life Production Factors using Kohonen Self-Organising Map. *Int. J. Ind. Syst. Eng.* **2018,** *28*, 193–215.

Van Engelen, J. E.; Hoos, H. H. A Survey on Semi-Supervised Learning. *Mach. Learn.* **2019,** *109*, 373–440.

Vanani, I. R.; Amirhosseini, M. IoT-Based Diseases Prediction and Diagnosis System for Healthcare. *Internet of Things for Healthcare Technologies*; Springer: Berlin/Heidelberg, Germany, 2021; pp 21–48.

Wang, M. H.; Chen, H. K.; Hsu, M. H.; Wang, H. C.; Yeh, Y. T. Cloud Computing for Infectious Disease Surveillance and Control: Development and Evaluation of a Hospital Automated Laboratory Reporting System. *J. Med. Internet Res.* **2018,** *20*, e10886.

Wang, S.; Pathak, J.; Zhang, Y. Using Electronic Health Records and Machine Learning to Predict Postpartum Depression. *Stud. Health Technol. Inform.* **2019,** *264*, 888–892.

Wiens, J.; Shenoy, E. S. Machine Learning for Healthcare: On the Verge of a Major Shift in Healthcare Epidemiology. *Clin. Infect. Dis.* **2018,** *66*, 149–153.

Wilkinson, J.; Arnold, K. F.; Murray, E. J.; van Smeden, M.; Carr, K.; Sippy, R.; de Kamps, M.; Beam, A.; Konigorski, S.; Lippert, C.; et al. Time to Reality Check the Promises of Machine Learning-Powered Precision Medicine. *Lancet Digit. Health* **2020,** *2*, e677–e680.

Wu, T.; Wu, F.; Redoute, J. M.; Yuce, M. R. An Autonomous Wireless Body Area Network Implementation towards IoT Connected Healthcare Applications. *IEEE Access* **2017,** *5*, 11413–11422.

Xu, J.; Glicksberg, B. S.; Su, C.; Walker, P.; Bian, J.; Wang, F. Federated Learning for Healthcare Informatics. *J. Health Inform. Res.* **2021,** *5*, 1–19.

Xu, J.; Xue, K.; Zhang, K. Current Status and Future Trends of Clinical Diagnoses via Image-Based Deep Learning. *Theranostics* **2019,** *9*, 7556–7565.

Yahaya, L.; Oye, N. D.; Garba, E. J. A Comprehensive Review on Heart Disease Prediction Using Data Mining and Machine Learning Techniques. *Am. J. Artif. Intell.* **2020,** *4*, 20.

Yan, Y.; Zhang, J. W.; Zang, G. Y.; Pu, J. The Primary use of Artificial Intelligence in Cardiovascular Diseases: What Kind of Potential Role does Artificial Intelligence Play in Future Medicine? *J. Geriatr. Cardiol.* **2019,** *16*, 585–591.

Ye, J.; Chow, J. H.; Chen, J.; Zheng, Z. Stochastic Gradient Boosted Distributed Decision Trees. In *Proceedings of the 18th ACM Conference on Information and Knowledge Management—CIKM'09*, Hong Kong, China, November 2–6, 2009; pp 2061–2064.

Yuvaraj, N.; SriPreethaa, K. R. Diabetes Prediction in Healthcare Systems Using Machine Learning Algorithms on Hadoop Cluster. *Clust. Comput.* **2017,** *22*, 1–9.

Zame, W. R.; Bica, I.; Shen, C.; Curth, A.; Lee, H. S.; Bailey, S.; Weatherall, J.; Wright, D.; Bretz, F.; Van Der Schaar, M. Machine Learning for Clinical Trials in the Era of COVID-19. *Stat. Biopharm. Res.* **2020,** *12*, 506–517.

Zoabi, Y.; Deri-Rozov, S.; Shomron, N. Machine Learning-Based Prediction of COVID-19 Diagnosis Based on Symptoms. *NPJ Digit. Med.* **2021,** *4*, 1–5

CHAPTER 6

Implementation of Compact Dual-Band Printed Antenna for Biomedical Applications

RACHNA PRABHA[1] and SAWAN KUMAR[2]

[1]*Department of Electronics Engineering, Indian Institute of Technology (ISM), Dhanbad, Jharkhand, India*

[2]*Bharat Sachar Nigam Limited, Hoshiarpur, Punjab, India*

ABSTRACT

This chapter proposed the design of a compact dual-band antenna having inclusive applications in the field of medical applications. The antenna design for medical applications involve the electromagnetic energy into the human body or into other biological systems. These proposed antennae experimentally measure the measurements with low specific absorption rate (SAR) which is used for the application of health care systems designed with defective ground structures. The suggested antenna is designed on the FR4 substrate. This proposed antenna provides dual slotted bands at 1.65 and 2.5 GHz which are characteristics at industrial, scientific, and medical (ISM) bands and medical implant communication system (MICS). These antennas are designed in the volume of 18 mm × 24 mm × 1.57 mm, which is very compact and are applicable for health care systems having microstrip feedlines. There are generally three key ways that antenna technology is being used in modern medicine, that is, Transmission of information, diagnosis, and treatment are the first three. We already use wireless capsule endoscopy for information transmission in our daily lives. Radio frequency identification (RFID) tags used in wearable or implanted monitoring devices

Smart Electronic Devices and Systems for Biomedical and Healthcare Applications. Upendra Kumar, Rajesh K. Kesharwani, & Raj K. Keservani (Ed.)

DOI: 10.1201/9781003560852-6

have received extensive research and reporting. Antenna technology is needed for magnetic resonance imaging (MRI) devices for diagnosis. The use of radiometry or microwave computed tomography (CT) is another promising noninvasive diagnostic method. The heat produced by antenna radiation has been employed in medical applications. Thermal therapies that are frequently used to treat cancer include hyperthermia and coagulation. Three different types of antennas are introduced in our research work. First, a relatively compact antenna for a measuring system that can be implanted is proposed. A potential antenna for such a system is an H-shaped cavity slot antenna. The antenna's numerical and experimental properties are illustrated. Additionally, SAR distributions created in a coil inside a bird cage are seen in the abdomen of a pregnant lady. The introduction of slotted antennas and array applicators made up of several slotted antennas for minimally invasive microwave thermal therapies follows a brief discussion of thermal therapy and microwave heating. Then, a few technical examples of actual antennas are presented which are the applications of the coaxial–slot antennas such as physiological state treatment for brain tumors and intracavitary hyperthermia for bile duct carcinoma.

6.1 INTRODUCTION

Nowadays, a research domain focuses on wireless medical implantable devices. The main factors of medically applicable antenna are resonant frequency, bandwidth, efficiency, and specific absorption rate (SAR). An antenna has a compact bandwidth and good radiation efficiency are important factors in the communication link between an implanted device and an exterior instrument for biotelemetry. The present antenna focuses on the design of compact dual U-slot antenna for wireless medical implantable devices. The medical applicable factor, specific absorption rate (SAR), will measure the electromagnetic power density which will affect the human body (Kim et al., 2004). The main constraints of medically applicable antennas are compact size, propagation losses, biocompatibility, and safety issues. This type of antenna is placed easily in the human body (Bahrami et al., 2016). The planar inverted-F antenna or compact dual-band antenna is used with some recent technologies, which is operated at radio frequency (RF) bands for medical applications (Declercq and Rogier, 2010). The planar inverted-F antenna is also applicable for medical applications and easily links with communication devices (Ha and Jun, 2011). For the body signal processing, telemetry applications are proposed in the antenna when designed (Prudhvi Nadh et al., 2015).

The effects of SAR on body is reduces by various new technologies such as ferrite loading, radiated conductors, and electromagnetic band gap energy (Li et al., 2018). These antennas are wireless implantable devices which are covering the human body placed below the 2 mm in the skin friction. The proposed antenna is simulated on finite element method (FEM)-based Ansys high-frequency structure simulator (HFSS) software.

6.1.1 EFFECTS OF ANTENNA RADIATED POWER AND EFFICIENCY ON THE HUMAN BODY

The loss of human body tissues has an impact on the radiation efficiency radiated power of the proposed antenna. The following equation gives the radiation efficiency of implanted antennas:

$$\eta = \frac{Prad}{Pin}$$

where the input power is p_{in} and the three power components are the absorbed power, radiated power, and reflected power combined to make up the input power for implantable antennas ($P_{\text{ref}} + P_{\text{abs}} + \text{P}_{\text{rad}}$). The implantable antennas near field coupling cause the absorbed power to be greater than the reflected power, resulting in low radiated efficiency and power. The following equation gives the absorbed power:

$$P_{abs} \frac{w}{2} \int_v \in_0 \in_r \ |E|^2 \ dv$$

The SAR similarly rises as the absorbed power. Antenna radiation efficiency arises along with the absorbed power. SAR is provided by (Bhattacharjee et al., 2016):

$$SAR = \frac{P_{\text{L}}}{\rho} = \frac{\sigma |E|^2}{2\rho}$$

where the mass density (kg/m^3) and the electric field (V/m) are both present. It is possible to represent the radiation efficiency using radiation resistance and loss resistance (Hammouda and Allam, 2014):

$$\eta = \frac{R_{rad}}{R_{rad+R_L}}$$

where R_{rad} and R_{L} are the respective resistances to radiation and loss in ohms. The antenna's gain would then be

$$G = \eta D$$

where D stands for the power's directivity as it is radiated. The ratio of an antenna's radiation strength coming from a specific direction to its overall average radiation intensity is known as an antenna's directivity. The average directivity-

$$D = \frac{U(\theta,\varphi)}{U(\theta,\varphi)_{\text{avg}}} = \frac{4\pi U(\theta,\varphi)}{P_{\text{t}}}$$

where P_t is the total radiated power; an increase in the absorbed power also increases the specific absorption. The efficiency of the antenna radiation decreases as the antenna radiation resistance. Increasing the antenna's radiation resistance. The antenna's radiation resistance is heavily influenced by the power it radiates. As a result of this connection between the antenna radiating element and the body tissues, the antenna's radiation efficiency falls inside human body tissues. Considering the biocompatibility issue might be advantageous for the antenna's insulation.

6.1.2 EFFECT OF ANTENNA BANDWIDTH ON THE HUMAN BODY

Due to their small sizes, implanted antennas are anticipated to have a limited bandwidth. However, losses and detuning within the body cause bandwidth to grow. All of the power that is radiated because of tissue absorption and reflection does not reach the receiver. For implanted antennas, the absorbed power is more than the radiated reflected power which is in a larger bandwidth at the expense of a reduced radiation efficiency (Merli, 2011). Various methods, including impedance matching and bio-encapsulation, can be used to reduce these losses. These methods could reduce losses, which would lead to a reduced bandwidth. However, frequency inside the human body inevitably affects implanted antennas with little bandwidth. However, frequency detuning inside the human body inevitably affects implanted antennas with little bandwidth. Therefore, it is necessary to approach this problem with caution.

6.1.3 EFFECT OF ANTENNA RADIATION PATTERN ON THE HUMAN BODY

The antenna's radiated field plot as a function of position at an exact distance is the definition of the radiation pattern of the antenna (Huang and Boyle, 2008). An implanted radiation pattern of the antenna is entirely distinct from a typical antenna in free space. The human body's lossy medium could widen of the radiation patterns. The antenna position and radiation patterns are related. If mounting circumstances and placements vary, an implanted antenna's radiation pattern would similarly vary in the same medium.

6.2 TECHNIQUES TO INCREASE RADIATION EFFICIENCY OF IMPLANTABLE ANTENNAS

The effectiveness of antenna radiation can be increased in a variety of ways. Suppressing the antenna's surface waves is one method. This method is used in Rachna (2017) as depicted in Figure 6.1. An increase in efficiency is provided by a ring antenna supplied by a coplanar waveguide (CPW). To accomplish this, a small part of the substrate near the patch is removed. In doing so, the radiated surface waves would be removed, resulting in a steady radiation efficiency. Applying high permittivity biocompatible superstrate is another way to increase the radiation efficiency of implanted antennas. By improving radiation efficiency, would be inside the human body. In (Oraizi and Rezaei, 2011), this method is used.

6.2.1 EMBEDDED ANTENNAS

The growth of the healthcare sector has brought a lot of interest to biomedical telemetry. This technique allows for the remote monitoring of patients' anatomical indicators without the need for routine hospital visits and examinations. IMDs (implantable medical devices) are crucial for wireless telemetry patient monitoring. IMDs are made up of sensor nodes, and an antenna is one of those nodes. For inculcating antennas to operate continuously a number of variables including abbreviated, patient safety, biomedical, efficiency, lower radiated frequency of operation, and dual-band radiation must be taken into account. In the design of implantable sensors, choosing the antenna is a difficult problem because it affects how well the entire implant performs. A design and analysis of implantable antennas for biomedical applications is provided in this paper. The pattern needs, strategies and difficulties discussed in the extended literature are summarized in this chapter, along with comments on the benefits and drawbacks of various techniques. A variety of implanted antenna examples are provided, and comparisons are drawn. The analysis of antenna implanted antenna performance in the human body and the consequences of the body's inherent looseness is provided. The possibility of enhancing antenna performance parameters is investigated. The various communication scenarios for path loss estimates are then provided. Figure 6.1 (Yan and Vandenbosch, 2016) depicts a schematic of the embedded antenna design process, specifications, difficulties, implantable antenna kinds, and impacts on the human body.

6.2.2 IMPLANTABLE ANTENNAS AND SENSORS

In the literature, various embedded sensors are presented. In (Jiang et al., 2014), a wireless embedded sensor for neural recording called nanoneuro RFID is presented. It is based on magnetoelectric (ME) antennas. Because ME antennas may be reduced to their size, an array of them is used in this sensor. The ME antennas can also gather energy, lengthening the lifetime of the sensor. In the literature, various implanted sensors are presented. In (Scarpello et al., 2012), a wireless implantable sensor for neural recording called nanoneuro RFID is described. It is based on ME antennas. Because ME antennas may be reduced due to their size, an array of them is used in this sensor. The ME antennas can also receive energy increasing the sensor's lifetime (Hamouda et al., 2018) describes the use of an implantable antenna for the detection of mouth cancer using a planar inverted F antenna (PIFA). The computer-simulation technology (CST) Voxel model Katja is used to create the human mouth. There is a noticeable shift in the resonance frequency when malignant tissue is discovered. An implantable RFID sensor tag for continuous glucose monitoring is discussed in (Kumar and Gupta, 2016). The ferrite antenna, RFID tag, potentiostat, glucose sensor, temperature sensor, analog-to-digital converter (ADC), and a digital baseband for signal processing are all components of this sensor system, which is powered by a logical connection. The glucose sensor can measure between 0 and 30 mm. Implantable sensors heavily rely on the antennas, so it is obvious that they play a crucial role in these applications.

6.2.3 REQUIREMENTS AND CHALLENGES IN THE DESIGN OF IMPLANTABLE ANTENNAS

The functioning of the proposed antenna within the human body significantly differs from that of conventional antennas operating in free space due to the lossy nature of human body tissues. Some major design provisions are universal and are therefore included in this section despite the fact that their needs change according to the application.

6.2.4 MINIATURIZATION

Because the implantable antenna accounts up a significant part of an implantable device, its reduction is a critical element. As according (Ruaro et al., 2017) implantable antennas perform in the medical implantable communication service (MICS) band (402–405 MHz), the medical device radio communication service (MedRadio) band (401–406 MHz), the ISM band (ISM) (902–928 MHz), and the 2.4–2.45 GHz band. A variety of reduction techniques are being used to reduce the antenna.

6.3 APPLICATION OF A HIGH PERMITTIVITY DIELECTRIC SUBSTRATE

Using a high permittivity dielectric substrate or superstrate will allow the implanted antenna's size to be lowered. The shorter wavelength produced by the substrate with increased permittivity would lower the resonant frequency (Tahi and Javed, 2015). To minimize the antenna's size, Yan and Vandenbosch, (2016) use a Rogers 6002 substrate with a relative permittivity of 10.2. MgTa1.5Nb0.5O6 is a substrate with an extremely high relative permittivity of 28 and is utilized in for size reduction. The electrical characteristics of the tissue fluid (r = 45.2 and = 0.61 S/m at 402 MHz) were taken into consideration when the antenna was put inside the fluid for testing. In the first scenario, a 14% size reduction is made, whereas a 29% size reduction is made in the second case. The antenna's input power may be transformed into surface waves if a substrate or superstrate with a very high permittivity is used, which will reduce the antenna's radiation efficiency. Because each layer of human body tissue has a different dielectric constant, it is also necessary to take this into account. Depending on the kind of tissue layer in which the implanted antenna is placed, the resonance frequency of the device will change to a higher or lower value.

6.3.1 ENLARGING THE RADIATOR CURRENT PATH

Another method for shrinking the size of implanted antennas is to extend the radiator's current path. The implanted antenna's resonance frequency is lowered by a longer current channel (Tahir and Javed, 2015). Two radiators are stacked on top of one another, either vertically or horizontally, as part of the effective radiator stacking technique to lengthen the current route of the radiator (Fan et al., 2018). This method can result in a 33% size reduction.

6.3.2 CAPACITIVE/INDUCTIVE LOADING

To match the impedance at the appropriate operating frequency, loading techniques might be applied. Inductive or capacitive loading would reduce the imaginary portion of the impedance and aid in size reduction. Yan and Vandenbosch, (2016) employed the inductive loading approach. The capacitive loading method is utilized in (Stoppa and Chiolerio, 2014) to reduce the size. The antenna size can be reduced by 72% by using inductive/ capacitive loading.

6.3.3 HIGHER OPERATING FREQUENCY USAGE

Using a higher operating frequency is the simplest technique to minimize the antenna's size. The resonant frequency of an antenna shifts to a higher value with decreasing antenna size and vice versa. The issue with higher frequency operation is that the losses are greater, which has an impact on the system's overall performance. Since implantable antennas operate in the MICS, MedRadio, and ISM frequency bands (Esselle, 2018), to reduce the elevated degree of losses, the MICS band is increasingly frequently used.

6.3.4 ENGINEERING GROUND PLAN

Impedance matching is crucial to the effective operation of the antenna. The impedance matching and consequently the antenna size change if the ground plane structure is altered (Nag et al., 2018). Making slots in the ground plane to change the return current's route is one technique to accomplish that. This slows the current because the phase shift caused by the current shifting from one edge of the slot to the other results in a lower antenna size (Simorangkir et al., 2018).

6.3.5 PATIENT SECURITY

Inside the human body are implantable medical devices. Therefore, it is important to consider ruling out electromagnetic exposure-related tissue damage. As detailed in the following, several safety measures are used to protect patients.

6.3.5.1 SPECIFIC ABSORPTION RATE (SAR)

The quantity of RF energy absorbed by human body tissues is measured as the SAR. It serves as a gauge to guarantee the protection of biological tissues from electromagnetic radiation. Two of the 13 SAR standards have received widespread acclaim. IEEE C95.1-1999 specifies that the SAR must be less than 1.6 W/kg averaged across a tissue volume of 1 g (Bakogianni and Koulouridis, 2016). According to the IEEE C95.1-2005 standard, the SAR must be less than 2 W/kg on average over a tissue volume of 10. where T_p is the length of the pulse. The biological tissue's temperature may rise as a result of the electromagnetic energy it has absorbed. The temperature of the bodily tissue next to the implanted device should not rise by more than a couple of degrees centigrade (Prudhvi Nadh et al., 2015).

6.3.5.2 EFFECTIVE ISOTROPIC RADIATION POWER (EIRP)

The human body can suffer damage from an implantable antenna with a high EIRP, and it will also interfere with surrounding radio equipment. For an implanted antenna operating in the MedRadio band, the standard limit of EIRP is −16 dBm, while for the ISM band, it is −20 dBm (Rachna, 2017). The input power should be restricted if the implantable antenna is utilized for data telemetry to prevent tissue harm. If the implantable antenna serves as a receiver, these requirements should be followed by the external source power.

6.4 MATERIALS AND CONSTRUCTION TECHNIQUES FOR IMPLANTABLE ANTENNAS

Due to its intended use, implantable antenna construction requires the highest level of wariness. A few of the numerous manufacturing techniques that have been mentioned in the literature includes photolithography, microfluidic antennas, embroidered antennas on fabric, contained polymer composites, and 3D-printed antennas (Mohamadzade et al., 2019). Polydimethylsiloxane (PDMS) was embedded in the conductive fabric in (Simorangkir et al., 2018). Contrary to standard fabrication methods, fabricating the antenna with conductive fabric and PDMS (which simultaneously serves as the substrate and a protective encapsulation) enables easier realization and a more durable flexible antenna structure. The flexibility and robustness are achieved by fabricating an ultrawideband antenna with the same method. Metamaterials are used in the design of implantable antennas when conventional materials are unable to provide the requisite performance. Based on the intended usage of the antenna, various materials are used to create implantable antennas. Their uses range from microwave hyperthermia and medical imaging to wireless pacemaker charging (Wang et al., 2007; Gong and Wang, 2009). In (Bhattacharjee et al., 2018), a proposed implantable wideband antenna is inspired by metamaterials. The antenna's overall dimensions are 0.420.420.0072, with a guided wavelength of 38.1 mm, and it operates at a frequency of 2.3 GHz. The antenna's actual size is roughly 16 × 16 × 0.275 mm^3. Characteristic modal analysis is used to estimate the maximum shrinkage and the bandwidth enhancement. To accomplish the requisite downsizing, the antenna makes use of complementary split ring resonators (SRRs). The antenna's biocompatibility is ensured using silicon and gold as the substrate layer and radiating constituent. The antenna has a 93.5% fractional bandwidth. The antenna performance is validated using a simplified body model. An implantable dual-band circularly polarized antenna was presented. The operational frequencies of 1.65 and 2.5 GHz the antenna's gain was measured to be −16.1 and −7.81 dBi, respectively, while providing an impedance bandwidth of 35.8 and 17.8%, respectively. (Khan et al. 60) demonstrated a triband implantable metamaterial antenna for biotelemetry applications. The antenna is designed on the Medical Implant Communications Services frequency of 402–405 MHz as well as the ISM bands of 902–928 MHz and 2.4–2.45 GHz. While dual slots are also carved in the bottom layer to modify the lower frequency of operation, the SRR metamaterial radiator is integrated into open-ended stubs in the top layer. Using CST Microwave Studio software, the antenna was created and optimized. In (Gong and Wang, 2009), a dual-band implantable antenna with short-circuited SRR was presented. Both the ISM band (2.4–2.45 GHz) and the MICS band (402–405 MHz) are used by the antenna. Miniaturization is accomplished by short-circuiting the antenna's ground plane and radiating component. Spirally coupled SRRs serve as the radiating element. An 18 mm × 24 mm antenna is used. The antenna was made with the aid of CST Microwave Studio. Using an anatomical human body model, the performance of the antenna was examined. Although the suggested antenna has several useful aspects, its bandwidth is somewhat limited.

6.4.1 PERFORMANCE ANALYSIS OF PROPOSED ANTENNAS IN A HUMAN BODY MODEL

The performance of the antenna must still be analyzed using the anatomical body model since its simulation results are more predictive of actual outcomes, despite the simplified body model's accurate antenna simulation results.

Every proposed antenna is generated using a human voxel model. The whole-body simulation takes a lot of time and memory to finish. To minimize simulation time, only the upper half of the body is simulated. When the designed antennas are simulated in anatomical body models, the simulation time and memory requirements are shown in Table 6.2

6.5 ANTENNA CONFIGURATIONS

In the presented paper, dual U slots are constructed as a single patch element and connected with a single feed line. The improvement of bandwidth and miniaturization of antenna size has been improved by the cut the rectangular slot in the ground plane in the proposed antenna. In the presented paper the dual frequency has been designed with the first resonant frequency of 1.65 GHz and a second resonant frequency of 2.5 GHz. The bandwidth of the presented dual-band antenna is, respectively, 45 and 162 MHz, and the value of return loss of the presented antenna respectively −15.83 and −16.33 dB. The design specifications of the presented antenna having a thickness of substrate 1.57 mm and FR4- epoxy material have been used.

TABLE 6.1 Design Parameters of Front View of Antenna (mm) (Rachna, 2017).

L_f	L_1	W_f	L_2
6.3 mm	12.5 mm	1.5 mm	6.5 mm
L_3	W_2	W_3	W_1
6 mm	6.3 mm	11.5 mm	1.5 mm
D_2	D_1	W_5	L_5
6.5 mm	6 mm	18 mm	2 mm
Total volume : 18 mm × 24 mm × 1.57 mm			

6.6 SIMULATION AND RESULTS

6.6.1 RETURN LOSS

The simulated figure shows the presented dual-band antenna which has first band covering the MICS having a first resonant frequency (1.65 GHz) and 45 MHz bandwidth and having second resonate frequency (2.5 GHz), and bandwidth of 160 MHz covering ISM band. So, the proposed simulated antenna is designed for the dual-band which has the center point such as the return loss value of both the dips is below 10 dB. The value of S11 (return loss), respectively, −15.83 and −16.33 dB having resonant frequencies 1.65 and 2.5 GHz. So, this proposed antenna is applicable for implantable medical applications.

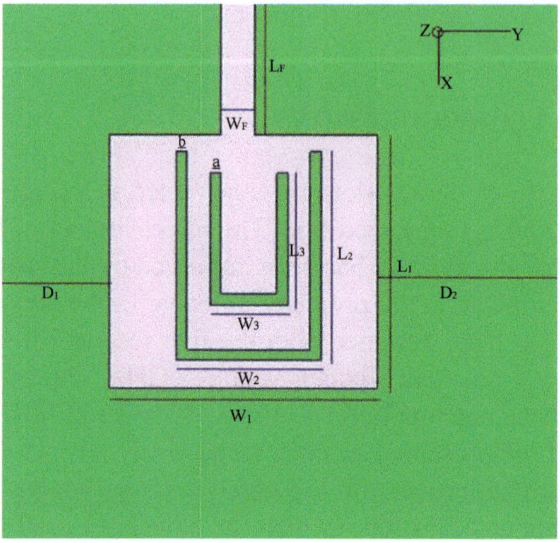

FIGURE 6.1 Simulated design of front view of antenna.

Source: Reprinted from Rachna, 2017

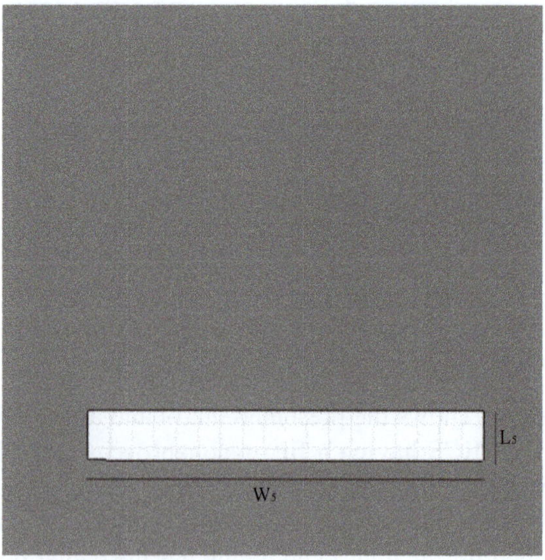

FIGURE 6.2 Simulated design of back view.

Source: Reprinted from Rachna, 2017

TABLE 6.2 Design Parameters of Back View of Antenna (mm) (Rachna, 2017).

W_5	L_5	Total volume
26	2	18 × 20

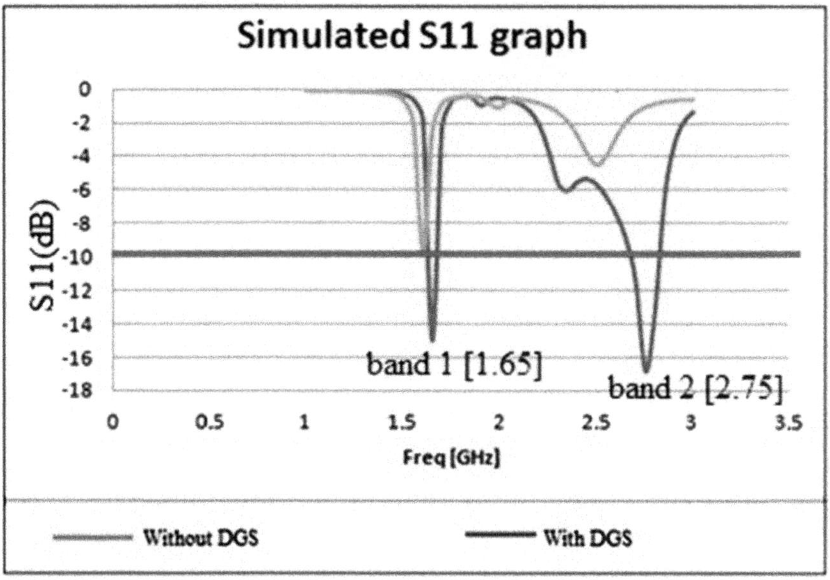

FIGURE 6.3 Simulated result of return loss.

Source: Reprinted from Rachna, 2017

6.6.2 CURRENT POSITIONS

The direction of current flow and the flow of intensity have been distributed such as surface current distributions. The surface current distribution of the lower U-slots' rectangular patch has an effect on the lower resonant frequency of 1.65 GHz and the surface current distribution of upper U-slots shows the resonating frequency of 2.5 GHz which affects the maximum frequency radiations.

6.6.3 GAIN

In the presented antenna the reduced ground structure has been used for the efficient gain. The simulated gain of the proposed antenna is 3.2 dB at 1.65 and 3.34 dB at 2.5 GHz.

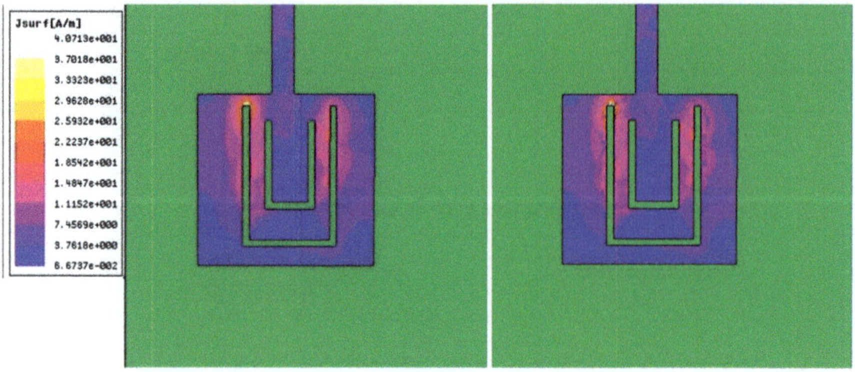

FIGURE 6.4 The surface current dispositions at resonate frequency (a) 1.65 and (b) 2.5 GHz.

Source: Reprinted from Rachna, 2017.

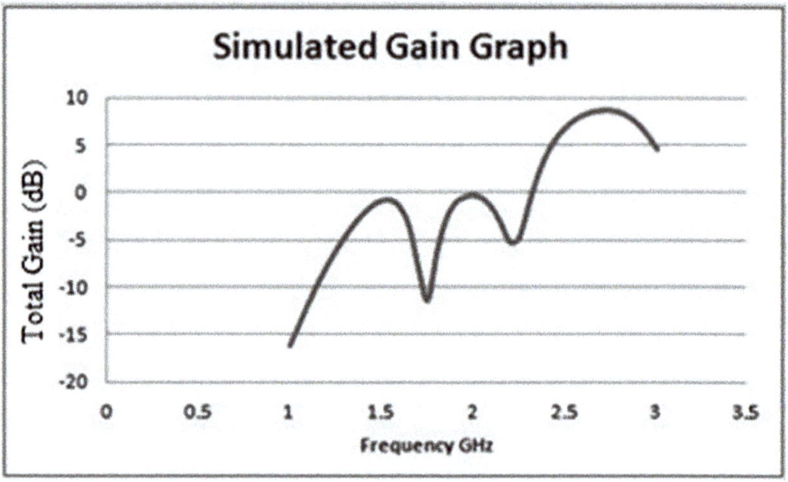

FIGURE 6.5 Simulated gain value at 1.65 and 2.5 GHz.

Source: Reprinted from Rachna, 2017

6.6.4 RADIATION PATTERN

In the antenna design, the radiation pattern is an important parameter for the dependence of the strength of the radiated wave. The radiated field strength of the simulated antenna in different directions has been noticed. The simulated proposed antenna has a unidirectional radiation pattern at frequency 1.65 GHz and an omnidirectional one at a frequency 2.5 GHz which will demand a miniaturized dual-band implantable antenna for medical applications.

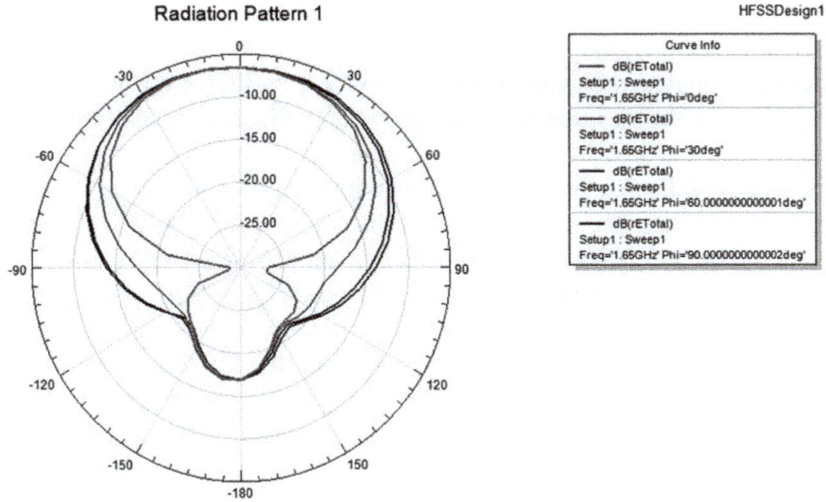

(a)

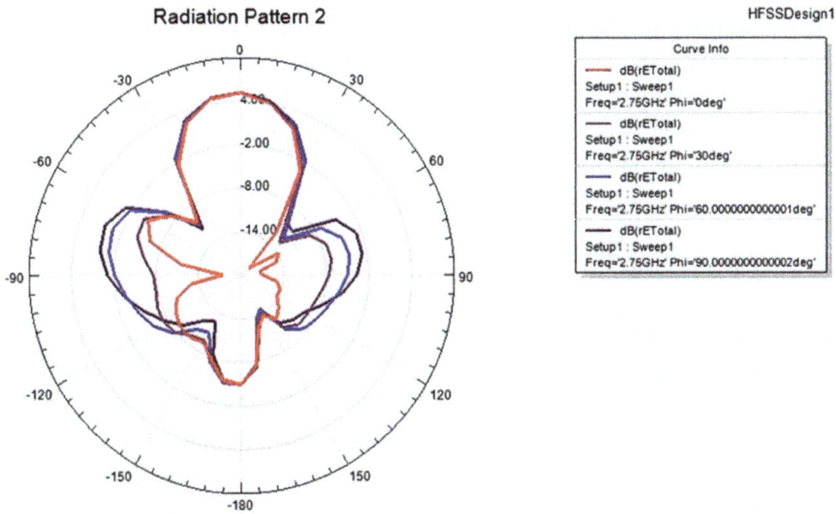

(b)

FIGURE 6.6 Radiation pattern at 1.65 and 2.5 GHz.

Source: Reprinted from Rachna, 2017

6.7 ANALYTICAL BODY MODEL PERFORMANCE EVALUATION OF RECTANGULAR PRINTED ANTENNA

The rectangular patch antenna was designed after the fat layer of an anatomical body model. A streamlined fat layer phantom and an anatomical body model were used to simulate the return loss plot of the rectangular patch antenna in Figure 6.3. At 402, 915, and 2.4 GHz, the rectangular patch antenna resonates in an anatomical body model and fat layer phantom. The bandwidth of the rectangular patch antenna in the anatomical body model is 206, 150, and 150 MHz while the bandwidth of the simplified fat layer phantom is 108, 170, and 250 MHz. The fat layer phantom is simplified at these frequencies. The gain of the antenna of the rectangular patch antenna at 402 MHz, 915 MHz, and 2.4 GHz in a simplified fat layer phantom is −47.7, −37.2, and −25.5 dBi, respectively, while it is −50.9, −42.2, and 23.2 dBi in an anatomical body model shows the polar plot of the radiation pattern from of the rectangular patch antenna on a simplified fat layer phantom and anatomical body model. The reduced fat layer phantom's primary lobe of radiation does have a 3 dB smaller angular diameter in the E and H planes than the anatomical body model for frequencies of 402, 915, and 2.4 GHz. For the simplified fat layer phantom, the primary lobe of radiation at 915 MHz is oriented in the E and H-plane at 0 and 1 degrees, whereas for the anatomical body model, it is at 0° and 1°. The E and H-plane radiation lobes have 157° and 158° inclinations, respectively, at 915 MHz. The simplified fat layer phantom has a primary radiation lobe in the E and H planes of 148° and 143°, whereas the anatomical body model does have a main radiation lobe of 41° and 4° at 2.4 GHz.

6.8 CONCLUSION

This paper shows the advanced design of the implantable antenna which placed in the human body. The parametric analysis has been studied for the antenna parameters of medically applicable antennae. Several studies have been done with aim of achieving the dual-band antenna which resonates with the operating frequencies 1.65 and 2.5 GHz. The operating resonating bandwidth is 45 MHz at 1.65 GHz, and 162 MHz at 2.5 GHz which covers the ISM bands and medical implant communication systems (MICS) which have the desired value of SAR. Implantable medical devices have gained a significant amount of attention in biotelemetry lately due to their benefits, including wireless capsule endoscopy, blood sugar level monitoring, blood glucose level monitoring, cardiac pacemakers, and many others. They can follow the physiological data continuously. The fact that these gadgets are made up of so many tiny components, which makes their design an extremely difficult undertaking. The antenna is a crucial component among all of these since it serves as both a transmitting and a receiving antenna. Implantable sensors with a reduced size are preferred for biotelemetry. The antenna is particularly challenging to keep compact while retaining performance compared to the other implantable sensing elements. As a result, designing an implanted antenna is a difficult task. Three implantable antennas were created for use with medical implants. A circular patch antenna was the initial amazing step forth. To replicate the antenna, various implant depths and multilayer tissues were utilized. The antenna continued performing well in the skin and muscle phantom layers even though the frequency was changed to higher bands in the fat phantom layer, which can be tuned. The results of every single antenna simulation were favorable. Some calculations were done using average antenna absorption rates of 1 and 10 g. According to SAR analysis, the antenna SAR values are within acceptable limits. For use in biotelemetry, a rectangular microstrip patch antenna has, indeed, been developed. The antenna used many bands. These frequency bands can be used for data transmission, wake–up signals, and wireless power transfer applications. A reasonable match was discovered after analysis of the simulated results for both free space and implanted devices. The antenna worked in multilayer tissues successfully at various implant depths. Gain and radiation of the antenna improve with deeper implants efficiency improved. This antenna's SAR values complied with the required norms. A small antenna was created. This weighs little and has a tiny overall volume. The originality of this antenna is that it retains its operating frequency despite the varied conducting qualities of the body tissues and the fact that it is quite small. The antenna functions at the same frequency in multilayer tissues (skin, fat, and muscle). This antenna has a greater bandwidth of 170 MHz and a lower ISM resonance frequency of 915 MHz, in contrast to small-size antennas that frequently have narrow bandwidths and higher operating frequencies. Even at a deeper implantation inside the layers of phantom tissue, the antenna displayed good function. The antenna is the greatest fit for implanted sensors due to its low SAR values. The comparative examination of every developed antenna has demonstrated that the designed antennas are smaller and perform better than earlier designed antennas that were published in the open literature. The circular patch antenna offers a 15% overall size reduction and a 9% overall performance improvement when compared to other antennas of a comparable design, according to Table 6.1. According to comparison Table 6.2, rectangular patch antenna design achieves a size reduction of about 29% while improving overall performance by 15%. Table 6.2 shows that when compared to similar antennas described in the literature, the compact-size antenna has an overall performance improvement of 12% and a size reduction of about 47%. Different implantation situations were used to analyze the designed antenna performance. Different form phantoms with the same volume were used to approximate the little antenna. Although the gain and radiation efficiency were altered, the antenna return loss and bandwidth remained unaffected. Different insulating layers were used to analyze the performance of the rectangular patch antenna. Different layer thicknesses have a minor impact on antenna performance at lower frequencies, however, for higher ISM bands, the antenna's bandwidth was enhanced while the return loss was decreased. To test the developed antennas' compatibility with sensors, imitations of circuit components were employed. The outcomes of the simulation demonstrated that the antennas worked effectively with the circuit elements. The created antennas additionally shown good agreement between the simulation outcomes of the anatomical body model and reduced body model.

KEYWORDS

- **dual band**
- **defected ground structure**
- **medical applications**
- **specific absorption rate**

REFERENCES

Agarwal, K.; Guo, Y. X.; Salam, B. Wearable AMC Backed Near-Endfire Antenna for On-Body Communications on Latex Substrate. *IEEE Trans. Compon. Packag. Manuf. Technol.* **2016,** *6* (3), 346–358.

Agneessens, S.; Rogier, H. Compact Half Diamond Dual-Band Textile HMSIW On-Body Antenna, *IEEE Trans. Antennas Propag.* **2014,** *62* (5), 2374–2381.

Ahlbom, A. Guidelines for Limiting Exposure to Timevarying Electric, Magnetic, and Electromagnetic Fields (up to 300 GHz). *Health Phys.* **1998**, *74*, 494–521.

Alomainy, A.; Hao, Y. Modeling and Characterization of Biotelemetric Radio Channel from Ingested Implants Considering Organ Contents. *IEEE Trans. Antennas Propag.* **2009,** *57* (4), 999–1005.

Bahrami, H.; Mirbozorgi, S. A.; Ameli, R.; Rusch, L. A.; Gosselin, B. Flexible Polarization-Diverse UWB Antennas for Implantable Neural Recording Systems. *IEEE Trans. Biomed. Circuits Syst.* **2016**, *10*, 38–48.

Bakogianni, S.; Koulouridis, S. An Implantable Planar Dipole Antenna for Wireless Medradioband Biotelemetry Devices. *IEEE Antennas Wirel. Propag. Lett.* **2016**, *15*, 234–237.

Bharadwaj, R.; Parini, C.; Alomainy, A. Experimental Investigation of 3-D Human Body Localization using Wearable Ultra-Wideband Antennas. *IEEE Trans. Antennas Propag.* **2015,** *63* (11), 5035–5044.

Bhattacharjee, S.; Maity, S.; Chaudhuri, S. R. B.; Mitra, M. Metamaterial-Inspired Wideband Biocompatible Antenna for Implantable Applications. *IET Microw. Antennas Propag.* **2018,** *12* (11), 1799–1805.

Bhattacharjee, S.; Maity, S.; Metya, S. K.; Bhunia, C. T. Performance Enhancement of Implantable Medical Antenna using Differential Feed Technique. *Int. J. Eng. Sci. Technol.* **2016,** *19* (1), 642–650.

Declercq, F.; Rogier, H. Active Integrated Wearable Textile Antenna with Optimized Noise Characteristics. *IEEE Trans. Antennas Propag.* **2010,** *58* (9), 3050–3054.

Declercq, F.; Rogier, H.; Hertleer, C. Permittvity and Loss Tangent Characterization for Garment Antennas based on a New Matrix-Pencil Two-Line Method. *IEEE Trans. Antennas Propag.* **2008,** *56* (8), 2548–2554.

Esselle, K. P. Antennas for Wireless Medical Devices. In *Antennas for Small Mobile Terminals*; Artech House, 2018; pp 183–202.

Fan, Y.; Huang, J.; Chang, T.; Liu, X. A Miniaturized Four-Element MIMO Antenna with EBG for Implantable Medical Devices. *IEEE J. Electromagn. RF Microw. Med. Biol.* **2018,** *2* (4), 226–233.

Gong, Y.; Wang, G. Superficial Tumor Hyperthermia with Flat Left-Handed Metamaterial Lens. *Prog. Electromagn. Res.* **2009,** *98* (5), 389–405.

Ha, S. J.; Jun, C. W. Reconfigurable Beam Steering using a Microstrip Patch Antenna with a U-Slot for Wearable Fabric Applications. *IEEE Antennas Wirel. Propag. Lett.* **2011,** (10), 1228–1231.

Hammouda, O.; Allam, A. Utilizing Implanted Antennas to Detect the Presence of Oral Cancers. *Res. J. Eng. Sci.* **2014,** *3* (7), 22–27.

Hamouda, Z.; Wojkiewicz, J. L.; Pud, A. A.; Koné, L.; Bergheul, S.; Lasri, T. Magneto-Dielectric Nanocomposite Polymer based Dual-Band Flexible Antenna for Wearable Applications. *IEEE Trans. Antennas Propag.* **2018,** *66* (7), 3271–3277.

Huang, Y.; Boyle, K. In *Antennas: From Theory to Practice*; John Wiley & Sons, 2008.

Islam, S. M.; Esselle, K. P.; Bull, D.; Pilowsky, P. M. In *Bandwidth Enhancement of an Implantable RFID Tag Antenna at 900 MHz ISM Band for RF Telemetry*, International Symposium on Communications and Information Technologies (ISCIT), Gold Coast, QLD, Australia, 2012; p 741.

Jiang, Z. H.; Brocker, D. E.; Sieber, P. E.; Werner, D. H. A Compact, Low-Profile Metasurface-Enabled Antenna for Wearable Medical Body-Area Network Devices. *IEEE Trans. Antennas Propag.* **2014,** *62* (8), 4021–4030.

Kim, J.; Rahmat-Samii, Y. Implanted Antennas Inside a Human Body: Simulations, Designs, and Characterizations. *IEEE Trans. Microw. Theory Techn.* **2014,** *2*, 1934–1943.

Kiourti, A.; Nikita, K. S. Miniature Scalp-Implantable Antennas for Telemetry in the MICS and ISM Bands, Design, Safety Considerations and Link Budget Analysis. *IEEE Trans. Antennas Propag.* **2012,** *60* (8), 3568–3575.

Kumar, V.; Gupta, B. On-Body Measurements of SS-UWB Patch Antenna for WBAN Applications. *AEU-Int. J. Electron. Commun.* **2016,** *70* (5), 668–675.

Li, H.; Guo, Y.; Liu, C.; Xiao, S.; Li, L. A Miniature-Implantable Antenna for Medradio-Band Biomedical Telemetry. *IEEE Antennas Wirel. Propag. Lett.* **2015,** *14*, 1176–1179.

Li, H.; Sun, S.; Wang, B.; Wu, F. Design of Compact Single-layer Textile MIMO antenna for Wearable Applications. *IEEE Trans. Antennas Propag.* **2018,** *66* (3), 3136–3141.

Masood, R.; Person, C.; Sauleau, R. A Dual-Mode, Dual-Port Pattern Diversity Antenna for 2.45- GHz WBAN. *IEEE Antennas Wirel. Propag. Lett.* **2017,** *16.*

Memarzadeh-Tehran, H.; Abhari, R.; Niayesh, M. A Cavity-Backed Antenna Loaded with Complimentary Split Ring Resonators. *AEU-Int. J. Electron. Commun.* **2016,** *70* (7), 928–935.

Merli, F. Implantable Antennas for Biomedical Applications. PhD Dissertation, EPFL University, Lausanne, Switzerland, 2011.

Mohamadzade, B.; Hashmi, R. M.; Simorangkir, R. B.; Gharaei, R.; Ur Rehman, S.; Abbasi, Q. H. Recent Advances in Fabrication Methods for Flexible Antennas in Wearable Devices. *Sensors* **2019,** *19* (10), 2312.

Moore, R. Effects of a Surrounding Conducting Medium on Antenna Analysis. *IEEE Trans. Antennas Propag.* **1963,** *11* (3), 216–225.

Nag, A.; Simorangkir, R. B.; Valentin, E.; Björninen, T.; Ukkonen, L.; Hashmi, R. M.; Mukhopadhyay, S. C. A Transparent Strain Sensor based on PDMS-Embedded Conductive Fabric for Wearable Sensing Applications. *IEEE Access* **2018,** 71020–71027.

Nikolayev, D.; Zhadobov, M.; Karban, P.; Sauleau, R. In *Increasing the Radiation Efficiency and Matching Stability of In-Body Capsule Antennas*, 10th European Conference on Antennas and Propagation (EuCAP), Davos, Switzerland, 2016; pp 1–5.

Oraizi, H.; Rezaei, B. Improvement of Antenna Radiation Efficiency by the Suppression of Surface Waves. *J. Electromag. Analys. Appl.* **2011,** *3* (03), 79.

Prabha, R. Design of Dual U Slotted Printed Antenna with DGS. *Int. J. Innov. Sci. Res. Technol.* **2017,** *2* (5).

Prudhvi Nadh, B.; Madhav, B. T. P.; Siva Kumar, M. Design and Analysis of Dual Band Implantable DGS Antenna for Medical Applications, 2015.

Rachna, P. Design of Dual U Slotted Printed Antenna with DGS. *Int. J. Innov. Sci. Res. Technol.* **2017,** *2 (5).*

Rais, N. H. M.; Soh, P. J.; Malek, M. F. A.; Vandenbosch, G. A. Dual-Band Suspended-Plate Wearable Textile Antenna. *IEEE Antennas Wirel. Propag. Lett.* **2013,** *12*, 583–586.

Ruaro, A.; Thaysen, J.; Jakobsen, K. B. Wearable Shell Antenna for 2.4 GHz Hearing Instruments. *IEEE Trans. Antennas Propag.* **2017,** *64* (6), 2127–2135.

Saeed, S. M.; Balanis, C. A.; Birtcher, C. R.; Durgun, A. C.; Shaman, H. N. Wearable Flexible Reconfigurable Antenna Integrated with Artificial Magnetic Conductor. *IEEE Antennas Wirel. Propag. Lett.* **2017,** (16), 2396–2399.

Sakthi, A.; Nair, R. S.; Raju, S. In *Design and Equivalent Circuit Analysis of Textile Antenna for WLAN and WBAN Application*, Region 10 Conference (TENCON), 2016.

Scarpello, M. L.; Kazani, I.; Hertleer, C.; Rogier, H.; Ginste, D.V. Stability and Efficiency of Screen-Printed Wearable and Washable Antennas. *IEEE Antennas Wirel. Propag. Lett.* **2012,** (11), 838–841.

Shakib, M. N.; Moghavvemi, M; Mahadi, W. N. L. Design of a Tri-band Off-Body Antenna for WBAN Communication. *IEEE Antennas Wirel. Propag. Lett.* **2017,** *16*, 210–213.

Simorangki, R. B.; Kiourti, A.; Esselle, K. P. UWB Wearable Antenna with a Full Ground Plane Based on PDMS-Embedded Conductive Fabric. *IEEE Antennas Wirel. Propag. Lett.* **2012,** *17* (3), 493–496.

Simorangkir, R. B.; Yang, Y.; Hashmi, R. M.; Björninen, T.; Esselle, K. P.; Ukkonen, L. Polydimethylsiloxane-Embedded Conductive Fabric: Characterization and Application for Realization of Robust Passive and Active Flexible Wearable Antennas *IEEE Access* **2018,** 48102–48112.

Song, L.; Rahmat-Samii, Y. A Systematic Investigation of Rectangular Patch Antenna Bending Effects for Wearable Applications. *IEEE Trans. Antennas Propag.* **2018,** *66* (5), 2219–2228.

Steinhaus, B. M.; Smith, R. E.; Crosby, P. In *The Role of Telecommunications in Future Implantable Device Systems Engineering in Medicine and Biology Society*, Proceedings of the 16th Annual International Conference of the IEEE Engineering Advances: New Opportunities for Biomedical Engineers, 1964; (2).

Stoppa, M.; Chiolerio, A. Wearable Electronics and Smart Textiles: A Critical Review. *Sensors* **2014,** *14* (7), 11957–11992.

Tahir, F. A.; Javed, A. A Compact Dual-Band Frequency-Reconfigurable Textile Antenna for Wearable Applications. *Microw. Opt. Technol. Lett.* **2015,** *57* (10), 2251–2257.

Tak, J.; Woo, S.; Kwon, J.; Choi, J. Dual-Band Dualmode Patch Antenna for On-/Off-Body WBAN Communications. *IEEE Antennas Wirel. Propag. Lett.* **2016,** *15*, 348–351.

Tsao, C. Radiation Resistance of Antennas in Lossy Media. *IEEE Trans. Antennas Propag.* **1971,** *19* (3), 443–444.

Wang, G.; Fang, J.; Dong, X. Resolution of Near-Field Microwave Target Detection and Imaging by using Flat LHM Lens. *IEEE Trans. Antennas Propag.* **2007,** *55* (12), 3534–3541.

Wittig, T. SAR Overview [Online]. *https://www.cst.com/Content/Events/UGM2007/05-Witting.pdf.*

Xiao, Z.; Tan, X.; Chen, X.; Chen, S.; Zhang, Z.; Zhang, H.; Wang, J. An Implantable RFID Sensor Tag Toward Continuous Glucose Monitoring. *IEEE J. Biomed. Health Inf.* **2015,** *19* (3), 910–919.

Xu; L. J.; Guo, Y. X.; Wu, W. Dual-Band Implantable Antenna with Open-End Slots on Ground. *IEEE Antennas Wirel. Propag. Lett.* **2012**, *11*, 1564–1567.

Yan, S.; Vandenbosch, G. A. E. Radiation Pattern-Reconfigurable Wearable Antenna based on Metamaterial Structure. *IEEE Antennas Wirel. Propag. Lett.* **2016,** *15*, 1715–1718.

CHAPTER 7

Biomedical Applications Based on Mobile Phone Cell Network Systems

GAYATRI TIWARI[1], SAURABH KUMAR[2], and SHALINI SHARMA[3]

[1]Department of Electronics & Communication Engineering, Institute of Engineering & Technology, Lucknow, Uttar Pradesh, India

[2]Department of Electrical Engineering, Engineering Degree Division, Institute of Engineering and Rural Technology, Prayagraj, Uttar Pradesh, India

[3]Department of Electronics & Communication Engineering, University of Lucknow, Lucknow, Uttar Pradesh, India

ABSTRACT

The study focuses on the usability and efficacy of wireless mobile medical devices in light of the development of mobile wireless technology in telebiomedical applications. Future medical advancements in a variety of fields will support wireless networking. By updating carers on a patient's health in real time, a patient who is in a remote location may be handled remotely. The primary purpose of the electronic medical device was to track physiological parameters such as body temperature, oxygen saturation, heart rate, blood pressure, and blood sugar levels. However, with the advancement of wireless technologies, e-healthcare systems have seen a range of changes and advancements. Telemedicine is a distant medical practice that allows collaboration between different individuals and supports their collaborative efforts to diagnose or treat disease using IT and telecommunications, Telebiomedical with wireless

Smart Electronic Devices and Systems for Biomedical and Healthcare Applications. Upendra Kumar, Rajesh K. Kesharwani, & Raj K. Keservani (Ed.)

DOI: 10.1201/9781003560852-7

networks for mobile communication uses the app. Medical databases and servers are distant computers that help doctors and hospitals evaluate and monitor vital signs and provide adequate facilities to patients in real time. The term "telebiomedical environment" refers to the use of mobile phones (handheld devices) and other wireless systems and communication devices to provide information to users or customers of medical preventive healthcare services. In the monitoring and delivery of healthcare services, mobile phones are becoming increasingly vital. Because of their superior computing functions, expanded preferences, and diversified capabilities, they are frequently referred to as pocket computers. Mobile healthcare (mHealth) apps are more viable and inventive because of their advanced sensors and extensive software applications. This chapter focuses on the latest exploration of existing mobile phone-based sensor applications in biomedical healthcare. This chapter highlights the benefits of cell phones and recent advancements in patient monitoring, teaching, and management, there are also certain significant difficulties and obstacles to be addressed, including data security and privacy, acceptance, reliability, and cost.

7.1 INTRODUCTION

We are hearing about electronics in our daily life, Semiconductors have undoubtedly changed the world beyond anything that could have been imagined before them (Kumar et al., 2013). People have always required to communicate and process data, it is acknowledgment to the semiconductors that these two important tasks have become easy and take up infinitely less time than, for example, the time of vacuum tubes. Without semiconductor materials and integrated circuits (chips) they cannot be done as small, light-weight, high speed, and low power consumption (Singh et al., 2013). There is a vast variety of applications of semiconductor materials in different sectors of modern electronics and communications. Semiconductor devices integrate into complex circuits and are easily manufactured in micro-electronic circuits (Kumar et al., 2017). The majority of electric and electronic instruments and systems, including those in communications, data processing, consumer products, biomedicine and mobile health, and industrial control equipment, present a strong chance for them to do so in the future (Bochicchio et al., 2019). When we travel back in time 100 years, we encounter obstacles from the past and picture ourselves facing them. Medical electronics and equipment are crucial technologies that have advanced significantly in this period.

The entrance of future development and their explosions is truly challenging. As we look to a greater extent of development of the past and imagine this challenge 100 years ago when we know that no one predicted the advancements in medical devices and medical electronics that we have today (Easttom and Mei, 2019). One such example of the equipment is stethoscope. This first medical instrument provided information to the clinician in the form of moderately amplified sound waves, due to the design of the device. However, these sound waves yet had to be interpreted by the clinician, who had to recognize not only the major sounds, but the complex nuances in timing and low-amplitude interstitial sounds between the sound bursts. Thus, the science of using a stethoscope implied that you could interpret sounds in terms of tone and intensity over time.

The basic physiology of electrically active cells such as nerve and muscle cells, it also led to the development of electrophysiological instrumentation that is used in the diagnosis of disease (Mannatunga et al., 2020). Compared to Einthoven's initial chain galvanometer, the electrocardiograph (ECG) of today is substantially simpler. It allows cardiologists to better understand heart rate, which is the temporal relationship of a series of heartbeats, in terms of frequency and morphology the currents of secondary damage to the heart muscle, the effects of drugs on the heart, and the size and position of the heart cavities (Vullings, 2019). It also provides recordings of the electrical activity of the heart as seen from various angles inside the body. The remainder of this essay is structured as follows: The latest global Internet of Things (IoT) breakthroughs in e-health and mHealth, as well as its monitoring, are discussed in Section 7.5. The idea of an implanted antenna for use in biological applications was proposed in Section 7.6. Section 7.7 provides a brief overview of the use of data analysis and digital signal processing in remote mHealth patient monitoring systems. In Section 7.8, the issues and prospects for software-driven biomedical devices are outlined and analyzed. The paper is concluded in Section 7.9.

7.2 RECENT INTERNET OF THINGS FOR MHEALTH CARE AND MONITORING

The development of the IoT, which connects various medical devices, equipment, sensors, and healthcare providers to provide high-quality medical services in remote or nonremote locations, has been the subject of extensive research in the last few decades in the fields of health services, pharmaceuticals, and their technological advancement (Vijayakumar and Bhuvaneswari, 2020). By lowering healthcare expenses, facilitating quicker access to healthcare services, and enhancing operational efficiency, this technology has enhanced patient safety. According to (Ray et al., 2021), the current study provides updated, solid proof of the prospective healthcare uses of IoT (especially HIoT) based technology. From the perspective of enabling technology, healthcare services, and applications in resolving different healthcare concerns, it has been observed that the application of HIoT has advanced in this area. The healthcare sector has experienced considerable expansion in recent years and has made significant contributions to both employment and revenue.

The IoT is described as an exchange of data over a network that belongs to physical devices that use connectivity by enabling the server. These devices are not necessarily intricate technological equipment and their advancement (Mohsen et al., 2020). However, as they go, procedures become more efficient and health professionals can perform their jobs more successfully. Healthcare and technology-focused companies are more likely to make significant IoT investments (Gupta et al., 2021). Nowadays, the majority of technological tools and gadgets have some kind of connectivity, from biosensor-equipped wearables to Wi-Fi or Bluetooth-enabled X-ray machines. Medical gadgets that are IoT-enabled offer vital information that helps healthcare professionals do their duties. IoT in healthcare simplifies routine but crucial procedures to enhance patient outcomes and lessen difficulties and obstacles for medical professionals. Medical gadgets with integrated IoT may perform tasks including remote patient monitoring, treatment progress observation, and immunization monitoring.

7.2.1 WORLD WIDE DEVELOPMENTS IN E-HEALTH AND MHEALTH

Mobile technology is used in many different industries, but its effects on healthcare have not yet been completely realized. This isn't a result of a lack of curiosity. Downloadable health applications number 85,000, and in 2018–2019, more than $8.5 billion in "digital health" sponsorships were made. Novel tiny sensors are being developed to continually detect biomarkers that have been realistically quantified in a clinic (e.g., sweat4, tear fluid5) (Gupta et al., 2021). We emphasize the use of these sensors and applications for universal mobile health (or mHealth) data collection that is customized for health. Data gathering is just the beginning toward developing mHealth solutions for better health. This is accurate whether you are evaluating the consequences of mHealth on heart failure patients' admission rates to hospitals or lifestyle changes. Health results such as weight and blood pressure, or adherence to advised rehabilitation exercises. To enhance the number of mHealth initiatives that succeed in their goals, whether it is developing a technique for incorporating mHealth technology into clinical workflows or accurately forecasting patient outcomes from mHealth data (Mohsen et al., 2020). That also suggests a chance for group efforts to advance operations for the entire organization. To emphasize the impacts of sharing digital resources in particular, this is necessary. Promote "sharing models, software, datasets, and other digital resources whenever possible to accurately predict health outcomes from mHealth data" as a result. Sharing data is necessary, but it would not be enough to advance mHealth applications into planned research and clinical use (Zaman et al., 2020). By process software and hardware models are required to conclude the data, therefore, is crucial that these must be shared as well. Ideally, they should be released under an open–source license that allows for revision, modification, and reuse for new applications.

7.2.2 ARCHITECTURE OF HEALTHCARE IOT (HIOT)

The introduction of healthcare apps that use IoT frameworks thanks to the convergence of IoT and cloud computing has revolutionized the area of medicine. These frameworks make it possible for data to be seamlessly transferred from various medical devices and sensors to a specialized healthcare network created with patient data in mind. Valuable insights may be gleaned by analyzing data gathered from sensors including blood pressure monitors, body temperature sensors, oxygen saturation monitors, and pulse rate trackers.

In this situation, the publisher is in charge of continually transmitting the patient's data to a broker across a network. The broker processes and stores the data on the cloud while serving as a middleman. Authorized users, healthcare professionals, or carers may then access and view the processed data via gadgets such as smartphones, PCs, or tablets. This makes it possible to easily access the patient's information and monitor them remotely.

In this case, the subscriber enjoys the benefit of ongoing patient information monitoring. They can spot and evaluate any changes or anomalies in the patient's health. The subscriber receives feedback about any discovered abnormalities or deterioration in the patient's health after the publisher and broker process the data.

In summary, this framework facilitates remote monitoring, real-time data transmission, and analysis of patient data, enabling healthcare professionals to gain valuable insights and promptly respond to any concerning changes in the patient's health condition.

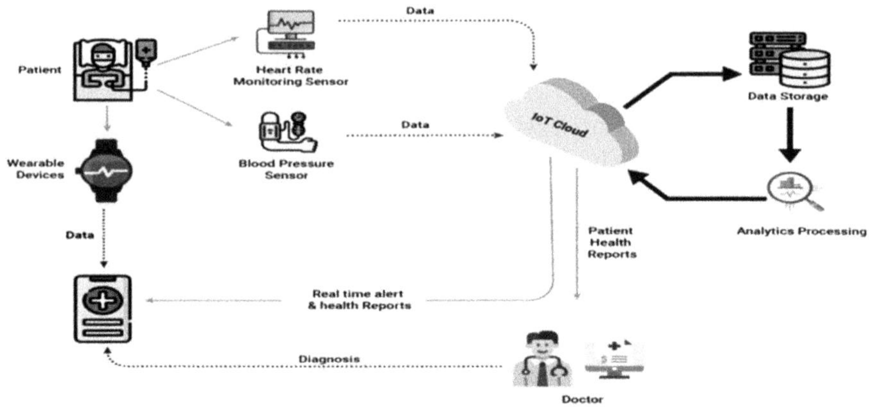

FIGURE 7.1 Architecure of HIoT.

Source: https://data.embeddedcomputing.com/uploads/articles/wp/12298/5efb903f31440-Picture1.png

7.3 HIoT TECHNOLOGIES

A HIoT system's technology selection is essential since it may greatly improve its capabilities (Ray et al., 2021). Identification technology, communication technology, and location technology are the three primary categories of cutting-edge technologies that have been implemented for integrating health-care applications with an IoT system (Zaman et al., 2020). The utilization of apparel IoT devices in healthcare offers both benefits and challenges. It provides healthcare providers with new opportunities to monitor patients and enables patients to monitor themselves (Zaman et al., 2020).

7.3.1 IDENTIFICATION TECHNOLOGY

Patient information obtained from the authorized node (sensor), which may be situated far away. The ability to accurately identify the nodes and sensors that are present in the healthcare network will enable this. Each authorized entity is given a unique identity (UID) as part of the identification process, which enables quick identification and clear data transmission (Vijayakumar and Bhuvaneswari, 2020). Every resource connected to the healthcare system, such as a hospital, doctor, nurse, carer, medical gadget, etc., often comes with a digital UID.

7.3.2 COMMUNICATION TECHNOLOGY

In an healthcare Internet of Things (HIoT) network, linking various qualities requires communication technology. Short-range and medium-range communication technologies can be distinguished between them. A body area network (BAN) or other small–area network of items are connected using short-range communication technology. Radiofrequency identification (RFID), wireless fidelity (Wi-Fi), ZigBee, near-field communication (NFC), and Bluetooth are typical examples of these technologies. Through tags and readers, RFID identifies and tracks items using radio waves. Devices may connect and interact wirelessly inside a local area network thanks to Wi-Fi. ZigBee is a low-power wireless protocol suitable for short-range communication in home automation and industrial applications. NFC allows for close-range communication between devices, commonly used in contactless payments and data sharing between smartphones. Bluetooth facilitates wireless communication between devices, often used for connecting peripherals and IoT devices. Medium-range communication technologies, on the contrary, support communication over larger distances, typically between a base station and a centralized node of the BAN. Examples of medium-range technologies in HIoT applications include cellular networks (4G/5G), satellite communication, and other long-range wireless protocols. Short-range communication technologies are preferred in HIoT applications due to their low power consumption, cost-effectiveness, and suitability for localized data exchange. However, the choice of communication technology depends on factors such as range requirements, power consumption, data transfer rates, and specific application needs.

7.3.3 LOCATION TECHNOLOGY

Real-time location systems (RTLS) or location technologies are extensively used in healthcare networks to accurately identify and track the real-time position of objects. These systems are also utilized to monitor the movement of medication and ensure effective resource distribution in healthcare services (Zaman et al., 2020). The global positioning system (GPS), widely recognized and popularly known, is a prominent technology used for tracking. To offer accurate object placement, GPS depends on a constellation of satellites. GPS can pinpoint an object's location as long as there is an open line of sight between it and at least four satellites. GPS may be used in the world of the HIoT to track the whereabouts of patients, ambulances, healthcare professionals, carers, and other important people in real time.

It is crucial to remember that GPS is primarily intended for outside use. Obstacles in the way of the environment or specific infrastructure may make it difficult for the item and the satellite to communicate. A local positioning system (LPS) network can be used to get around these restrictions. LPS employs an array of predeployed receivers to track objects by utilizing radio signals emitted by the moving objects. Several short-distance communication technologies including RFID, Wi-Fi, ZigBee, and others can be utilized for LPS implementations. Among these technologies, ultra-wideband (UWB) radio stands out due to its higher temporal resolution, which enables more precise tracking and positioning capabilities.

7.4 REMOTE HEALTH MONITORING AND TELEHEALTH

Returning the patient to his "physical environment" is important, and there is one way to lessen the strain on healthcare professionals who, in many nations and at various times, are unable to keep up. A scarcity of healthcare staff, particularly during and after seasons when more diseases affect us, is a recurrent yearly catastrophe in various countries due to budgetary issues (Gupta et al., 2021). Figure 7.2 illustrates how remote health monitoring—made possible by the IoT—partially addresses the problem of the spread and aging population and other factors contributing to the increase of chronic illnesses. When patients reside in remote places, remote health monitoring is also beneficial. Today, a wide variety of (specialized) wearables, biosensors, and other medical equipment are available that allow for remote health monitoring.

7.4.1 THE CONNECTED AND SMART HOSPITAL—A BROAD RANGE OF APPLICATIONS

IoT applications abound in hospitals and other healthcare settings, fostering a connected and intelligent environment. The administration of medical facilities and equipment is one of the major uses of IoT. Hospitals may manage and monitor assets, assuring their availability and reducing loss, by using RFID and IoT-enabled equipment. This increases operational effectiveness while simultaneously lowering equipment acquisition costs. Predictive maintenance is another valuable use case of IoT in healthcare facilities. Through the use of IoT sensors and connectivity, hospitals can proactively monitor the condition of medical devices and infrastructure. Real-time data collected from these devices enables early detection of potential issues, allowing for timely maintenance and minimizing equipment downtime. This approach enhances reliability and ensures critical healthcare systems are consistently available to patients.Through IoT, scheduling and personnel coordination in the healthcare industry may also be optimized. Hospitals can more efficiently schedule operations, surgeries, and appointments by linking healthcare equipment and gadgets to a centralized system. Effective resource allocation is made possible by integrating real-time data on equipment availability, patient status, and personnel availability. This lowers the wait times and boosts overall effectiveness.

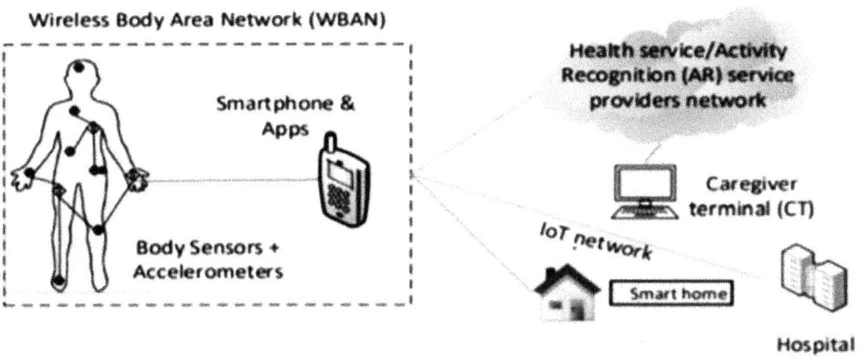

FIGURE 7.2 Schematic of remote health monitoring wireless body area network (WBAN) system.

Source: Reprinted with permission from Anupma Gupta et al., 2021. Copyright © 2021, Indian Academy of Sciences

7.4.2 THE INTERNET OF HEALTHCARE THINGS AND THE DUAL ROLE OF WEARABLES

As we explore the potential for a true Internet of Healthcare Things (some refer to this reality of the connecting of health-related devices and equipment as the Internet of Medical Things (IoMT); some name it the medical IoT or MIoT). According to Vijayakumar and Bhuvaneswari (2020), the IoMT is already a reality and has been for some time. When seen holistically, however, key concerns including health device interoperability, integrated systems connecting healthcare providers and patients, secure standards, real-time health system (RTHS) integration, and others must be addressed.

7.4.3 INTERNET OF THINGS MEETS ROBOTICS IN HEALTHCARE

Hospital robots such as Panasonic HOSPI, which among other things handles deliveries (medication, beverages, etc.), rehabilitation robots, and surgical robots used in precision or remote surgery are all well-known examples of robots utilized in medicine today. In today's society, the advent of the society-wide Web has helped advance IoT science among everyday people. The likelihood of this linked technology delivering significant advantages in everyday life is quite high. Before the IoT was ever implemented, the mHealth system had been anticipated. These interconnected, reliable systems play a significant role in mHealth technology. The two things that patients most desperately need money and time are both reduced as a result. By utilizing several biomedical sensors, mHealth is developed in this work as a scaled-down live healthcare system (Mohsen et al., 2020). These sensors can capture health-related information in numerical and waveform formats for the carer and the hospital server.

7.4.4 INTERNET OF THINGS MEETS ROBOTICS IN HEALTHCARE

Prior to the current progress in IoT technology, doctors could use medical equipment to do real-time analysis. Additionally, it has helped healthcare facilities serve a larger population at once and do it at a low expense. As healthcare needs increase and technology is improved by offering answers to diverse healthcare issues, more services are launched daily. Figure 7.3 illustrates how concepts and services have altered the healthcare sector.

These are increasingly important while constructing an HIoT system. A set of healthcare solutions are provided by each service in an HIoT context. IoT-integrated artificial intelligence (AI) utilized to support senior citizens AAL is a specialized subfield of assisted living. AAL provides a method for monitoring patients in real-time and ensuring that they will get human services, such as aid in the event of a medical emergency. The primary goal of AAL is to assist older persons in living independently at home in a convenient and secure environment. With the use of cutting-edge AI technologies, big data analysis, machine learning (ML), and their application in the healthcare sectors are made feasible (Zaman et al., 2020).

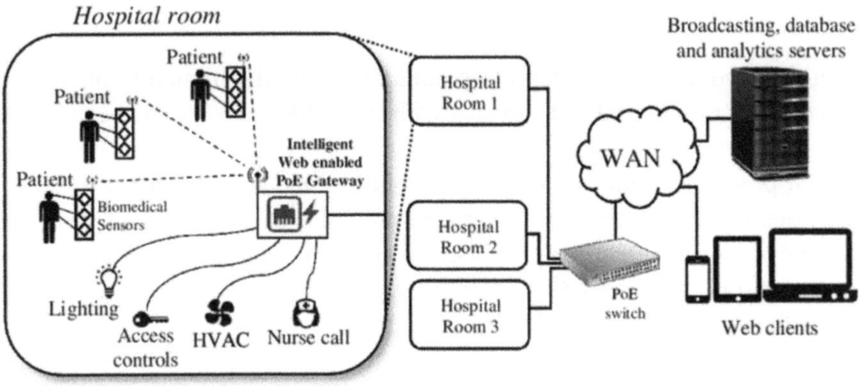

FIGURE 7.3 IoT architecture for healthcare application.

Source: Reprinted with permission from Grandos et al., 2015. Copyright © 2015, Institute for Computer Sciences, Social Informatics and Telecommunications Engineering

7.4.5 MOBILE IoT

The combination of mobile computers, sensor technology, communication technologies, and cloud computing to track patient health data and physiological parameters is referred to as mobile IoT or MIoT. In essence, it creates a communication link between local area networks and mobile networks, such as 4G and 5G, to enable effective Internet-based healthcare services (Gupta et al., 2021). By collecting and analyzing patients' data, it enables healthcare practitioners to quickly diagnose and provide appropriate treatment, leveraging the capabilities of mobile devices in the field of HIoT.

7.4.6 COMMUNITY-BASED HEALTHCARE SERVICES

A healthcare network that serves the requirements of a local community, such as a private clinic, a small residential neighborhood, or a hotel, is referred to as community-based healthcare monitoring. Its main goal is to keep an eye on the patients who live in that particular area's health status. To supply comprehensive and collaborative healthcare services, this network connects several networks and makes it possible for joint activities (Easttom and Mei, 2019).

An IoT-based cooperative medical network was set up to deliver healthcare monitoring services in outlying locations. Different authentication and authorization procedures were used to guarantee a secure connection between these networks.

7.4.7 COGNITIVE COMPUTING

Cognitive and analytical computing simulates how the human brain solves problems. The fusion of IoT devices with brain-like sensors has been made possible by the development of sensor technology and ML. These sensors make it possible to analyze huge data volumes and reveal hidden patterns. Additionally, they improve the processing of medical data and make it possible for sensors to change their surroundings on their own. Sensors work together with other smart devices in a cognitive IoT network to offer effective healthcare services.

7.4.8 ADVERSE DRUG REACTION AND BLOCKCHAIN

The negative effects that occur as a result of taking medicine are referred to as an adverse drug response (ADR). After a single dosage or after prolonged treatment, several responses might happen. When two different drugs are used at the same time, negative side effects might also occur. ADRs might differ from person to person and are not reliant on the condition or medicine.

Each drug is uniquely identified at the patient's terminal in an IoT-based ADR system using a unique identification or graphic code, such as a QR code or barcode. It is possible to use intelligent pharmaceutical equipment to confirm that a medicine is compatible with the patient's body. Each patient's allergy profile is recorded and stored by the electronic health records system. In an HIoT network, data sharing between different healthcare providers and medical equipment is essential.

Data fragmentation, however, poses a substantial obstacle to safe data exchange. Information gaps between healthcare professionals working on a specific patient's care may result from this fragmentation. The course of treatment might be hampered by inadequate information. Blockchain technology is used to connect data repositories inside the network and resolve the issue of data fragmentation to address this issue. This guarantees the secure exchange of private medical information and improves communication between patients and physicians.

7.4.9 CHILD MHEALTH INFORMATION

The goal of child health information (CHI) is to increase public knowledge of a kid's health and well-being. Its main objective is to inform and empower kids and the adults who look after them about several facets of a child's general well-being, such as nutrition, emotional and mental health, and behavior. The development of a platform that can track and manage a child's health; thanks to the IoT which has made it easier to attain this goal (Guo et al., 2020). As a result, digital gadgets now make it simple to obtain healthcare services. The healthcare sector has been transformed by the integration of smart sensors, cloud computing, and communication technology. IoT does, however, also provide several difficulties and problems that need more studies. The cost of servicing and maintaining equipment, power consumption, data privacy and security, standardization, scalability, identity, self-configuration, ongoing monitoring, discovery of new illnesses, and environmental effects are some of the major issues.

7.5 IMPLANTABLE ANTENNA FOR BIOMEDICAL APPLICATIONS

Enhancing medical diagnosis, treatment, and academic research are all made possible by advances in biomedical engineering. It has cleared the door for several medical applications, including endoscopy, deep brain simulations, insulin pumps, and glucose monitoring. These technologies profit from implanted devices such as pacemakers and remote monitoring systems that wirelessly communicate critical information (Bashir et al., 2019). Implantable medical devices are being explored in great detail for purposes such as monitoring blood pressure and body temperature, tracking dependent people or animals, and wireless diagnostic data transfer (Kirtonia et al., 2019).

Antennas are inserted into the human body to read brain activity, measure blood sugar levels, and support implants including cochlear implants, oral implants, and retinal prosthetics to provide continuous health monitoring. However, because of the strict restrictions, executing these implant applications presents substantial problems (Mishra et al., 2020).

In addition to on-body antennas, in-body antennas also need to be characterized using tissue-like materials. Bahrami et al. conducted a study comparing the performance of single and dual-polarized antenna structures for brain implants. They utilized a monopole antenna and single-arm spiral antennas with CPW feed for single and dual-polarization, respectively. The implantable antennas were coated with a biocompatible superstrate layer made of flexible Kapton polyimide. To achieve impedance matching at WASM and UWB bands, they designed a truncated shape of the transmission line and a staircase ground plane. The antenna's efficiency was evaluated based on near-field link characteristics such as specific absorption rate (SAR), figure of merit, and fidelity factor.

7.5.1 BIOMEDICAL TELEMETRY

In child wireless telemetry, signals from the body are measured and transmitted to a base station—such as a phone or computer—located far from the body through RF transmission. Then, for diagnosis and patient guidance, this information is sent via the Internet to physicians or other pertinent parties. When compared to open space antennas, designing antennas for biomedical telemetry is particularly difficult since antenna performance can vary greatly when placed close to the human body. Biotelemetry uses a range of frequency bands, including:

- 402–405 MHz for medical implant communication services (MICS).
- Medical device radiocommunications service (MedRadio): 401–406, 413–419, 426–432, 438–444, and 451–457 MHz.
- Industrial, scientific, and medical (ISM) band: 433.05–434.79, 902–928, 2400–2483.5, and 5725–5850 MHz.
- WMTS bands (608–614, 1395–400, and 1427–1432)
- 3100–10,600 MHz for the UWB.
- 2360–4000 MHz for the medical body area network (MBAN)

Depending on where the sensor node is located on or within the human body, different types of biotelemetry antennas can be divided into these categories. These categories include implanted antennas, body-centric wearable antennas, and ingestible antennas for uses such as capsule endoscopy. Figure 7.4 shows several antennas used in the realm of medicine.

7.5.1.1 A INGESTIBLE ANTENNAS CAPSULE ENDOSCOPY

It's a well-known application for in-body to off-body data transmission. Biological statistical information in the form of images/videos is gathered by the camera and transmitted by the capsule antenna while passing through the digestive system. Compared to the wired endoscopy, it is capable of tracing the hidden location of the track and providing an effective visualization of the entire convoluted track with the help of a wireless capsule endoscope. Antenna resonance should remain stable when bent across the wall of the capsule. Each organ of the human body is constructed of heterogeneous tissue layers with varying electric properties (Bashir et al., 2019).

Moreover, conformal structures supply more space for other components in the capsule. Wide impedance bandwidth: It interferes with the electromagnetic properties of the antenna and causes frequency detuning. Good impedance matching is required with the continuous operation of ingestible antenna wideband performance with Orientation matching: While moving through the gastrointestinal (GI) track, the orientation of the capsule changes randomly causing a time-varying orientation mismatch with the receiver antenna and reducing the reliability of data transmission. Therefore, an antenna that fulfills the need for orientation matching is necessary for capsule endoscopy.

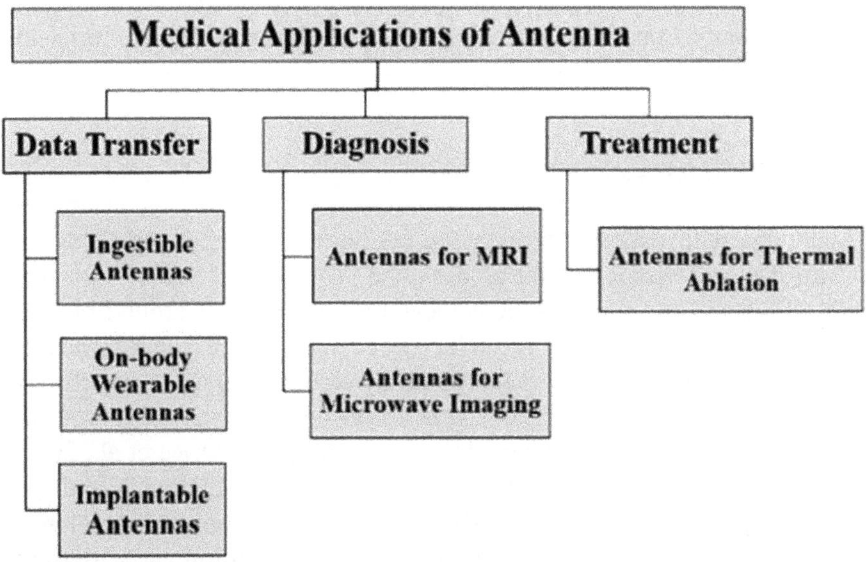

FIGURE 7.4 Application of antenna in medical sector.

Source: Reprinted with permission from Anupma Gupta et al., 2021. Copyright © 2021, Indian Academy of Sciences

7.5.1.2 CHARACTERIZATION WITH HUMAN BODY TISSUE

The researchers employ a variety of anatomical models to characterize the antenna impedance with the human body. The male body phantom employed for the numerical simulation is called the Gustav model. Single and multilayered canonical structures are modeled to reduce the complexity of the calculation, which is time consuming and memory-intensive. It is a geometric construction with great complexity. Antenna testing on actual human and animal bodies is prohibited because of ethical and safety concerns (Mishra et al., 2020). Thus, to verify the results of numerical simulations, tissue-equivalent phantom gels, and liquids are fabricated for experiments. The phantom gel consists of deionized water, sugar, salt, and gelatin. Animal tissue is also used for the verification of antenna performance.

7.5.1.3 FREQUENCY DETUNING AND IMPEDANCE MATCHING

Managing bodily tissue affects the antenna's surface current and throws off its impedance. Electromagnetic waves are complicated and lossy medium in the human body. Impedance matching and frequency detuning a complicated and lossy medium for electromagnetic waves is the human body.

7.5.1.4 HIGH GAIN AND EFFICIENCY

An antenna with wide beam emission normal to the body surface or null toward the body tissue should be designed. The desired radiation pattern, high efficiency, and gain can be achieved by using a full ground plane: a unidirectional radiation pattern can be obtained with a full ground structure. Antenna gain and efficiency that reduces by significant power absorbed by body tissue. The body tissue is damaged by absorbed power and it limits the transmission range of the antenna to a few meters (Chatterjee et al., 2017). It is possible to get the desired radiation pattern, high planning, and gain by employing a whole ground plane. With a complete ground structure, a unidirectional radiation pattern may be generated. Therefore, it is advisable to develop an antenna with a broad beam of radiation that is either null or normal to the surface of the body.

7.5.1.5 AMC BACKED YAGI ANTENNA

AMC-backed Yagi antenna is now in design on a flexible latex substrate. First, the T-resonator technique and quality factor were used to evaluate the electric characteristics of latex at 2.68 GHz. Then, a meander-type Yagi antenna was created, and a driven dipole was supplied by a balun to transition an unbalanced signal into a balanced one.

7.5.1.6 SUBSTRATE INTEGRATED WAVEGUIDE (SIW)

The wearable antennas are the main focus of surface-integrated waveguide (SIW) technology. Entire ground planes and arrays of shorting pins or shorting walls were employed in SIW technology. By using SIW antenna structures, excellent isolation between the antenna and bodily tissue may be accomplished using SIW technology. When antennas were examined for structural deformation, impedance matching, and steady frequency were noted.

7.5.1.7 SLOTTED PATCH WITH METAMATERIAL

Using denim fabric, Ashyap et al. created a U-shaped patch antenna that operates at 2.68 GHz. To improve antenna performance and resilience in biological environments, 296 arrays of metamaterial structures have been introduced. After the placement of the metamaterial, significant improvements in the front-to-back ratio (14.5 dB), high gain (7.5 dBi), and bandwidth (14.5%) were made. Three-layer simulation model, phantom gel for measurement, and the Gutsav model.

7.5.1.8 FRACTAL PATCH WITH DEFECTED GROUND

In the 2.68 GHz ISM band, a wearable antenna was suggested by Arif et al. The antenna and its output were made smaller by the authors using meandering slits and the Koch fractal geometry. Wider bandwidth was intended for the deflected ground structure, which has many slits and truncated corners. Antenna performance was seen at many body tissue locations, including the arm, thigh, leg, and back.

7.5.1.9 DIPOLE WITH METALLIC REFLECTOR

For wearable technology, Al-Ghamdi et al. developed a three-layered dipole structure. To produce mechanical strength and resilience, a flexible substrate made of a composite of nitrile butadiene rubber (NBR) is frequently utilized. The front-to-back ratio is increased and the impact of electromagnetic radiation on bodily tissue is diminished thanks to the placement of a metallic reflector beneath the substrate. A low radiation efficiency of 15.7% at 2.68 GHz is a problem for the antenna.

7.5.1.10 MINIATURIZATION

The in-body antenna should be miniaturized to fit in exceedingly limited space on the chip. Several size reduction techniques widely used by researchers includes homocentric complementary split ring shape radiator, annular ring shape radiator, capacitive coupling, offset fed meandered patch, and circular PIFA with slotted ground.

7.5.1.11 PATIENT SAFETY

To ensure the patient's safety, the maximum incident power that may be applied to the implanted antenna is restricted. SAR is frequently used to calculate the heating impact of tissues. Near-field electromagnetic radiations raise the temperature of human tissue. Due to the antenna and limits the bodily tissue's ability to absorb power.

7.5.1.12 POWER REQUIREMENT

Medical implants use energy continually, shortening their useful lives. Implanted devices that provide continuous monitoring should only be utilized when necessary. You can utilize a device that operates on two bands, one for transmission and the other for wake-up. A dual band stacking antenna in the shape of a spiral is developed for inductively coupling electricity from an external connection.

7.5.1.13 CIRCULAR POLARIZATION

As a result of different body motions and postures, time-varying orientation mismatch makes it impossible to ensure the reliability of a communication link. In addition, low efficiency, excessive power loss in bodily tissue, and multipath reflections in an indoor setting during transmission are all problems with implanted antennas. For high-quality in-body biotelemetry, a high-gain antenna with circular polarization is, therefore, required. In literature, various techniques are employed to create orthogonal modes, including many slots with met material, truncating corners, adding shorting pins, and ring shapes ground with Z-shaped radiators.

7.5.1.14 ANTENNAS FOR DIAGNOSIS

To assess how well bodily tissue is operating, antenna metrics including impedance, reflection coefficient, and near-field radiation are employed. Some applications of antennas include the detection of a paralyzed person's muscle movement, the identification of tumors using magnetic resonance imaging (MRI), and microwave imaging. MRI is a well-known method for non-invasively describing living things. To stimulate and receive a magnetic resonance (MR) signal, radiofrequency (RF) transmitter and receiver antennas are crucial and vital components (Farshchi et al., 2006). The electromagnetic pulses that are sent into the bodily tissue by the RF transmitter antenna are its primary purpose. The reception antenna is an extremely sensitive tool that can quickly capture the RF signals released by a patient's body during an examination and save this data in the computer system.

7.5.1.15 DESIGN OF ANTENNA FOR MRI

A patch antenna with a cylindrical form that operates in the UWB band between 7.8 and 15 GHz was created by Emran and Elvi. To obtain a wide bandwidth, radiators with slots in a triangular flared design were employed. The inside wall of the cylindrical antenna was developed with an in-phase frequency selective surface to increase gain and front-to-back ratio. Body-tissue-related electromagnetic information was removed from the antenna's S-parameters. The transmission of the MR signal via a patch antenna. To get very sensitive parallel signals from the whole brain, the antenna was merged with a 24-channel receive-only array.

7.5.1.16 MICROWAVE IMAGING (MI)

According to Arulmozhivarman et al. (2013), MI operates on the theory of change in a backscattered variable with changing electrical characteristics of bodily tissue. It is necessary to have an antenna that is more aware of the presence of various dielectric media in its near-field area. The antenna also has to be able to function across a broad frequency range while retaining its quality over the full range. A number of antennas that are used in the near-field zone are positioned in a certain configuration and close to the breast tissue. The transmitting antenna is delighted sequentially at a varying angle to impinge low-power microwave signal to the breast, reflected signal is then collected by the receiver antenna and initialization using various signal clarifying techniques and used for image formation.

7.6 DIGITAL SIGNAL PROCESSING APPLICATION FOR REMOTE PATIENT MONITORING SYSTEM USING LIVE EEG DATA TRANSMISSION

One of the most complex systems in the cosmos is the human brain. One of the methods used to measure brain waves nowadays is electroencephalography (EEG). Aberrant brain waves have been connected to particular brain diseases, and this approach to interpreting brain signals helps researchers better comprehend the intricate inner workings of the brain. The technology that enables the generation, transformation, and interpretation of information is known as signal processing. At certain times, our brain has varied reactions. To properly study how the brain functions, human neural impulses are used for several purposes, including creating, converting, and analyzing the recorded signal. This process is known as "brain signal processing." The study of brain signals is crucial in the treatment of various brain abnormalities. The brain contains billions of nerve interconnected neurons that connect through electrical. The brain generates a huge amount of electrical activity as a result of millions of neurons transmitting signals at the same time, which can be monitored through precise medical devices along with an EEG, that analyzes electrical impulses throughout sections of the brain.

EEG records would be a significant method for identifying the brain's functioning status and identifying specific abnormalities (Dalton et al., 2012) Because of its pulse form, the collection of electrical impulses is usually referred to as a neural activity pattern. Many people around the world are currently suffering from severe brain disorders. The patient's physical status or condition can be determined by examining his EEG data. As a result, doctors feel compelled to periodically review a patient's EEG data. However, either entirely hospitalizing the patient for the sake of the readings or requesting frequent visits to check on data is not realistic.

That is where the designed methodology enters the picture. It allows physicians to study the patient's EEG data while having to communicate with them directly. The individual checks into a database and the appropriate equipment is used to take accurate readings. The collected EEG data are subsequently saved in a database that is maintained by the physician. Physicians can access the system and access the information that is organized by patient.

The process of integrating EEG activity that is time-locked to the presence of an input of some type (visual, somatosensory, or aural) is known as eliciting potentials (EPs). Event-related potentials (ERPs) are EEG responses that are time-locked to the more generating. This approach is used in a study on psychophysiology, educational sciences, psychology, and neuroscience. The brain has hundreds of billions of neurons that maintain a steady electrical charge.

Plasma carrier proteins transfer ions over neuron membranes, causing them to become electrically charged (or "polarized"). Neurons exchange ions with the external environment regularly, for instance, to establish resting potential and transmit action potentials (Tian et al., 2010). In a waveform, numerous ions that are pushed out of multiple neurons at the same time may push their neighbors, and then their neighbors, and so on. Volume conduction is the term for this phenomenon. Whenever a pulse of ions enters the scalp's electrodes, researchers either push or drag ions on the electrodes' metal. A voltmeter can calculate the distance in push or pull voltages among any two electrodes because metal efficiently transmits electrons' push and pull. The EEG is created by recording these voltages throughout time.

7.6.1 EEGLAB

EEGLAB is an EEG signal processing application with a client frontend. Such a system can assess occurrence as well as other sorts of brain reactions. It also offers many signal viewing options in the spectral and temporal domains. Through an EEGLAB database, EEGLAB stores data, occurrences, and data in a centralized layout. EEGLAB has an immersive graphic user interface (GUI) that allows users to interpret the greater EEG information freely and effectively. It is imperative to measure data once a database has been put into the program. The following are the key characteristics of EEGLAB: (1) GUI: The GUI in EEGLAB is not as sophisticated as in consumer uses. The command line is used for a majority of the operations. (2) Integrating progressive scan information: This includes integrating MATLAB arrays, Biosemi. BDF files, and other data structures. The analysis and reporting functionality in EEGLAB scrolls. (3) Specified EEG data layout: It is for individuals who want to use EEGLAB and its operations in MATLAB.

EEG structures have been simplified so that users may analyze user data efficiently. (4) EEGLAB's open-source plugin facility: EEGLAB also has a "plug-in" feature that searches for operations imported into a given plug-in directory and displays them inside the EEGLAB interface. (5) Immersive calculating procedures: In general, interactive calculating operations are used for this function. A colorful image is plotted using pop erpimage (6) and erpimage (7). (8) Artifact removal: EEGLAB has an alternative for artifact rejection. One of the benefits of EEG is its speed, as it can capture distinct patterns of cerebral activity that occur in fractions of a second. There are several reasons why EEG may be used to assess the strength and location of electrical impulses, including the following:

- Monitoring alertness.
- Monitor cognitive activities
- Control anesthesia
- Investigate epilepsy and its drug effect
- Locate damaged areas following a head injury
- Monitor brain development
- Test effects of drugs
- Investigate sleep disorders

7.6.2 FREQUENCY BANDS OF THE EEG SIGNAL

Electrical signals are categorized based on their frequency and magnitude, as previously stated. The individual's psychological functioning will be abnormal if either of the pulses' wavelengths is abundant or insufficient. Five spectrum bands can be identified in the raw EEG signal:

- Delta (<4 Hz)
  Stages 3 and 4 of sleep are when the lowest delta frequencies, which represent the unconscious mind, are present.
- Theta (4–8 Hz)
  Theta brainwaves are related to creativity, memories, emotions, daydreaming, sensations…They are strong when the person is focused (internal focus: praying, meditating…).
- Alpha (8–12 Hz)
  Normal alpha levels are correlated with a good mental imagination, a good mood, and efficiency in accomplishing tasks. They are present when an individual is alert and not processing any information. They can be increased by closing the eyes or decreased by thinking.

- Beta (>12 Hz)
  Beta activity "is typically distributed symmetrically on both sides and is most prominent frontally." Since beta activity has a relatively wide range, it is divided into:
  Low beta (12–15 Hz)
  Midrange beta (15–18 Hz)
  High beta (>18 Hz)
- Gamma (>30 Hz)
  Every portion of the brain contains this frequency, which is necessary for processing information when the brain gathers data from various sources.

7.6.2.1 EXISTING SYSTEM DESIGN

To record their EEG data under the current method, a patient must go to a medical facility and await his turn.

As is shown in Figure 7.5, the primary components here are the database and the transmission medium or technology. The proposed system shall have a database which is managed at the doctor's end. Individual patients are given access credentials which they use to log into the system and record their EEG samples.

The system is primarily based on four modules:

i. Data acquisition system
ii. Administration
iii. Data storage
iv. Transmission

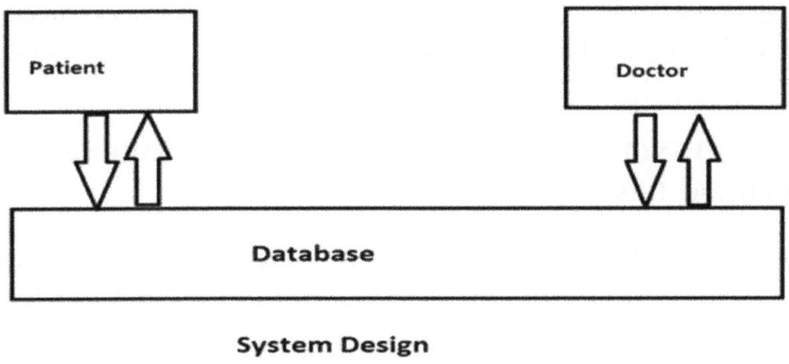

FIGURE 7.5 Block diagram.

7.6.2.2 MODULES

The system was implemented using Java on the Netbeans IDE. The interface and design of the web pages were all developed using the IDE. Mysql was used as the database to store the login credentials of doctors and patients.

7.6.2.2.1 Data Acquisition System

This module involves the required hardware to measure the EEG data along with the other hardware such as the user's computer. This module is restricted to the patient's computer alone. The patient makes use of the system to record his EEG data once he/she is logged in to the system. The patient's computer is part of a network that connects to a database managed by the doctor or the hospital.

7.6.2.2.2 Data Storage

The data server is a standalone program that offers the following features: EEG data storage, access interfaces for uploading and viewing applications, and a web interface for users to browse and examine the recordings that are accessible. Between the doctor's private network and the outside world, the data server serves as a bridge.

7.6.2.2.3 Data Transmission

This module covers the transmission of data from the patient's computer to the database which is managed at the doctor's end. Transmission control protocol (TCP) is used here to ensure the data is transferred successfully.

7.7 CHALLENGES AND OPPORTUNITIES IN SOFTWARE-DRIVEN BIOMEDICAL DEVICES

The environment for developing medical devices is rapidly changing. Medical device makers now have to think about more than merely striking a balance between creative gains and safety (Parihar et al., 2019). The emergence of linked, software-driven medical devices, tightening regulations on medical devices, and disruptive new market entrants are the three primary issues that medical device firms will have to deal with in the upcoming years. There are big issues at play given that the medical device business is expected to grow rapidly to $800 billion by 2030. Here is a closer look at some of those significant issues that must be taken into consideration when creating modern medical devices.

7.7.1 FIRST CHALLENGE: THE RISE OF CONNECTED, SOFTWARE-DRIVEN BIOMEDICAL DEVICES

Up until recently, medical device software had only a few functionalities. Like so many other items, medical gadgets' primary features and functions gradually began to be dominated by software.

Medical devices have become significantly more complex as a result of the marriage of high-end hardware and cutting-edge software, which translates to greater development labor for those responsible for creating the goods (Bairöz et al., 2020). The environment of medical device development has changed substantially with the addition of connectivity, which, although highly powerful and convenient, may also pose several hazards such as viruses and hackers.

The complexity of the product design adds to the risk that teams creating modern medical devices must manage in addition to keeping up with the quick rate of innovation and market needs. Medical device development cannot be managed through spreadsheets and documentation, therefore enterprises need to innovate if they want to keep up with the competition.

7.7.2 SECOND CHALLENGE: INCREASINGLY STRICT BIOMEDICAL DEVICE REGULATION

Regulators are being compelled to reevaluate current policies and guidelines due to the growing complexity of medical devices, and for good reason. According to Stericycle's Recall Index, the average size of a medical device recall in the first quarter of 2019 was 823,126, which was the largest quarter since 2006. The main reason for all recalls was poor quality. Compliance with both industry and government requirements, whether it be ISO 14971 or ISO 13485, continues to be a challenge for medical device companies as regulatory authorities strive to keep up with the pace of the market.

This increases the expense of medical device compliance and makes it more challenging to understand regulatory standards. For instance, combing through the paperwork created by both the development and design groups, including outputs from various software tools, is one of the most challenging and time-consuming components of demonstrating medical device compliance (Kothari, 2019). Finding the proper documentation to demonstrate FDA compliance and methodical risk management is difficult, even in the best–case situation for documentation.

7.7.3 THIRD CHALLENGE: NEW MARKET ENTRANTS MAKING A DISRUPTIVE SPLASH

There is good cause to be concerned when well-known brands including Apple and Amazon choose to begin competing in your business. These well-known brands profit from immediate brand awareness, fast distribution, and seemingly endless resources. For example, the FDA granted a Class II de novo clearance for Apple's Watch Series 4's ECG and notice of abnormal heart rhythm features.

Despite the fact that goliath firms entering the medical device market can be daunting, it's crucial to keep in mind that the majority of them are still getting used to FDA regulation. This raises additional hurdles for the FDA's examination of every single software solution and poses new dangers for tech businesses with histories outside of traditional FDA regulation, which has led to a reform of that system.

This situation makes it more important than ever the medical device companies to evaluate their processes and ensure that their products are made correctly the first time.

7.8 BIOINFORMATICS IN MHEALTH

Internal antenna the subject of bioinformatics involves the use of computing to derive information from biological data. Through the creation of algorithms and software, it encompasses the gathering, storing, retrieving, manipulating, and modelling of data for analysis, visualization, or prediction (Singelmann et al., 2019).

In this multidisciplinary area, new computational techniques are New computational techniques are created in this multidisciplinary discipline to analyze biological data and produce biological discoveries. For instance, the methods of sequencing and annotating an organism's whole DNA set are two standard operations in genetics and genomics. Neuroimaging methods are used in neurosciences to examine brains in vivo and comprehend the inner workings of the nervous system. These methods include computerized tomography (CT), positron emission tomography (PET), functional magnetic resonance imaging (fMRI), and diffusion tensor imaging (DTI).

Our comprehension of complicated illnesses like cancer or neurodegenerative and mental disorders will be improved; thanks to the use of ML in biological and neuroimaging data. The development of automated diagnostic tools and precision medicine, which targets individualized medical treatments taking into account individual variability, lifestyle, and environment, can be attributed to advancements in this sector (Bugnon et al., 2020).

Before the advent of ML algorithms, bioinformatics algorithms had to be explicitly written by hand, which proved to be quite challenging for issues such as protein structure prediction. Through the use of automatic feature learning, which is made possible by ML techniques such as deep learning, the algorithm can learn how to combine various features of the input data into a more abstract set of features from which to carry out further learning. When trained on huge datasets, these systems are capable of making rather complicated predictions thanks to their multi-layered approach to understanding patterns in the input data. The amount and quantity of biological datasets that are now readily accessible have increased dramatically in recent years, making it possible for bioinformatics researchers to employ these ML methods. Six biological fields—genomics, proteomics, microarrays, systems biology, stroke diagnostics, and text mining—have all seen the use of ML.

7.8.1 ML TO RESOLVE ISSUES IN BIOINFORMATICS

The study of DNA and protein sequences in bioinformatics covers indicators of functionality and related issues, such as categorization of homologs, multiple sequence alignment, pattern finding, and evolutionary analysis.

ML techniques are favored because sequence analysis can address all of these issues. Protein structures are three-dimensional representations of data; issues with them include the following:

- Examining protein structures for indicators of functionality.
- Prediction of structure (possessing secondary and tertiary protein structures).

- Structure alignment.
- Graph-theoretic techniques are used to handle numerous related issues, such as the construction and interpretation of massive-range networks, which are shown on graphs and include many biological networks, such as the gene regulatory network, protein-protein interaction networks, etc.
- Neural networks are a kind of soft computing; they provide network systems the capacity to learn. One input layer, one or more hidden layers, and one output layer make up the neural network's design.
- Additionally, when dealing with biological data, categorization becomes a challenging problem that cannot be accomplished using standard techniques of analysis. For this reason, artificial neural networks are frequently utilized in bioinformatics as an ML tool.

7.8.2 APPLICATIONS OF NEURAL NETWORKS IN BIOINFORMATICS

With the exponential rise of biological data, it is important to focus on the effective administration and preservation of information as well as the extraction of pertinent information from this data. In addition, suitable computational techniques must be used to turn this diverse data into insightful knowledge. These computational techniques, or ML techniques, allow us to understand more detailed data and deliver knowledge in the form of testable models that allow us to forecast the behavior of the system. The following are some examples of how neural network applications in bioinformatics may be used to extract information from data in various biological areas using ML methods:

- In recognizing the coding area of genes.
- In identifying faults with genes.
- Detection of features, categorization, and sequence.
- Image and signal processing.
- Signal identification and analysis from regulatory locations.
- Expression of genomic and genetic information.

Currently, bioinformatics has many medical applications, including the ability to predict or visualize protein structure from amino acid sequences, to help with the development of novel drugs, and to track the medical progress of patients based on their DNA sequences.

7.8.3 BIOMEDICAL AND MHEALTHCARE

ML and AI are widely used in healthcare settings to improve patient care and improve the quality of care. In the near future, hospitals could use ML technology to get real-time data from multiple health systems in different countries, to increase treatment effectiveness.

7.8.3.1 DRUG DISCOVERY AND MANUFACTURING

ML is often used in the first phases of the drug development process. Precision medicine and next-generation sequencing are two examples of research and development technologies. They have been useful in locating alternative options for the treatment of complicated disorders.

7.8.3.2 MEDICAL IMAGING AND DIAGNOSIS

ML and deep learning were used in a breakthrough technology called computer vision. The technology has been accepted across a broad spectrum of applications. One example is the Microsoft InnerEye project, which is building innovative tools for quantifiable investigation of 3D medical images.

7.8.3.3 PERSONALIZED MEDICINE

To help with individualized therapy, predictive analysis may be applied to patient data. Currently, doctors are constrained to a few diagnoses or are required to predict a patient's likelihood of developing a disease based on their health and limited genetic information. However, given the huge strides machine learning is making in medicine, which uses patient data to provide a wide range of therapy alternatives that may soon change.

7.8.3.4 STROKE DIAGNOSIS

To diagnose, treat, and forecast problems from numerous neurological illnesses, ML employs model recognition. Acute ischemic stroke (AIS) therapy has advanced significantly in recent years. ML can be used to forecast motor impairments in stroke patients. Support vector machines (SVM) and 3D conventional neural networks (CNNs) are the techniques most often utilized.

7.8.4 ML TOOLS USED IN BIOMEDICAL AND MHEALTH

Processing enormous datasets produced by recently developed technologies into useful information is one of the most urgent problems in biomedical and mHealth as a whole (Alaoui et al., 2021). However, ML in bioinformatics is playing a crucial part in bringing about this revolution as we move into the era of AI and big data.

7.8.4.1 DEEPVARIANT

This deep learning technology is employed in the analysis of genetic data. Compared to earlier traditional approaches, it can more precisely anticipate frequent genetic variants. One of the first biological tools to do so is DeepVariant, which uses Google's processing power and ML to offer a scalable, cloud-based solution that can handle even the most demanding genomics datasets.

7.8.4.2 ATOMWISE ALGORITHMS

The San Francisco-based biotech business Atomwise created the first deep learning-based algorithm that aids in the conversion of molecules into 3D images. This conversion aids in the atomic-level analysis of the three-dimensional structures of proteins and other molecules. Additionally, it essentially predicts chemicals that might interact with a certain protein. Drug discovery is the principal use for the algorithms.

7.8.4.3 CELL PROFILER

Only one aspect of each image was measured by the program. However, owing to ML techniques, that has changed. Scientists may prepare and photograph hundreds of samples per day using the CellProfiler program. The program can count the luminous cells in a field. It is capable of highlighting hundreds of cell characteristics by utilizing deep learning techniques. From the extracted data, machine learning using deep learning may create a model. Other biological data may be analyzed using the model.

7.9 CONCLUSION

Always advancing controls and the unused complexities displayed by associated gadgets make the administration of these moving complexities troublesome. Electronic Gadgets for Biomedical and mHealth can offer assistance by decreasing taking after troubles such as effectively illustrating traceability, overseeing chance investigation, keeping up review trails and trade information, progress item advancement with reuse and pattern administration, streamlining compliant audits and endorsements, overseeing plan confirmation and approval. 2020 was a year that was persistently depicted as "unprecedented" and "unparalleled." The framework permits an incredible level of comfort for specialists and patients in cases where EEG recordings are required at visit intervals for conclusion. The framework spares time, and cash and is exceptionally helpful. Patients do not have to travel to a clinic to record their EEG information. The specialists moreover advantage from this framework by permitting them the extravagance of checking up on their patient's readings from their domestic or through the Android application. Storing and transmitting EEG information in a database could be a challenge due to the analog nature of the signals. Once completely created, the framework is a perfect instrument for clinics and ancient-age homes that bargain with a few patients enduring sicknesses for which it is imperative to screen EEG information signals habitually. The framework can also be overhauled or made strides to cover and back other therapeutic readings such as EEG and blood weight. As we begin a modern year, it's difficult to say what might anticipate us. One thing that ready to be beyond any doubt of is that advancement in pharmaceutical, science, and innovation appears no sign of abating down. The advertisement for restorative gadgets proceeded to boom as imaginative, associated innovations got to be more coordinated into inpatient, outpatient, and domestic care. To combat the mounting complexity, groups must get a more grounded handle on their process and avoid holes in biomedical and mHealth improvement just such as the rise of associated, software-driven therapeutic gadget; therapeutic gadget quality issues are expanding, and advanced advertise participants expanding the pace of competition.

KEYWORDS

- **diagnosis**
- **mHealth**
- **healthcare**
- **Internet of Things**
- **biomedical**
- **machine learning**

REFERENCES

Alaoui, E. A. A.; Tekouabou, S. C. K.; Hartini, S.; Rustam, Z.; Silkan, H.; Agoujil, S. Improvement in Automated Diagnosis of Soft Tissues Tumors using Machine Learning. *Big Data Mining Analytics* **2021,** *4* (1), 33–46.

Arulmozhivarman, P.; Reddy, G. R.; Tatavarti, R. In *Low Cost EEG Signal Acquisition System*, ISSNIP Biosignals and Biorobotics Conference: Biosignals and Robotics for Better and Safer Living, 2013; pp 1–3.

Bağiröz, B.; Doruk, E.; Yildiz, O. In *Machine Learning In Bioinformatics: Gene Expression And Microarray Studies*, Medical Technologies Congress, 2020; pp 1–4.

Bashir, Z.; et al. In *A Miniaturized Wide Band Implantable Antenna for Biomedical Application*, UK/ China Emerging Technologies, 2019; pp 1–4.

Bochicchio, M. A.; Vaira, L.; Mortara, A.; De Maria, R. In *A Preliminar Analysis and Comparison of International Projects on Mobile Devices and mHealth Apps for Heart Failure*, 5th Experiment International Conference (exp.at'19), 2019; pp 280–285.

Bugnon, L. A.; Yones, C.; Milone, D. H.; Stegmayer, G. Deep Neural Architectures for Highly Imbalanced Data in Bioinformatics. *IEEE Trans. Neural Netw. Learn. Syst.* **2020,** *31* (8), 2857–2867.

Chatterjee, B.; Saini, L. M.;Gandhi, T. K. In *Non-invasive Wireless EEG Monitor*, International Conference on Energy, Communication, Data Analytics and Soft Computing, 2017; pp 3537–3542.

Dalton, A., et al. Development of a Body Sensor Network to Detect Motor Patterns of Epileptic Seizures. *IEEE Trans. Biomed. Eng.* **2012,** *59* (11), 3204–3211.

Easttom, C.; Mei, N. In *Mitigating Implanted Medical Device Cybersecurity Risks*, IEEE 10th Annual Ubiquitous Computing, Electronics & Mobile Communication Conference, 2019; pp 0145–0148.

Farshchi, S.; Nuyujukian, P. H.; Pesterev, A.; Mody, I.; Judy, J. W. A TinyOS-enabled MICA2-BasedWireless Neural Interface. *IEEE Trans. Biomed. Eng.* **2006,** *53* (7), 1416–1424.

Guo, J.; Zhou, H.; Yang, L.;Chen, X. In *Research on Digital Copyright Blockchain Technology*, 3rd International Conference on Smart BlockChain (SmartBlock), 2020; pp 1–5.

Gupta, N.; Juneja, P. K.; Sharma, S.; Garg, U. In *Future Aspect of 5G-IoT Architecture in Smart Healthcare System*, 2021 5th International Conference on Intelligent Computing and Control Systems (ICICCS), 2021; pp 406–411.

Hussain, F.; Hammad, M. In *Smart Healthcare Facilities via IOT in the Healthcare Industry*, 3rd Smart Cities Symposium (SCS 2020), 2020; pp 386–391.

Kirtonia, P.; Hosain, M. K.; Rahman, T. In *A Circularly Polarized Implantable Wideband Antenna for Bio-Telemetry Applications*, International Conference on Electrical, Computer and Communication Engineering, 2019; pp 1–4.

Kothari, D.; Patel, M.; Sharma, A. K. In *Implementation of Grey Scale Normalization in Machine Learning & Artificial Intelligence for Bioinformatics using Convolutional Neural Networks*, 6th International Conference on Inventive Computation Technologies, 2021; pp 1071–1074.

Kumar, S.; Kumar, R.; Singh, N. In *Comparative Analysis of Battery with Paper Battery for Renewable Energy Storage*, International Conference on Challenges in Sustainable Development from Energy & Environment Perspective, Gorakhpur, India, 2017; pp 222–229.

Kumar, S.; Singh, R.; Singh, D.; Prasad, R.; Yadav, T. Electronics-Waste Management. *Int. J. Environ. Eng. Manag.* **2013,** *4* (4), 389–396.

Mannatunga, K. S.; Ali, S. H. M.; Crespo, M. L.; Cicuttin, A.;Samarawikrama, J. G. High Performance 128-Channel Acquisition System for Electrophysiological Signals. *IEEE Access* **2020,** *8*, 122366–122383.

Mishra, P. K.; Raj, S.; Tripathi, V. S. In *A Novel Skin-Implantable Patch Antenna for Biomedical Applications*, IEEE 7th Uttar Pradesh Section International Conference on Electrical, Electronics and Computer Engineering, 2020; pp 1–5.

Mohsen, S.; Zekry, A.; Abouelatta, M.; Youssef, K. In *A Self-Powered Wearable Sensor Node for IoT Healthcare Applications*, 2020 8th International Japan-Africa Conference on Electronics, Communications, and Computations (JAC-ECC), 2020; pp 70–73.

Parihar, J.; Kansal, P.; Singh, K.; Dhiman, H. In *Assessment of Bioinformatics and Healthcare Informatics*, Amity International Conference on Artificial Intelligence, 2019; pp 465–467.

Ray, P. P.; Dash, D.; Salah, K.; Kumar, N. Blockchain for IoT-Based Healthcare: Background, Consensus, Platforms, and Use Cases. *IEEE Syst. J.* **2021,** *15* (1), 85–94.

Singelmann, L.; Swartz, E.; Pearson, M.; Striker, R.; Vazquez, E. A. In *Design and Development of a Machine Learning Tool for an Innovation-Based Learning MOOC*. IEEE Learning with MOOCS, 2019; pp 105–109.

Singh, R.; Kumar, S.; Chauhan, S. Energy:-Conservation, Management, Efficiency & Storage. *Int. J. Appl. Eng. Res.* **2013,** *8* (7), 80–85.

Tian, Z.; Qiu, F. In *Research on Physiological Parameters Monitor Based on Multi-Functional Device*, International Conference on Biomedical Engineering and Computer Science, 2010; pp 1–4.

Vijayakumar, K.; Bhuvaneswari, V. In *A Ubiquitous First Look of IoT Framework for Healthcare Applications*, 2020 International Conference on Emerging Trends in Information Technology and Engineering (ic-ETITE), 2020; pp 1–7.

Vullings, R. In *Fetal Electrocardiography and Deep Learning for Prenatal Detection of Congenital Heart Disease*, 2019 Computing in Cardiology (CinC), 2019; pp 1–4.

Zaman, F.; Aloqaily, M.; Sallabi, F.; Shuaib, K.; Othman, J. B. In *Application of Graph Theory in IoT for Optimization of Connected Healthcare System*, GLOBECOM 2020 - 2020 IEEE Global Communications Conference, 2020; pp 1–6.

CHAPTER 8

Intelligent Visual to Textual Mapping in Automatic Medical Diagnosis and Treatment

PREM SHANKER YADAV[1*], DINESH KUMAR TYAGI[1] and SANTOSH KUMAR VIPPARTHI[2]

[1]*Department of Computer Science and Engineering, Malaviya National Institute of Technology, Jaipur, Rajasthan, India*

[2]*School of Artificial Intelligence and Data Engineering, Indian Institute of Technology, Ropar, Rupnagar, Punjab, India*

*Corresponding author

ABSTRACT

Real-time scene description impacts instant logic for blind or disabled persons. It becomes more critical if related to health. Scene-to-text conversion on biomedical-related images can assist and impel the diagnosis process. Medical image description sorts correct diagnosis and treatment in the right direction, presenting connections in objects that exist in the images useful in clinical findings. Correct findings lead to satisfactory treatment and help early health recovery with time and overall expenditure savings.

We have discussed various important approaches and methodologies, together with relevant datasets, applications, limits, future prospects, and finally summarize the application values of biological image description.

8.1 INTRODUCTION

Today, the world is permeated with technologically controlled era with autonomous, intelligently managed, and controlled equipment. Machine and neuron-based technology is infusing self-learning capabilities into devices

Smart Electronic Devices and Systems for Biomedical and Healthcare Applications. Upendra Kumar, Rajesh K. Kesharwani, & Raj K. Keservani (Ed.)

DOI: 10.1201/9781003560852-8

day by day. What was impossible yesterday is now a reality. Favorite food, clothes, and groceries are just one click away. Self-driving vehicles are becoming increasingly common. This increases the significance of scene understanding. Pictorial detailing into textual records creates huge impact in manufacturing industry of assistive equipment like security sector, automobile industry, and healthcare sector, etc. Our eating habits and way of life are making us sick. Disease prevention seems impossible. Everyone needs adequate health care. A quick and precise health diagnosis saves time and lowers total treatment costs. Its times need to produce automatic medical reports from medical-related images. Accurate results and diagnosis based on imaging-based tests such as X-rays and ultrasound reports, among others, will lead to successful therapy.

8.1.1 TERMINOLOGY: TERMS IMPORTANT TO KNOW

Here are some important explanations given to understand their root impact as we move further in this chapter.

i. **Objects:** Various things and items are considered as object (live or dead) in the scene.

FIGURE 8.1 Objects in scene (a) giraffe, (b) zebra, and (c) tree.

ii. **Caption:** According to Oxford Language, "the words written above or below a picture, photograph, etc. are to explain what it is about." The Merriam-Webster dictionary emphasizes explanatory commentary as well as pictorial illustration. From a technical point of view, the caption is textual information about the object-content in the image. Caption from Figure 8.1 is "a Giraffe is standing near zebra."

iii. **Image captioning:** According to, Huang et al., 2019; Alsharid et al., 2020, image captioning is "generating textual sentence(s) for images or videos."

A girl is stretched out in shallow water.
A girl wearing a red and multi-colored bikini is laying on her back in shallow water.
A little girl in a red swimsuit is laying on her back in shallow water.
A young girl is lying in the sand, while ocean water is surrounding her .
Girl wearing a bikini lying on her back in a shallow pool of clear blue water .

FIGURE 8.2 Image-text sample from the dataset Flicker8k.

Now we hope you can easily define image captioning as the "task of generating description from the image content, that is, what object(s) are there and their relationship."

8.1.2 NEED MEDICAL DESCRIPTION

Scene in descriptive form is always a challenging task as it involves deep learning and natural language processing. It has a significant impact on the life of a blind or disabled person; it even makes their daily routine smoother and easier. A description in meaningful form is the result of the detection of various objects and their relationships among them, their features, and their interpretations. Image-to-text information plays a critical role in classification, education, answering visual-based questions, helping visually impaired persons, and making medical observations.

The medical description is a favorable approach to diagnose the body without surgery or other hectic invasive procedures (Allaouzi et al., 2018). Proper diagnosis and treatment (Xiong et al. 2019) are performed on the basis of pathological details of organs and tissues (Zeng et al., 2020b).

Recently, technical advances enhance medical facilities in imaging as they are tedious to handle manually and time-consuming (Harzig et al., 2019). Manual diagnose takes a large amount of time which leads delay in findings and treatment on time, making treatment costly and worsening the patient's condition. Diagnosis speeded the complete healthcare chain by automatic medical image description. In medicine, there is no space for failure. Accuracy and complete report is the leading step to first fix the disease and if we are generating this from visual reports, detecting correct illnesses, this will save our labor costs, and fill the gap created due to lack of medical expertise.

8.1.2.1 MEDICAL IMAGE CAPTIONING

Current development in medicine domain using machine learning provides the ability to automatically comprehend and generate a textual description of medical photographs, which may be very helpful in medical analysis procedures such as diagnosis, therapy, and surgery.

We know that medical diagnosis is the process of concluding a disease after applying a standard procedure to its indications and symptoms. A diagnosis report is a complete document finalized by an expert to summarize the conclusion of the indications and symptoms from medical images (Monshi et al., 2020). Notably, this adheres to important procedures and makes use of a specific medical taxonomy that incorporates visual proof of findings.

Now medical image captioning, a medical image-based image captioning, easily supports the creation of diagnosis reports (Park et al., 2020). The textual creation of a diagnosis from a review of a patient's medical report for diagnostic purposes is known as diagnostic captioning (Pavlopoulos et al., 2021). This diagnostic captioning helps doctor to cure patients more effectively due to clinically significant information.

Here, we are presenting one sample of the medical image-to-text description. "In given image lifted polyps are described with their findings so the textual presentation of medical-related images considered medical captioning."

Tomar et al. (2022) proposed a novel approach to effectively segment biomedical images called feedback attention network (FANet) that integrates the feature map of the current training epoch with the previous epoch mask. Seven publicly available biomedical datasets are analyzed. By introducing this mechanism, we achieved an improvement over the existing state-of-the-art methods. Magnification of endoscopy provides improved captioning of early gastric cancer when applied with narrow-band imaging Gong et al. (2022). Aswiga and Shanthi (2022) have presented a medical image captioning model, architecture of Transfer Learning along with a decoder, Long Short-Term-Memory used as decoder. Transfer learning at input level is applied in three steps combining non-medical and medical images. A significant work is done by Mansour et al. (2022) to identify colorectal cancer (CRC) at the earliest. A fusion model named AIFM-CRC utilizes noise reduction and contrast-enhancing Gaussian filtering. SIFT-based handmade features and inception v4-based deep features are combined during feature extraction.

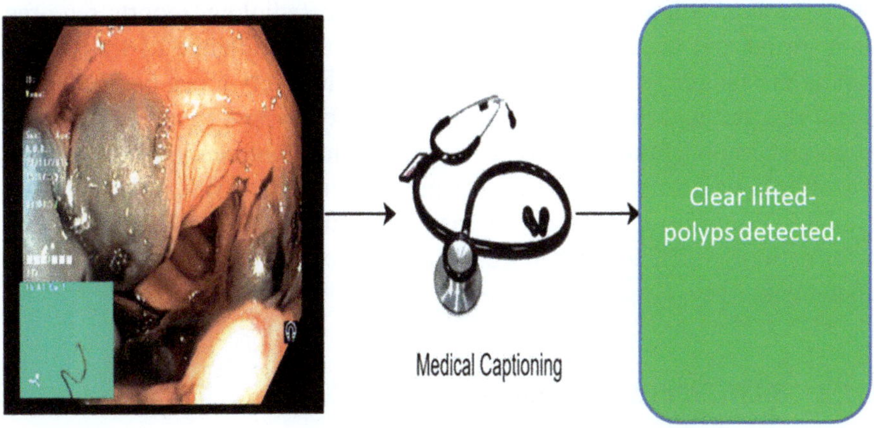

FIGURE 8.3 Description is generated from image having medical content.

The method for classifying medical picture modalities using visual and linguistic elements is proposed by Miranda et al. (2023). Convolutional neural networks are used in the suggested method to extract visual content from a medical picture. The textual characteristics from the picture captions are generated using word embeddings from biomedical word2vec models. The author uses support vector classifier to categorize medical content-based pictures. To merge visual and textual information, the paper is used in the late fusion strategy. The diagnosis of the majority of chronic kidney diseases is crucially dependent on histopathological analysis of glomerular immune fluorescence (IF) images but the analysis procedure is subject to inter-observer variability and always requires human judgments. In order to improve the root content extraction capabilities for glomerular IF images, a fusion model is developed.

In the research of Liu et al. (2023), hierarchically weighted features arranged in a specific manner were fused to an attention network. Second, a suggested intensity equalization (IE) technique brings all pictures to a consistent and acceptable intensity level, enhancing the quality of images and resulting IF descriptions. Thirdly, to provide descriptions that mimic doctors' reading preferences, deep-learning architecture with the fusion network at its core is used. Finally, to facilitate the model training, we provide a large dataset of glomerular IF pictures and their related textual descriptions. This dataset contains 11,506 image-text pairs.

Automatic radiology report production of Wang et al. (2023) has attracted a lot of interest recently as a means to avoid the tedious and difficult process of producing radiology reports. The creation of radiology reports requires both visual and linguistic aspects, similar to the challenge of translating vision into language. However, earlier approaches only focus on producing fluent reports, ignoring the paramount necessity of how to use and extract pictorial components efficiently. To overcome the aforementioned problems, we offer a unique architecture of a model combined with a CLIP-based module. Pre-trained encoders of vision language are applied for visual information, which is further given to the model to generate the textual document.

8.1.3 BASIC STEPS OF IMAGE CAPTIONING

- Pre-processing: Refine the raw image for subsequent phases by enhancing the quality using image processing tools relevant to captioning. We are applying various rotations, flips, augmentation, and conversions to increase the strength of the pre-processing.
- Feature extraction: This is the process to identify and extract those unique biomedical features that are relevant to caption generation. Significant features produce effective reports to impressive diagnoses. Figure 8.4 presents the importance of feature effectiveness in the correct generation of image content description. The classical mechanism requires selective parameters for feature extraction while deep-learning-based models automatically extract features. As good features are extracted effective caption will be generated. Generally, ResNet50, VGG16, and Inception3V are preferred for this phase. A number of models are architecture using various combinations of convolution, pooling, and flattening to get best performing model. Feature summarization, their reduction, and the use of transfer learning are other factors that highly participate in the final description.
- Textual description: Based on the extracted features, perform translation and decode them into sentences.

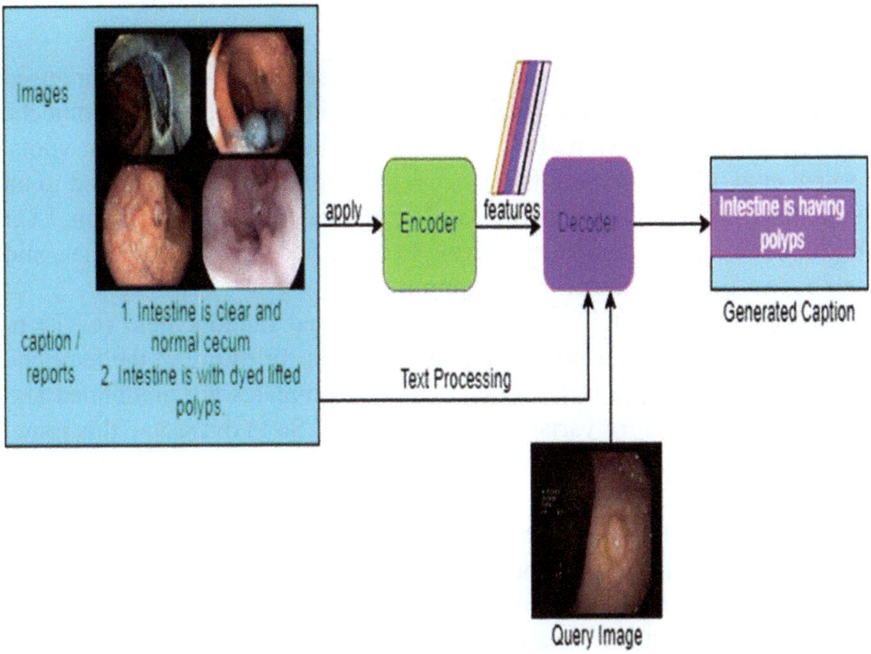

FIGURE 8.4 Generating description from image text pairs.

8.2 APPROACHES

Image captioning benefits self-driven vehicles, retrieving targeted images and caring for visually impaired persons. Evidently, without any human intervention, converting the visual content of an image into textual form relaxes the headache of the captioning mechanism. In most of the research studies natural images are involved in various applications and services but medical images are still challenging and digging area of interest as content is difficult to describe rather than natural images. Finding the actors, objects and their correlation can easily describe the content, however, specific domain knowledge and expertise are the main requirements to interpret the medial report since body organs and tissues create images. Any error can lead to the wrong description and further life risk. Though very few existing image captioning models are adopted in the medical domain, they do not present good results. Generally, text-to-pictorial conversion is studied in three main categories. These are also applied in medical description cases. We are going to discuss template, retrieval, and deep-learning-based approaches. In deep learning our focus is on encoder–decoder models.

8.2.1 TEMPLATE BASED

In template-based systems fixed templates with several vacant slots are used to produce captions. These methods start by identifying various objects, characteristics, and actions before filling up the template's vacant spots. Farhadi et al. (2010) present the concept of a triplet, which is applied to an image for fixed template slots to complete a proper textual description. Li et al. (2018) collect the sentences relating to identified objects, properties, and their connections to map efficient caption.

Kulkarni et al. (2013) introduce a conditional random field (CRF) to deduce the objects, their properties, and relations for fitting in the template slots. Though this approach accuracy is best, templates are predefined and fixed so problematic for variable length captions. So to overcome this issue, later on, parsing approach is preferred and applied to generate descriptions. Ayesha et al. (2021) adopted grammar literature and language rules and their constraints along with object detection and attributed relationships to distinguish accurately correct or incorrect leads in generated reports. Beddiar et al. (2022) generate small but grammatically more accurate descriptions within the constraints of variability.

Onita et al. (2020) authors used classification-based regression, targeting the effectiveness of generated word sentences. Wang et al. (2021) use medical-based combinations of patterns to produce sentences.

8.2.2 RETRIEVAL BASED

Similarity comparison plays a golden factor in this approach. Finding comparable photos in big datasets or archives and creating a new sentence. The most similar sentence is used in place of the new description, or the retrieved sentences are combined to generate an entirely new sentence. Figure 8.5 demonstrates the similarity phase after feature extraction. Various distance matrices like Hamming distance, Hellinger, Kendell tau, and Euclidean may be used as per requirement. Wang et al. (2019) applied Lucene Image Retrieval (LIRE) approach and used a classifier along with a topic modeling technique to retrieve medical-based images, making the clusters of various classes of images, the content of the images, and formulating the captions. The suggested method could be enhanced to extract the input image's most pertinent captions from captions of photos in its class.

Retrieval concept based on the approach of the same images will be the same descriptions. For query image, a set of similar images are identified from the huge database. Ayesha et al. (2021) discussed two viewpoints to generate new caption for novel image. First, map the most similar candidate description to the same description. Later, exploit the candidate description and merge them based on pre-specified rules and logic to generate description for a novel image.

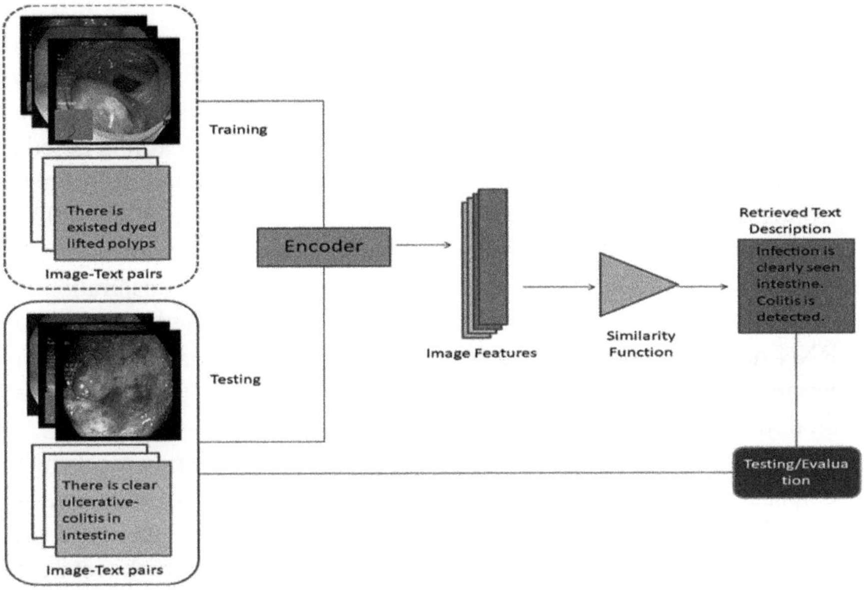

FIGURE 8.5 Retrieval architecture demonstrating feature extraction and similarity function.

Syeda-Mahmood et al. (2020) proposed a domain-centric retrieval system for achieving similar chest X-ray radiology report document from the available huge dataset. Feature extraction is the main phase as it leads to fine detailed descriptions of the scene. Labeling is properly integrated into features, and carried out using concept detection. In proposed algorithm group of phrases are critical factors, responsible for effective labeling and conversion into textual descriptions of the scene. Final report is retrieved using the labeling of similar images from the archive dataset.

Charalampakos et al. (2021) proposed a model integrated with classical model to consolidate retrieval approach effectively. The author found KNN and the previous work Kougia et al. (2021) impose a more realistic impact on the retrieval concept. Additionally, when creating new words, the authors used a visual attention method Pavlopoulos et al. (2021) to concentrate on certain vectors of the input text encoding.

8.2.3 ENCODER–DECODER MODEL

This approach consists of encoder and decoder module. Figure 8.6 demonstrates architecture properly. Encoder may be any formation of convolution, pooling, and flattening steps of deep learning, however, LSTM (Long Short-Term Memory), Bi-LSTM (Bidirectional-LSTM), and GRU (Gated recurrent units), etc. may be used at the decoder side.

Alsharid et al. (2020) used deep-learning approach to generate reports. It collects attributes from images and gives RNN (recurrent neural network) to map generated words into meaningful language. This is the most used attentive concept in image-to-text conversion because deep learning based is well performer as per current time. These methods include CNNs, fully linked networks, and encoder–decoder architectures Pavlopoulos et al. (2021). Vinyals et al. (2015) a deep-learning plus natural language processing-based model for pictorial description, is largely responsible for the inspiration for the methods currently used in the literature. The proposed method is built on an RNN network for the text generator and a CNN network for collecting important visual content (Encoder), which together make up the proposed method (decoder).

Encoder–decoder model is tremendously used with attention mechanism by emphasizing the specific part or region of the images. We are primarily interested in an area where the majority of the attributes are collected. Zeng et al. (2020) suggested employing encoder–decoder architecture to identify lesion sites, extract automatic pictorial attributes, diagnose pathological content, and finally generate medical report. Similarly, Yin et al., 2019; Tsuneda et al., 2021 used encoder-decoder-attention model to detect abnormal findings in generated medical reports. As usual, encoder is CNN and decoder is a hierarchical RNN. In this paper, lengthy captions are tried to generate using an attention mechanism.

Zeng et al. (2020) used fast RCNN as encoder followed by attention to features to get feature map from ultrasound images and finally, these feature maps are transferred to the LSTM to produce medical description. Harzig et al. (2020) proposed a dual LSTM model to justify positive and negative values in captioning. A multi-task approach is proposed Chelaramani et al. (2020) to identify and detect fundus based diseases.

Yang et al. (2021) have presented a combined encoder–decoder and attention network to characterize key fine aspects in ultrasound medical pictures and provide reports. Initially, multi-label classification is initiated in order to develop meaningful characteristics that are combined with visual

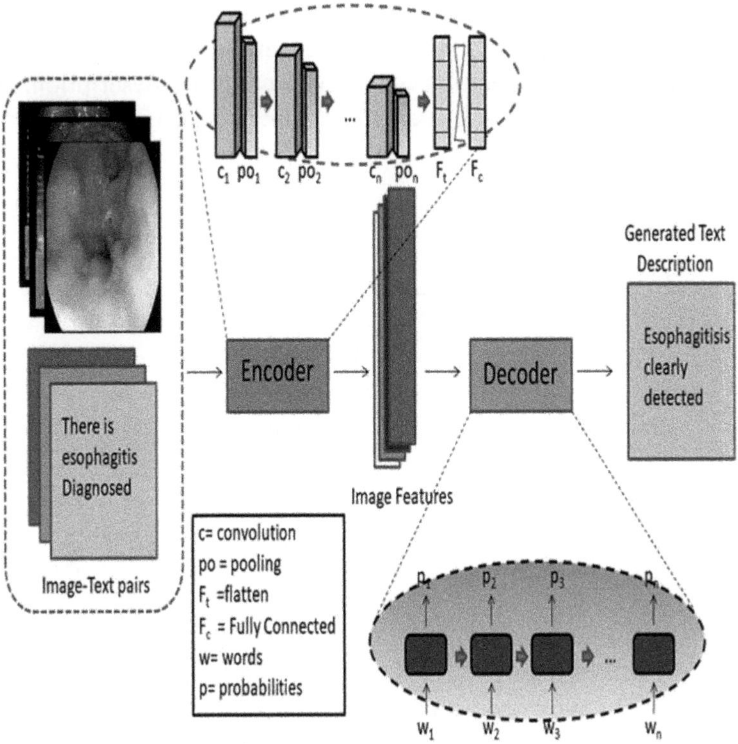

FIGURE 8.6 Encoder–decoder architecture.

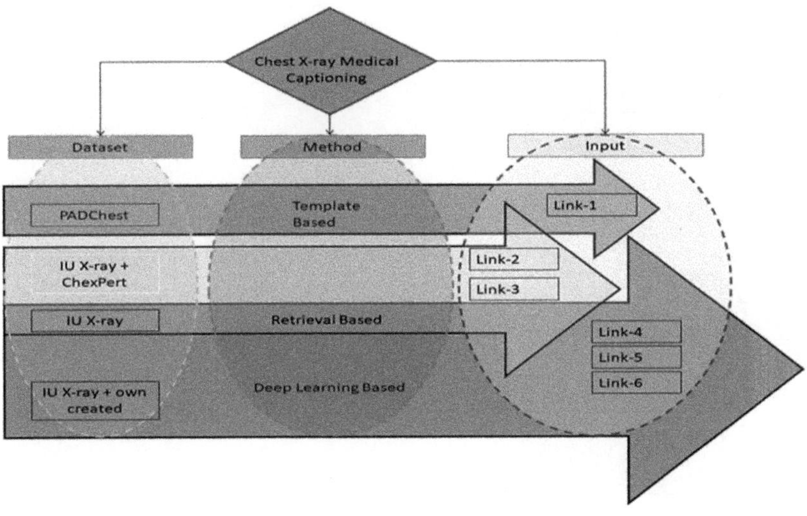

FIGURE 8.7 Relational references with methods having related datasets.

features and predict important local attributes. There is no change in LSTM side as decoder, only the attention mechanism is modified by adopting a sentinel gate to control attention while generating the description. Below categorical demonstrating is presented of papers dealing in chest medical captioning.

TABLE 8.1 Link–References.

S. No.	Link	References
1	Link-1	Onita et al. (2020)
2	Link-2	Li et al. (2018)
3	Link-3	Li et al. (2019)
4	Link-4	Zeng et al. (2020)
5	Link-5	Yin et al. (2019)
6	Link-6	Wang et al. (2019)

8.3 DATASETS

Many academics test their ideas on benchmark datasets in order to correctly assess how well the proposed approaches work in various image processing tasks as segment-division, class prediction and intend content retrieval, etc. Image captioning normally uses Flicker8k, Flicker30k, Coco datasets for natural images. Datasets like Sydney, UCF, and RSICD are recommended for remote sensing-based captioning.

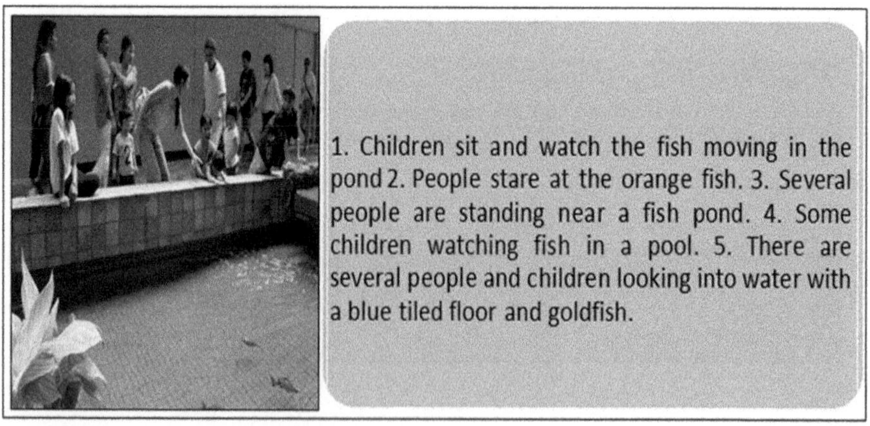

FIGURE 8.8 Image –text pair from Flicker8k dataset.

We show numerous datasets utilized in creating diagnostic reports based on the precise features of associated medical scene material. The report comprised one sentence or paragraph. Some publicly available medical datasets are presented.

8.3.1 THE INDIANA UNIVERSITY CHEST X-RAY

The Chest X-ray Collection at Indiana University (IU X-ray)-It is a collection of chest X-ray pictures-textual description. Textual descriptions are basically medical reports extracted from corresponding images Demner-Fushman et al. (2016). A collection of total 8121 related photos and 3996 reports was done by the Indiana Network for Patient Care Demner-Fushman et al. (2016). These images were then captioned manually and classified into normal and abnormal.

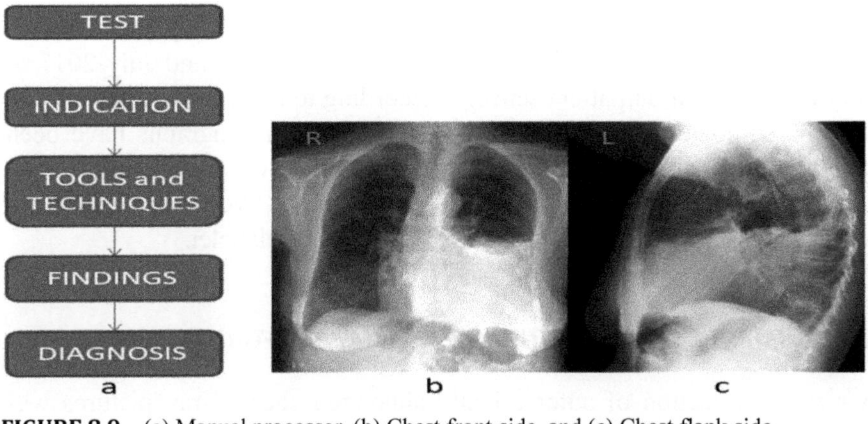

FIGURE 8.9 (a) Manual processor, (b) Chest front side, and (c) Chest flank side.

8.3.2 MEDICAL INFORMATION MART FOR INTENSIVE CARE

Medical information mart for intensive care is about to chest X-ray dataset. It is open sourced dataset. CheXpertlabeler is used to find explanations from radiology medico-reports (Johnson et al., 2019). It contains 227,943 reports and 371,920 chest X-rays that were acquired from the Beth Israel Deaconess Medical Center (BIDMC). The English-language documents were used to prevent someone's personal identity from being revealed, and the photos were examined to extirpate any patient-related information (Figure 8.10).

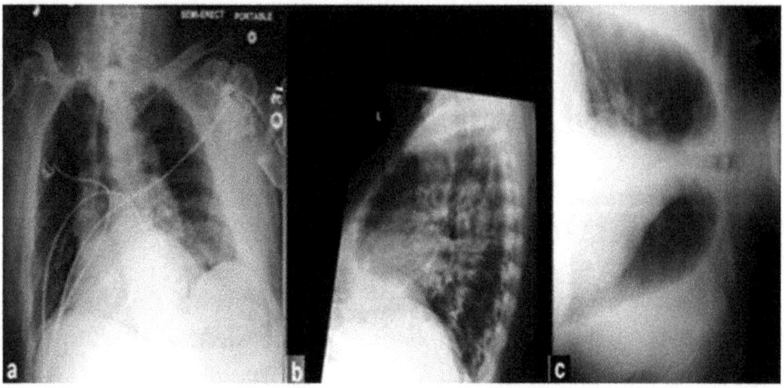

FIGURE 8.10 MIMIC-CXR. (a) Patient's imperfect positioning, (b) secondary collimation, and (c) wrong orientation.

8.3.3 CHEXPERT

A sizable collection of automatically annotated chest X-ray pictures was gathered from Stanford Hospital between October 2002 and July 2017 in both inpatient and outpatient settings, according to Irvin et al. (2019). From 65,240 candidates, total 224,316 multi-view chest radiographs have been classified as positive, negative, or uncertain based on the presence of 14 common chest radiographic findings. Annotations were extracted from unstructured radiological records using the CheXpertlabeler.

8.3.4 IMAGEL CROSS-LANGUAGE EVALUATION FORUM (ICLEF)

A sizable collection of automatically annotated chest X-ray pictures was gathered from Stanford Hospital between October 2002 and July 2017 in both inpatient and outpatient settings, according to Irvin et al. (2019). From 65,240 candidates, total 224,316 multi-view chest radiographs have been classified as positive, negative or uncertain based on the presence of 14 common chest radiographic findings. Annotations were extracted from unstructured radiological records using the CheXpertlabeler.

8.3.5 IMAGEL CROSS-LANGUAGE EVALUATION FORUM (ICLEF)

Image CLEF introduced a dataset for the purpose of concept prediction and caption generation in 2017. The PubMed Central (PMC) database had

184,614 photos, many of which were clinical and compound images. Due to the fact that compound pictures have an overall noise level of 10–20%, this dataset needed to be refined. Dataset contains total 232,305 image-text pairs.

8.3.6 BLADDER CANCER DATASET

The bladder cancer images and its diagnostic report (BCIDR) have 5000 photo-captions pairs that were gathered with the aid of pathologists in a publication Zhang et al. (2017). A pathologist explained observation in terms of five different feature categories, where each category explains how a cell appears. A conclusion made up of four classes follows the observations. Four non-experts in bladder cancer who performed the identical surgery expressed their findings in their own words while taking into account pathologist reports to ensure correctness. This gives each image a total of five ground truth reports. There are 30–59 words in each report.

8.3.7 PATHOLOGY EDUCATION INFORMATIONAL RESOURCE (PEIR) GROSS

The PEIR digital library is a publicly accessible image database. Jing et al. (2017) present the very first use of this dataset having 7442 image-text pairs of gross lesions divided into 21 pathology-class sub-categories. From the dataset, 6698 pairs are considered for training and 745 pairs for test instances. In this dataset, few captions are repeated. 632 captions are used for other images also. The image description sentence length is 17 words, and the vocabulary size is 4051.

8.3.8 ROCO DATASET

An open-access database dataset contains 81,000 radiology images using a variety of medical imaging modalities of biomedical literature collected from PubMed Central. Only related images were extracted and other photos were then removed using a binary classification of radiology and non-radiology. Pelka et al. (2018) effectively presented this multimodal medico photo-caption dataset.

8.4 ANALYSIS

Several criteria have been added to assess the effectiveness, and performance of the generated captions compared with classical picture descriptions. The captioning of medical images might use the same metrics. In fact, they enable us to assess the performance while generating new image description by matching equality ground truth and the generated captions. Additionally, most research used these indicators to compare their findings to prior state-of-the-art baselines Syedo-Mahmood et al. (2020). Few important performance metrics have been discussed below.

8.4.1 *BILINGUAL EVALUATION UNDERSTUDY*

A rapid, language-independent evaluation of automatic machine translation is called BLEU (Papineni et al., 2002) Additionally, BLEU can be utilized for speech recognition, picture captioning, and text summary. It enables us to compare candidate translations (made by machines) to the reference translations (made by humans), accounting for various aspects. When there are more matches, machine translation is thought to be better. BLEU score approaching one is considered better.

8.4.2 *RECALL ORIENTED UNDERSTUDY FOR GISTING EVALUATION-LONGEST COMMON SUBSEQUENCE (ROUGE-L)*

The metric subsequence ROUGE-L (Lin et al., 2004) can be applied to text summarization and machine translation. This is based on a common subsequence. Method calculates the longest common subsequence (LCS) and matches both the original summary and the predicted summary. Instead of matching words consecutively, with the use of LCS, we are able to determine in-sequence matches. Another benefit is that LCS incorporates common n-grams in-sequence automatically, eliminating the need to calculate the predefined sequence of n-grams.

8.4.3 *METRIC FOR TESTING EXPLICIT ORDERING IN TRANSLATION*

METEOR metric was primarily used in the machine translation literature. METEOR (Banerjee et al., 2005) was created to assess the consistency of automatically generated and human-translated sentences at the sentence level. The greatest score is calculated by word-by-word comparisons in candidate and reference sentences. It is now used in textual representation such as text summary, image description, speech recognition, and so on.

Consensus-based Evaluation of Image Description: The automatic consensus-based assessment metric CIDEr (Vedantam et al., 2015) was created primarily for the evaluation of picture descriptions. It evaluates the grammar, saliency, and the sentence's precision in contrast to a collection of manmade descriptions.

8.5 LIMITATIONS

Although several methods for the medico-images descriptions have been proposed in the literature, they are still constrained by a variety of issues. In this section, we list some of these restrictions, as follows:

- Specialists may produce different reports for the same image or write reports in a variety of styles that result in inconsistently labeled data (Han et al., 2021).
- Due to the possibility that medical data obtained might contain patient personal information and result in incomplete reports, privacy concerns may arise (Momshi et al., 2020).
- Because the terminology required to write medical documents is precise, complex, and distinct from normal speech (Zeng et al., 2020b), crowd-sourcing is not a viable choice for building datasets since it might cause mistakes to spread from the construction activity to the learning process. Medical picture annotation is subject to human mistakes as well.
- Incoherent or lengthy statements that do not adequately convey the content of the image are not yet clinically appropriate since they might be quite deceptive (Harzig et al., 2020).
- Since human review is still important though existing models are performing better Pavlopoulos et al. (2021). We need human experts to verify the result but finding qualified evaluators remains difficult and expensive.
- Model and disease generalization is still very difficult and challenging. The reason is that dataset samples of single modalities are very low and the majority of the known datasets are small and limited (Rodin et al., 2019), which has an impact on predictions made using other imaging modalities and prevents the model from generalizing.

- Problem of gradient vanishing in lengthy sentences. According to Xiong et al. (2019), lengthy phrases suffer substantially due to the gradient vanishing problem as the majority of produced reports are generated after the RNN module.
- Even while it is helpful to apply predefined and trained models to solve the problem of small data, there is a significant amount of domain transfer from classical natural photos to medical content images, which might lead to varying degrees of mistake.
- In most research works, we are using a deep-learning approach and it is very clear that labeled data should be largely required; it is still difficult to evaluate and compare with current state-of-the-art approaches (Ayesha et al., 2021).

8.6 CONCLUSION AND FUTURE DIRECTIONS

This chapter tries to provide a brief overview of the fundamentals in the field of image-to-textual conversion of study thus far. In the beginning, it introduces image-to-text terminology followed by basic steps of image captioning. The article then goes into the emphasis on the medical industry and the main goals that medical image captioning systems seek to achieve. We also examine the approaches already in use in various studies, which are divided into mainly three groups: template-based approaches, retrieval-based approaches, and encoder–decoder deep-learning approaches.

Performance analysis and popularly used datasets are also explained. Despite the advancements in image-to-textual descriptions in the literature, it was found throughout this study that its use in the medical field is still exceedingly difficult. This is because medical content in images and reports has different qualities than natural image descriptions. Clinical findings and the final reports are made up of many paragraphs that highlight just the clinically significant elements rather than the objects and their relationship in the image. Nowadays, it is highly promising that generative and retrieval-based models will increasingly be used in combinations.

Contrary to general picture captioning, where salient items (and activities occurring) normally have to be reported, prominent things presented (such as human organs) do not need to be noted if nothing clinically essential has to be stated about them. Medical photos differ far less from general image captions, and as a result, the diagnostic language that goes with them is sometimes extremely similar or even the same for various patients. These elements enable retrieval-based techniques to perform unexpectedly well in medical captioning, reusing diagnostic text from training examples with related pictures. Instead of producing them, sentences can instead be used often or as templates.

Additionally, deep-learning techniques are used as genuine medical imaging datasets are growing day by day. To further improve research in captioning, a dataset that includes images of various bodily components, such as the brain and breast, must be created. Additionally, since present measures are global and not 100% accurate, proper domain-specific assessment must be developed to cope with medical captioning.

KEYWORDS

- **feedback attention network**
- **colorectal cancer**
- **scale-invariant feature transform**
- **immune fluorescence**
- **conditional random field**
- **Lucene Image Retrieval**
- **long short-term memory**

REFERENCES

Allaouzi, I.; Ben Ahmed, M.; Benamrou, B. In *Automatic Caption Generation for Medical Images*, Proceedings of the 3rd International Conference on Smart City Applications (SCA'18), 2018.

Alsharid, M.; El-Bouri, R.; Sharma, H. A Curriculum Learning based Approach to Captioning Ultrasound Images. *Med. Ultrasound Preterm Perinat. Paediatr. Image Anal.* **2020,** *12437*, 75–84.

Aswiga, R. V.; Shanthi, A. P. A Multilevel Transfer Learning Technique and LSTM Framework for Generating Medical Captions for Limited CT and DBT Images. *J. Digit. Imaging* **2022,** *35* (3), 564–580.

Ayesha, H.; Iqbal, S.; Tariq, M.; Abrar, M.; Sanaullah, M.; Abbas, I.; Rehman, A.; Niazi, M. F. K.; Hussain, S. Automatic Medical Image Interpretation: State of the Art and Future Directions. *Pattern Recognit.* **2021,** *114*, 107856.

Banerjee, S.; Lavie, A. Meteor: In *An Automatic Metric for Mt Evaluation with Improved Correlation with Human Judgments*, Proceedings of the ACL Workshop on Intrinsic and Extrinsic Evaluation Measures for Machine Translation and/or Summarization, 2005; pp 65–72.

Beddiar, D. R.; Oussalah, M.; Seppänen, T. Automatic Captioning for Medical Imaging (MIC): A Rapid Review of Literature. *Artif. Intell. Rev.* **2022**.

Charalampakos, F.; Karatzas, V.; Kougia, V. In *Auebnlp Group at Imageclefmed Caption Tasks*, CLEF2021 Working Notes, CEUR Workshop Proceedings; CEUR-WS.org, Bucharest, Romania, 2021.

Chelaramani, S.; Gupta, M.; Agarwal, V.; et al. In *Multi-Task Learning for Fine-Grained Eye Disease Prediction*, 5th Asian Conference on Pattern Recognition, 2020; vol *12047*, pp 734–749.

Demner-Fushman, D.; Kohli, M. D.; Rosenman, M. B.; et al. Preparing a Collection of Radiology Examinations for Distribution and Retrieval. *J. Am. Med. Inform. Assoc.* **2016,** *23* (2), 304–310.

Farhadi, A.; Hejrati, M.; Sadeghi, M. A. In *Every Picture Tells a Story: Generating Sentences from Images*, European Conference on Computer Vision; Springer, 2010; pp 15–29.

Gong; L.; Wang, M.; Shu, L.; He, J.; Qin, B.; Xu, J.; Su, W. Automatic Captioning of Early Gastric Cancer using Magnification Endoscopy with Narrow-Band Imaging. *Gastrointest. Endosc.* **2022,** *96* (6), 929–942.

Han, Z.; Wei, B.; Xi, X.; et al. Unifying Neural Learning and Symbolic Reasoning for Spinal Medical Report Generation. *Med. Image Anal.* **2021,** *67*.

Harzig, P.; Chen, Y. Y.; Chen, F.; et al. In *Addressing Data Bias Problems for Chest X-Ray Image Report Generation*, 30th British Machine Vision Conference, 2019.

Harzig, P.; Einfalt, M.; Lienhart, R. In *Automatic Disease Detection and Report Generation for Gastrointestinal Tract Examinations,* 27th ACM International Conference on Multimedia, 2019; pp 2573–2577.

Huang, X.; Yan, F.; Xu, W. Multi-Attention and Incorporating Background Information Model for Chest X-Ray Image Report Generation. *IEEE Access* **2019,** *7*, 154808–154817.

Irvin, J.; Rajpurkar, P.; Ko, M.; et al. In *Chexpert: A Large Chest Radiograph Dataset with Uncertainty Labels and Expert Comparison*, Proceedings of the AAAI Conference on Artifcial Intelligence, 2019; pp 590–597.

Jing, B.; Xie, P.; Xing, E. On the Automatic Generation of Medical Imaging Reports, arXiv preprint arXiv:1711.08195, 2017.

Johnson, A. E.; Pollard, T. J.; Berkowitz, S. J.; et al. Mimic-cxr, a De-identifed Publicly Available Database of Chest Radiographs with Free-Text Reports. *Sci. Data* **2019,** *6* (1), 1–8.

Kougia, V.; Pavlopoulos, J.; Androutsopoulos, I. In *Auebnlp Group at Imageclefmed Caption 2020,* Proceedings of the Working Notes of CLEF 2021—Conference and Labs of the Evaluation Forum, Bucharest, Romania, September 21–24, 2021; CEUR Workshop Proceedings, 2021; vol *2936*, pp 1184–1200.

Kulkarni, G.; Premraj, V.; Ordonez, V. Babytalk: Understanding and Generating Simple Image Descriptions. *IEEE Trans. Pattern Anal. Mach. Intell.* **2013,** *35* (12), 2891–2903.

Li, C.; Liang, X.; Hu, Z. In *Advances in Neural Information Processing Systems*, NIPS, 2018.

Li, C. Y.; Liang, X.; Hu, Z.; Xing, E. P. In *Knowledge-Driven Encode, Retrieve, Paraphrase for Medical Image Report Generation*, Proceedings of the AAAI Conference on Artificial Intelligence, Atlanta, GA, USA, October 8–12, 2019; vol *33*, pp 6666–6673.

Li, C.; Liang, X.; Hu, Z.; Xing, E.; Li, C. Y.; Liang, X.; Hu, Z.; Xing, E. P. In *Hybrid Retrieval-Generation Reinforced Agent for Medical Image Report Generation*, Advances in Neural Information Processing Systems 31 (NIPS 2018), Proceedings of the Annual Conference on Neural Information Processing Systems 2018, NeurIPS 2018, Montreal, QC, Canada, December 3–8, 2018; Curran Associates, Inc.: New York, NY, USA, 2018; vol *31*.

Lin, W.; Chen, J. Improving Radiology Report Generation with Adaptive Attention. In *Multimodal AI in Healthcare*; Springer: Cham, 2023; pp 293–305.

Lin, C. Y. Rouge: A Package for Automatic Evaluation of Summaries. In *Text Summarization Branches Out*, 2004; pp 74–81.

Liu, H.; Peng, L.; Xie, Y.; Li, X.; Bi, D.; Zou, Y.; Lin, Y.; Zhang, P.; Li, G. Describe Like a Pathologist: Glomerular Immunofluorescence Image Caption based on Hierarchical Feature Fusion Attention Network. *Expert Syst. Appl.* **2023,** *213*, 119168.

Lyndon, D.; Kumar, A.; Kim, J.; et al. In *Neural Captioning for the Imageclef 2017 Medical Image Challenges*, 18th Working Notes of CLEF Conference and Labs of the Evaluation Forum, 2017; vol 1866.

Mansour, R. F.; Alfar, N. M.; Abdel-Khalek, S.; Abdelhaq, M.; Saeed, R.; Alsaqour, R. Optimal Deep Learning Based Fusion Model for Biomedical Image Classification. *Expert Syst.* **2022,** *39* (3), e12764.

Miranda, D.; Thenkanidiyoor, V.; Dinesh, D. A. Detecting the Modality of a Medical Image using Visual and Textual Features. *Biomed. Signal Process. Control* **2023,** *79*, 104035.

Monshi, M. M. A.; Poon, J.; Chung, V. Deep Learning in Generating Radiology Reports: A Survey. *Artif. Intell. Med.* **2020,** *106*, 101878.

Onita, D.; Birlutiu, A.; Dinu L. Towards Mapping Images to Text using Deep-Learning Architectures. *Mathematics* **2020,** *8* (9), 1606.

Papineni, K.; Roukos, S.; Ward, T.; et al. Bleu: In *A Method for Automatic Evaluation of Machine Translation*, Proceedings of the 40th Annual Meeting of the Association for Computational Linguistics, 2002; pp 311–318.

Park, H.; Kim, K.; Yoon, J. In *Feature Difference Makes Sense: A Medical Image Captioning Model Exploiting Feature Difference and Tag Information*, 58TH Annual Meeting of the Association for Computational Linguistics (ACL 2020): Student Research Workshop, 2020; pp 95–102.

Pavlopoulos, J.; Kougia, V.; Androutsopoulos, I. Diagnostic Captioning: A Survey, arXiv preprint arXiv: 2101.07299, 2021.

Pelka, O.; Koitka, S.; Rückert, J.; et al. In *Radiology Objects in Context (Roco): A Multimodal Image Dataset*, 7th Joint International Workshop on Computing and Visualization for Intravascular Imaging and Computer Assisted Stenting, CVII-STENT 2018, and the 3rd International Workshop on Large-Scale Annotation of Biomedical Data and Expert Label Synthesis, LABELS 2018, Held in Conjunction with the 21th International Conference on Medical Imaging and Computer-Assisted Intervention, MICCAI 2018, 2018; vol *11043*, pp 180–189.

Rodin, I.; Fedulova, I.; Shelmanov, A.; et al. In *Multitask and Multimodal Neural Network Model for Interpretable Analysis of X-Ray Images*, 2019 IEEE International Conference on Bioinformatics and Biomedicine, 2019; pp 1601–1604.

Syeda-Mahmood, T.; Wong, K.; Gur, Y. In *Chest X-Ray Report Generation through Fine-Grained Label Learning*, 23rd International Conference on Medical Image Computing and Computer-Assisted Intervention, 2020; vol *12262*, pp 561–571.

Tomar, N. K.; Jha, D.; Riegler, M. A.; Dagenborg, H. J.; Johansen, D.; Rittscher, J.; Halvorsen, P.; Ali, S. Fanet: A Feedback Attention Network for Improved Biomedical Image Segmentation. *IEEE Trans. Neural Netw. Learn. Syst.* **2022.**

Tsuneda, R.; Asakawa, T.; Aono, M. In *Kdelab at Imageclef 2021: Medical Caption Prediction with Effective Data Pre-processing and Deep Learning*, CLEF2021 Working Notes, CEUR Workshop Proceedings, CEUR-WS. org, Bucharest, Romania, 2021.

Vedantam, R.; Lawrence Zitnick, C.; Parikh, D. In *Cider: Consensus-Based Image Description Evaluation*, Proceedings of the IEEE Conference on Computer Vision and Pattern Recognition, 2015; pp 4566–4575.

Vinyals, O.; Toshev, A.; Bengio, S.; Erhan, D. In *Show and Tell: A Neural Image Caption Generator*, Proceedings of the IEEE Conference on Computer Vision and Pattern Recognition, Boston, MA, USA, June 7–12, 2015; pp 3156–3164.

Wang, X.; Guo, Z.; Xu, C. In *Imagesem Group at Imageclefmed Caption 2021 Task: Exploring the Clinical Signifcance of the Textual Descriptions Derived from Medical Images*, CLEF2021 Working Notes, CEUR Workshop Proceedings, CEUR-WS. org, Bucharest, Romania, 2021.

Wang; X.; Guo, Z.; Zhang, Y.; Li, J. In *Medical Image Labelling and Semantic Understanding for Clinical Applications*, International Conference of the Cross-Language Evaluation Forum for European Languages; Springer: Cham, 2019; pp 260–270.

Wang, X.; Zhang, Y.; Guo, Z.; Li, J. A Computational Framework Towards Medical Image Explanation. In *Artificial Intelligence in Medicine: Knowledge Representation and Transparent and Explainable Systems*; Springer: Cham, Switzerland, 2019; vol *11979*, pp 120–131.

Wu, Y.; Zhu, L.; Jiang, L.; Yang, Y. In *Decoupled Novel Object Captioner*, Proceedings of the 26th ACM International Conference on Multimedia, 2018; pp 1029–1037.

Xiong, Y.; Du, B.; Yan, P. In *Reinforced Transformer for Medical Image Captioning*, 10th International Workshop on Machine Learning in Medical Imaging, MLMI 2019 Held in Conjunction with the 22nd International Conference on Medical Image Computing and Computer-Assisted Intervention, 2019; vol *11861*, pp 673–680.

Yang, S.; Niu, J.; Wu, J.; et al. Automatic Ultrasound Image Report Generation with Adaptive Multimodal Attention Mechanism. *Neurocomputing* **2021,** *427*, 40–49.

Yin, C.; Qian, B.; Wei, J.; Li, X.; Zhang, X.; Li, Y.; Zheng, Q.; Yin, C.; Qian, B.; Wei, J.; et al. In Automatic Generation of Medical Imaging Diagnostic Report with Hierarchical Recurrent Neural Network, Proceedings of the 2019 19th IEEE International Conference on Data Mining (ICDM 2019), Beijing, China, November 8–11, 2019; pp 728–737.

Zhang, Z.; Chen, P.; Sapkota, M.; et al. In *Tandemnet: Distilling Knowledge from Medical Images using Diagnostic Reports as Optional Semantic References*, International Conference on Medical Image Computing and Computer-Assisted Intervention; Springer, 2017; pp 320–328.

Zeng, X.; Wen, L.; Liu, B.; et al. Deep Learning for Ultrasound Image Caption Generation based on Object Detection. *Neurocomputing* **2020,** *392*, 132–141.

Zeng, X.; Wen, L.; Xu, Y.; Ji, C. Generating Diagnostic Report for Medical Image by High-Middle-Level Visual Information Incorporation on Double Deep Learning Models. *Comput. Methods Programs Biomed.* **2020b,** *197*, 105700.

CHAPTER 9

Design and Analysis of OTA-OC CMOS Filter for Recording High Focal Oscillations-EEG Application

SWAGATA DEVI[1], SOURAV NATH[2], KOUSHIK GUHA[2], and KRISHNA LAL BAISHNAB[2]

[1]*Faculty of Engineering, Assam down town University, Sankar Madhab Path, Gandhi Nagar, Panikhaiti, Guwahati, Assam, India*

[2]*Department of Electronics and Communication Engineering, National Institute of Technology Silchar, Assam, India*

ABSTRACT

The neural recording arrangement in a biomedical signal acquisition system records the neural activities of the brain. The electroencephalogram (EEG) signals have weak amplitude and require proper amplification for further processing. The relatively small voltage in the signals results in their high susceptibility to collecting noise from supplies, electrodes, and added sources. The analog front-end (AFE) of a neural system demands a bio amplifier that will magnify the signal at low power and low noise. Also, it is followed by a low pass filter (LPF) that will allow the specific bandwidth of the signals to pass through and reject the unwanted signals. The acceptance of low power methods lessens heat dissipation. The high amount of heat has the capability to damage the surrounding tissues, which is undesirable. One of the advantages of low power and low supply devices is that the bulk-sized battery usage can be prevented. This will increase the battery longevity and prevent the trouble of frequent substitution of battery. In this article, the focus is on the design of an OTA-C filter that will operate by consuming low power in the circuit. The analog designers face challenges in lowering the power of the circuit,

Smart Electronic Devices and Systems for Biomedical and Healthcare Applications. Upendra Kumar, Rajesh K. Kesharwani, & Raj K. Keservani (Ed.)

DOI: 10.1201/9781003560852-9

in correspondence to its noise contribution and concerned bandwidth. These challenges are addressed in the proposed work. The low-frequency circuits necessitate huge capacitors and resistors, so they occupy a large chip area, but the design of implantable designs require less area. The OTA-C filters have gained popularity because they are fundamentally tuneable, resistor-less, and can be designed on small chips. The proposed OTA-C design is a third-order LPF and is capable of allowing only a specific range of frequencies, which belongs to the set of EEGs. These are known as high-frequency oscillations (HFOs) and are the biomarkers for the detection of neural diseases like epilepsy. The proposed filter is designed in a 180 nm SCL Cadence technology and operates at a 1.8 V. The neural amplifier achieves a mid-band gain of 54.2 dB and a −3 dB bandwidth in the range of 7.89–691 Hz.

9.1 INTRODUCTION

The biomedical technology in modern times is constructable on circuit integrated systems, also known as system-on-chip (SoC), which consists of discrete components. These SoCs are capable of acting as implantable bio-signal recording and detection systems. There are various discrete components present in these systems that are necessary to attain extremely precise bio-signals from the human body for further processing and classification of diseases. In this article, an analog LPF is implemented and it removes the out-of-band frequencies introduced by front-end circuits. In the AFE of the system, the operational transconductance amplifier (OTA) is required to amplify the signals because these bio-signals have very small magnitudes in the range of microvolts (μV) or millivolts (mV). They are continuous-time signals and are controlled by the bandwidth of the amplifier, however, there is a requirement for an analog LPF to remove all the anomalies and the shifted noise. Once the concerned signals are filtered out, they are converted into digital data for further signal processing. The current work focuses on EEG signal recording and acquisition for the detection of epilepsy disease. Epilepsy is one of the most familiar chronic neuro-biological diseases impacting around 65 million prolonged of all ages worldwide (Amudhan et al., 2015). This condition not only affects the health of the patient but also their quality of life by affecting education, social behavior, employment, and disorders, or mobility. There are various antiepileptic drugs prescribed to suppress or prevent seizures. However, around 30% of cases, predominantly those experiencing focal epilepsy, are repellent to epileptic drug therapy. In recent times, the fast ripple oscillations in EEG are considered as a potential electrophysiological component to measure pathological brain activity. The HFOs are ensuring EEG biomarkers of epileptogenicity (Kassiri et al., 2017). The HFOs are sub-categorized as ripples: 80–250 Hz, and fast ripples: 250–600 Hz.

In the AFE system of the neurostimulator designs, it is necessary that every building block should consume less power. To do this, an operational trans-conductance amplifier capacitor (OTA-C) network topology is a fascinating technique for accomplishing analog filter designs, which consume less power and are designed for low-frequency devices. The EEG signals have low band-width and Gm-C, OTA-C, switched capacitors are some of the common filter techniques that are generally implemented. In contrast to switching circuits, the OTA-C-based circuits do not unnecessarily charge or discharge the load capacitor due to the switching noise. The transconductance in the OTA structure is reduced by biasing the MOSFET transistors in a weak inversion region. The on-chip capacitance can be adjusted to a suitable value so as to occupy an acceptable silicon area and a greater time constant integrator can be calculated. Thus, the general design of the OTA-C filter in weak inversion reduces total consumed power and brings it down to a few tens of nanowatts. Therefore, the implementation of low-frequency filters is not easy as Gm comes in the range of nA/V and capacitor of value less than 100 pF. The foundries do not provide capacitor values of more than 50 pF.

However, a few of the works like Harrison and Charles (2003) designed an AFE to accommodate a bandwidth in the range of 1–30 Hz. The load capacitor has been replaced with a capacitor of larger value. This capacitor increases the die area of the circuit which is not preferable for implantable devices. Wattanapanitch et al. (2007) in their reported work placed a band-pass filter next to the gain stage. The bias current of the Gm-C OTA filter is dropped to decrease the power consumption. The input referred noise of the bandpass filter is unimportant, so to preserve the trade-off between noise and power in the device, it should be lowered as much as possible in the gain stage. Barsakcioglu et al. (2019) modified the AFE design reported by Harrison and Charles (2003) and introduced a second-order high-pass filter constructed on Bessel's function. While Della Sala et al. (2019) designed a sixth-order Butterworth filter with a pass band of HFOs. The filter is constituted of three Gm-C biquad filters which are functioning in subthreshold region to cut down power consumption. Similarly, Lee and Cheng (2009) used a fifth ladder category low pass Butterworth filter and generated a large time constant. The capacitor and resistor have a value of less than 10 pF and greater than 100 MΩ. The OTA-C designs working in subthreshold regions ensure low power and a Butterworth bandpass filter confirms attenuation in stop band and flat gain in pass band. Bihr and Ortmanns (2012) used an anti-aliasing filter and switched capacitor biquad filters with tuneable gain and cut-off frequencies to split spikes and LPF from one another and amplify them to full swing. The low frequency of biquad ensures either large capacitance or low transconductance. The switched capacitors have leakage problem, hence sample-hold circuits of switch-based topologies are not suitable for large time constants. The Gm-C filters have medium linearity, and the focus of the designers is mainly to reduce noise and power rather than linearity. As the biquad filters are comparatively power hungry and noisy, so the main aim of the research is to reduce power consumption and improve linearity. Various current reuse techniques are used to increase power efficiency in circuits. Therefore, it can be concluded that no report is registered for filtering HFOs to identify epileptogenicity. The HFOS have a low range of frequencies, and it is necessary to filter them out by consuming less amount of power. To do this higher order OTA-C filters are used and show efficient results in this work. In the next section of this article there is a thorough explanation of the design approach in a conventional OTA-C filter, while in section 3, the notion and network topology of OTA with less transconductance (Gm) is discussed. The execution of the proposed OTA-C filter in Cadence 180 nm SCL environment and its simulation outcomes are reviewed in section 4. The comparison of the design with previous works is also discussed here and to end this work conclusions are presented in section 5.

9.2 DESIGN APPROACH

The design of the basic AFE for the physiological signal acquisition system is shown in Figure 9.1. The analog processing block comprises two design challenges, firstly, the design of a low noise high gain preamplifier and secondly an OTA-C design that consumes low and has a small transconductance. Both these circuits should not bring in any kind of disfigurement in the signal, otherwise it will destroy the data within it. Power consumption is a critical factor, the complexity and steepness of the filter should be kept in mind. The higher the order of the filter, the steeper the cut-off slope, but the complexity of the circuit increases with the increase in the order. Therefore, the trade-off should be kept under consideration (Peng et al., 2017). The filter circuits allow a limited frequency band to pass through it and, therefore, to generate a large time constant, is a challenging job.

The cut-off frequency in Gm-C filters is dependent upon the transistors involved in designing the OTA and the monolithic capacitances. It is also necessary to create a small transconductance to prevent the use of larger capacitances in the design. The reason behind this is that the cut-off frequency of a Gm-C filter is directly related to the gain or transconductance of the OTA and inversely related to the load capacitance. However, one of the major limitations of the existing OTA is its controlled voltage swing which is obligatory for good linearity (Yen et al., 2004). This can be improved by introducing some passive components or OTAs in the existing design. A first-order filter is presented in Figure 9.2.

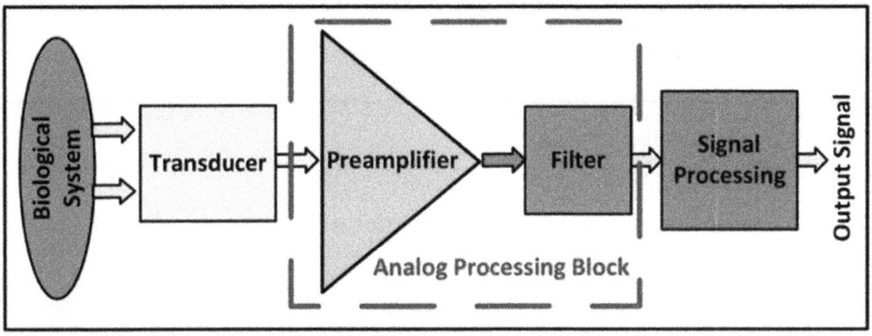

FIGURE 9.1 AFE for physiological signal acquisition system.

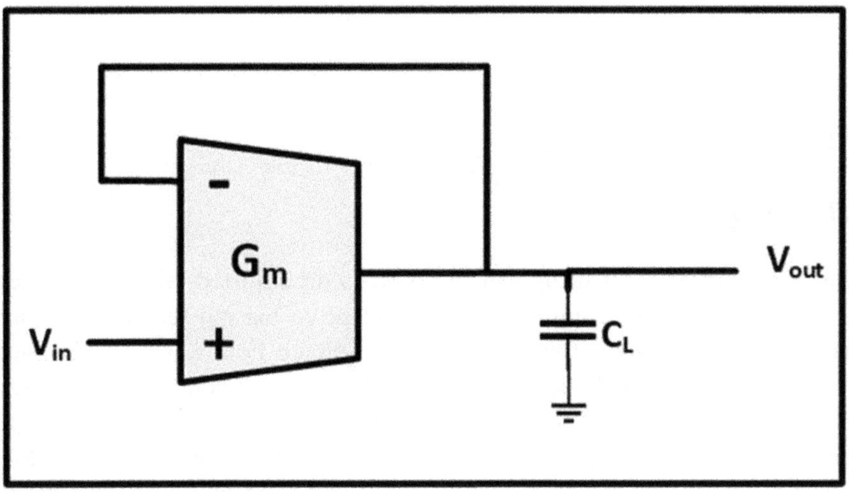

FIGURE 9.2 First-order OTA-C filter.

While the second-order filter comprises two OTAs and two grounded capacitors, which is displayed in Figure 9.3. The higher order filters can be designed by cascading first and second-order filters.

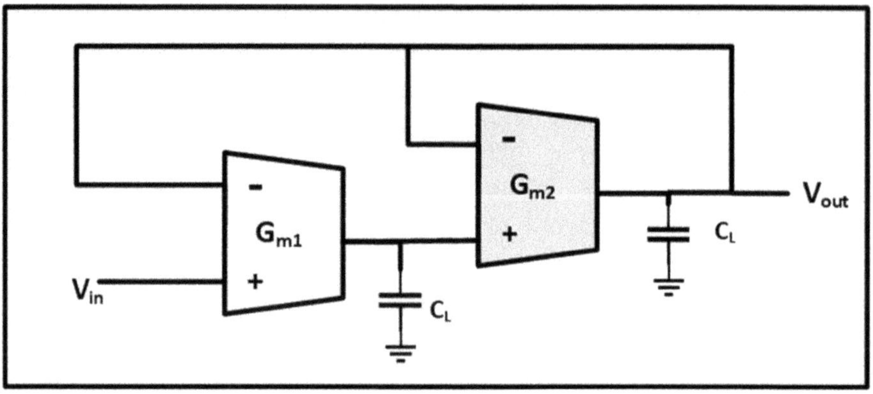

FIGURE 9.3 OTA-C filter of second order.

The transfer function for the second-order filter is shown in eq 9.1.

$$H(s) = \frac{V_{out}(s)}{V_{in}(s)}$$

$$H(s) = \frac{\frac{G_{m1}G_{m2}}{C_1C_2}}{s^2 + sG_{m2} + \frac{G_{m1}G_{m2}}{C_1C_2}} \tag{9.1}$$

9.3 DESIGN APPROACH

9.3.1 PREAMPLIFIER

The preamplifier OTA design in the AFE is an adjusted recycling folded cascode design, which implemented an adaptive biasing technology. This biasing design includes two pairs of Flip Voltage Followers (FVF). Each FVF has two similar devices, and the FVF are cross connected to design DC level control shifters. They demonstrate proficiency in channeling large currents as the load capacitor charges or discharges, owing to their low output resistances. Therefore, they remain unaffected by noise. The circuit diagram for the preamplifier OTA is displayed in Figure 9.4. M01–M02 and M03–M04 are the two pairs of FVFs, and consequence is the increment of gain, slew rate, and transconductance of the preamplifier.

The slew rate for the preamplifier design is given by eq 9.2:

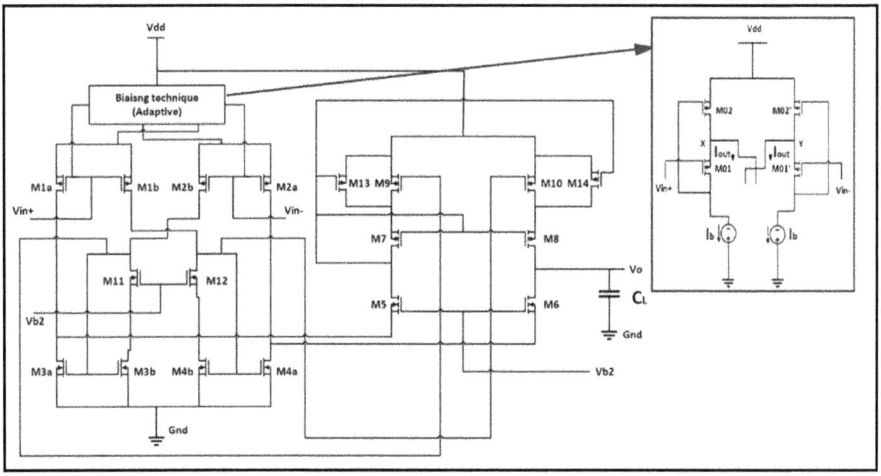

FIGURE 9.4 Schematic of preamplifier OTA.

$$SR_{IRFC,\ improvised} = \frac{2I_{out}(2k-1)}{C_L} \tag{9.2}$$

where current I_{out} can be expressed as $I_{in} + I_b$. When the voltage, V_{in}^{+} becomes high, immediately the transistor devices M_{a2} and M_{b2} goes off, while the device M_{b1} starts working in deep triode region. This happens because as the input transistors M_{a2} and M_{b2} goes off, also M_{a4} and M_{b4} are turned off. Then the current through M_{a4} will rise and M_6 goes off. The total circuit current will flow through M_{a3} and M_{b3} and further into M_5 by an element k. Then it further flows to M_9 and M_{10} by a factor of one. Then the whole current flows into capacitor C_L by multiples of k and (k-1). The entire circuit current depends on its input voltage. All the devices are biased in weak inversion region, so there is a reduction in consumed power and the circuit works at low supply voltage. The transconductance G_{m_OTA} is given by eq 9.3:

$$G_{m\_OTA} = (2k)\, g_{ma1} \tag{9.3}$$

where k is the ratio in which the input transistors are split and k has a value of 3 to avoid phase degradation. Also, there is a degradation in frequency as the value of currents is very low, and the flicker noise is dominant (Devi et al., 2021). The thermal noise, $4K_B T\gamma g_m \Delta f$, at low frequencies has very small magnitude because of small current in weak inversion region. So, the expression for dominant flicker noise expression is given by eq 9.4.

$$\overline{V_{n,}^{\ 2}} = 2\frac{K_{fp}}{2k\,fC_{OX}}[\frac{1+(2k-1)^2}{2k(WL)_{a1}} + \frac{K_{fn}}{K_{fp}}\frac{n_{a1}^{\ 2}}{n_{a3}^{\ 2}}\frac{k^2+k(2k-1)^2}{2k(WL)_{a3}} + \frac{n_{a1}^{\ 2}}{n_9^{\ 2}}\frac{(k-1)^2}{2k(WL)_9}]\Delta f \tag{9.4}$$

where, $\overline{V_{n,}^{\ 2}}$ is the measurable unit for the input referred noise, W and L represent the width and length dimensions of transistors, C_{ox} is the oxide capacitance, K_{fp} and K_{fn} are the flicker constants of p-mos and n-mos devices, and n is the subthreshold factor. The transistors Mc1 and Mc2 are functioning in deep triode region, to improve the phase degradation, and transistors M9 and M10 act as input driving transistors because of the cross over connection with Mb3 and Mb4. While M11 and M12 make it a single-ended architecture. The input referred to flicker noise has been reduced moderately (Akbari et al., 2015; Zijlmans et al., 2009; Gray and Meyer, 1982).

9.3.2 OTA-C FILTER DESIGN

The structure in Figure 9.5 is the schematic of OTA. The input differential transistor devices M_1 and M_2 are divided int: M_1 - MM_1, and, M_2 - MM_2. The essential level of transconductance for filters is particularly small, therefore, the required bias currents are also low. The effective transconductance G_m is derived from the input transistor pairs, and is given by $\frac{g_{m\_}M_{comp}}{(M+1)}$. M_{comp} is the compound transconductance value before splitting. The input transistors M_1, MM_1, and M_2, MM_2 are split by a factor of M in their sizes. Therefore, the overall transconductance is brought down by a factor of (M+1) (Gallegos and Huq, 2014).

The pairs of input transistor devices are p-mos devices because they have large gate area which minimizes the flicker noise of the design. The transistors M_p and M_n both are included in the OTA schematic to make the design a second-stage amplifier. This will increase the gain and hence the swing of the design. In this proposed OTA for filter design, other than transconductance reduction technique additional source degeneration technique is also used, and the total transconductance is given by eq 9.5 below:

$$G_{m\_filter} = \frac{g_{m\_M1}}{1+\frac{(M+1)g_{m\_M1}}{g_{o\_M14}}} \tag{9.5}$$

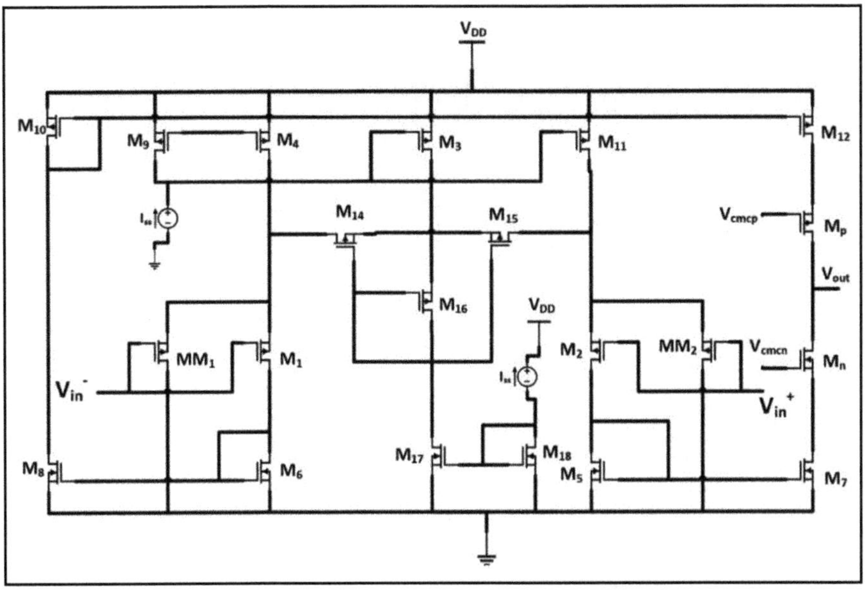

FIGURE 9.5 OTA schematic structure with current division and source degeneration.

The symbol g_m is transconductance of the MOSFET M_1 and g_o is the output transconductance. Therefore, G_m can be altered with the change in g_o, which is again controlled by bias current I_{ss}. M_{14} and M_{15}, both are biased in triode region so that they can act as source degeneration resistors, while the transistors M_3, M_{16}–M_{18} control the source gate voltage of M_{14} and M_{15}. The source resistances offered by these transistors are responsible for bringing down the transconductance by (1+M) (Saidulu et al., 2016). As mentioned, and discussed priorly, the realization of small transconductances is a difficult task and needs attention. The technique to produce very small controlled currents is tough. This problem can be solved by the use of large length transistors but can rise problems in its layout design. So, the current division and source degeneration techniques are used, which will help in increasing current levels but at very low transconductance levels (Mirzaei et al., 2013).

9.4 RESULTS AND DISCUSSION

A high-performance LPF is designed using OTA-C filter to reduce the bandwidth of the gain stage. The block diagram for the proposed AFE circuit is shown in Figure 9.6. The modeling, simulations and back-end design modeled and carried out in Cadence 180 nm SCL Virtuoso environment.

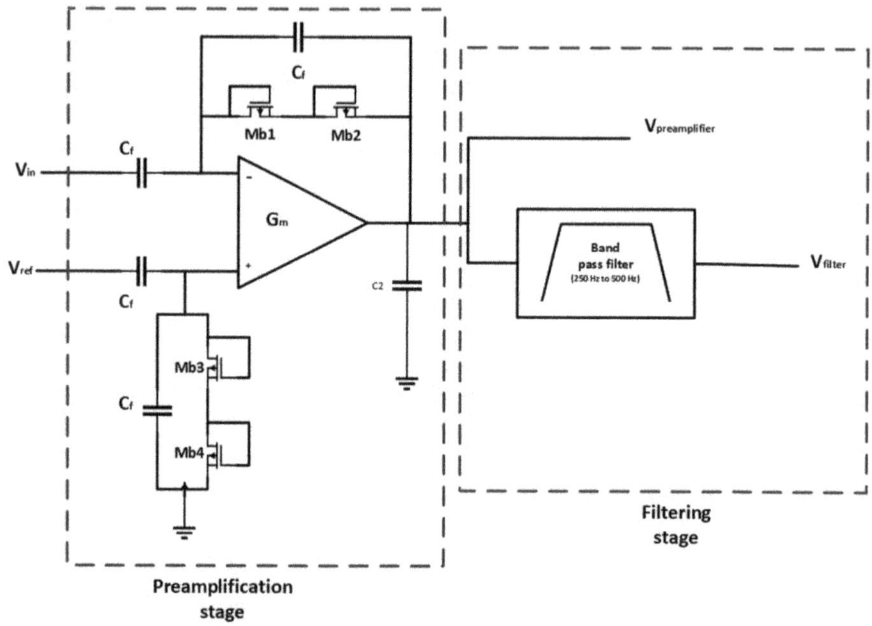

FIGURE 9.6 Block diagram for the proposed AFE.

The LPF added after the gain stage will process the HFOs. A first-order RC filter will offer attenuation of around 6 dB/octave, which is less for HFO processing. The brainwaves have a bandwidth of less than 600 Hz and more than 70 Hz. The oscillations within this range are called ripples, and they have the same amplitude. The oscillations that fall in the range of 80–200 Hz are categorized as ripples, while the oscillations beyond 200–600 Hz are the fast ripples or HFOs. The ripples should be essentially attenuated following up a minimum of 10–30 dB attenuation. The present-day medical intracranial EEG recordings for epileptic treatment and analysis require dB/octave or dB/decade filter roll-off. Thus, the filter order should be of third order or of order higher than that, and the upper roll-off for the filter must be capable of reducing the interferences above 600 Hz. The high-frequency power is located away from the concerned band of signal, while the roll-off obligation is not as rigorous as that for the ripple power. This explains why the order of the filter is determined by the lower roll-off for the required bandwidth. The schematic in Figure 9.7 is for a third-order OTA-C filter.

The graph interpretation for the third-order OTA-C filter implementation is shown in Figure 9.8. The three graphs represent the gain for preamplifier, the third-order filter, and the AFE. As seen, the wide bandwidth of the gain or preamplifier stage is brought down by the filter and is reduced to 691 Hz. The power plot is shown in Figure 9.9 and measures an average power of 3.2 μW.

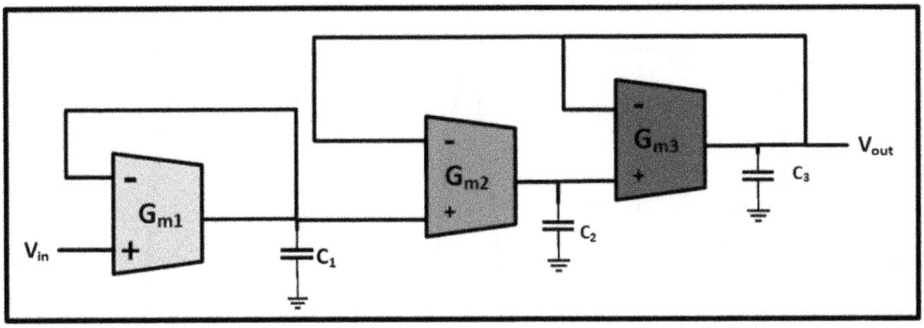

FIGURE 9.7 Third-order OTA-C filter.

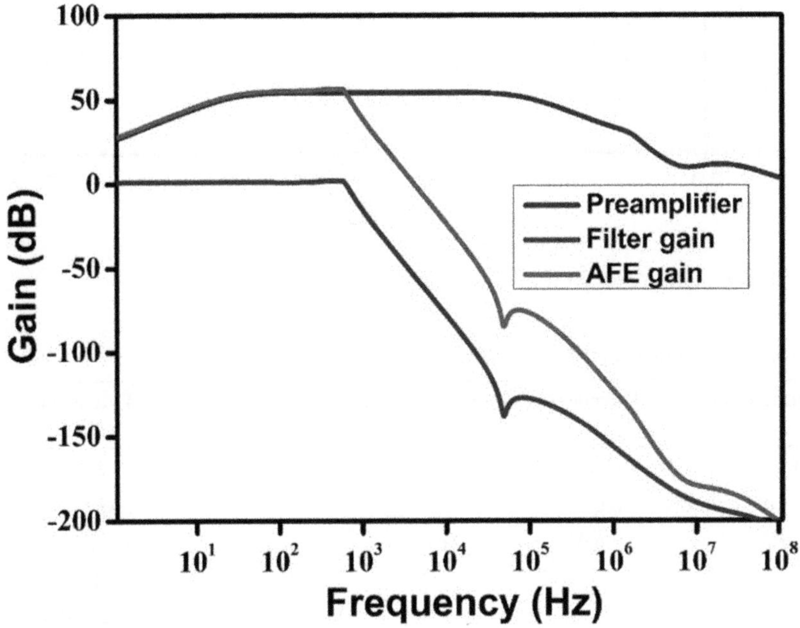

FIGURE 9.8 Third-order low pass G_m-C filter.

Table 9.1 summarizes the measured results for this design and compares them with the recent works. The EEG epilepsy AFE revealed a bandwidth limitation of 691 Hz with a used power of 3.2 μW and a measured input noise of 1.889 μV_{rms}. The evaluated mid-band gain for the amplifier is 54.2

dB, and single-pole roll-off takes place at 691 Hz. While the low-frequency roll-off happens at 7.89 Hz, owing to its input differential pair and the operation of current mirrors.

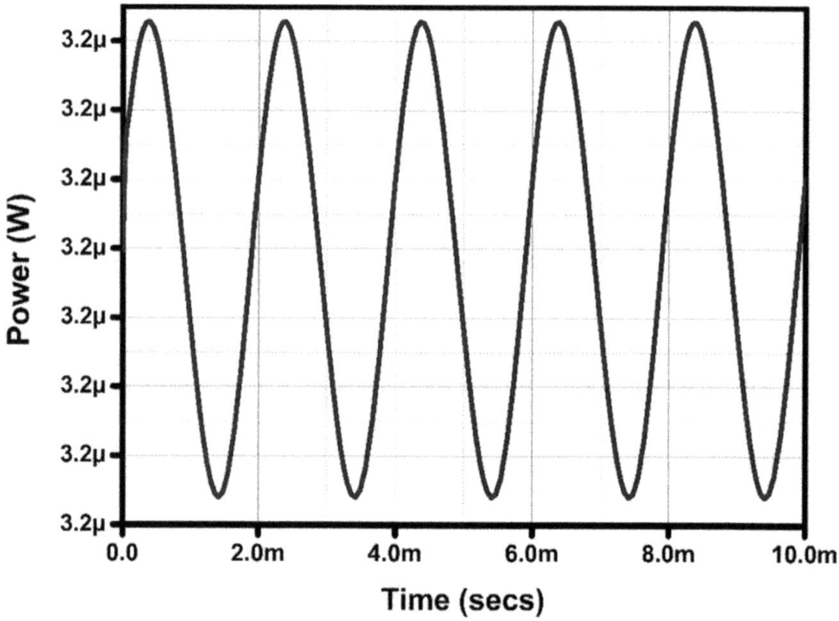

FIGURE 9.9 Power plot for the design.

TABLE 9.1 Comparison with Relevant Works.

Parameters	**Harrison et al.**	**Wattanapanitch et al.**	**Gallegos et al.**	**This work**
Gain	39.8 dB	40.9 dB	40 dB	54.2 dB
Input referred noise	1.6 μV_{rms}	1.66 μV_{rms}	4.39 μV_{rms}	1.889 μV_{rms}
Power	0.9 µW	3.21 µW	0.653 µW	3.2 µW
Type	–	–	Butterworth	Chebyshev
Topology	Capacitor	G_m-C filter	G_m-C filter	G_m-C filter
Order	–	–	3rd order	3rd order
Bandwidth	~30 Hz	392 mHz–295 Hz	1–154.21 Hz	691 Hz
Supply voltage	2.5 V	2.8 V	1.8 V	1.8 V
Technology	150 nm CMOS	500 nm CMOS	300 nm CMOS	180 nm CMOS
Application	EEG	Neural recording	EEG	Epileptic seizure

9.5 RESULTS AND DISCUSSION

This work concentrates on the design of a third-order filter with OTA-C structure. The transconductance in OTA design should be in the range of nA/V to avoid the use of large capacitances for low-frequency applications. The main target of this work is to design a low transconductance circuit that will operate at a low supply voltage to reduce the operating bandwidth of the AFE. The circuit-related simulations are performed using a 180 nm SCL CMOS model in a Cadence environment. A third-order Chebyshev OTA-C filter with good tunability is designed to provide variable gain and bandwidth. However, higher order filters can be designed to allow epileptic which are in the range of 80–200 Hz. The incoming signals must be attenuated necessarily to a minimum value of 10–30 dB. Therefore, higher order filter designs can be implemented furthermore for efficient results.

ACKNOWLEDGMENTS

We would like to give sincere thanks to those valuable authors whose references have been cited in this book chapter. We would like to gratefully acknowledge those who have motivated us to write down this book chapter and finally, we would like to give hearty thanks to all those contributors who were involved directly or indirectly in the preparation of this work.

KEYWORDS

- **neural amplifier**
- **EEG**
- **LPF**
- **bandwidth**
- **low power**
- **operational transconductance**
- **seizures**

REFERENCES

Akbari, M.; Nazari, M.; Sharifi, L.; Hashempiour, O. Improving Power Efficiency of a Two-Stage Operational Amplifier for Biomedical Application. *Analog Integr. Circuits Signal Process.* **2015,** *84*, 173–183.

Amudhan, S.; Gururaj, G.; Satishchandra, P. Epilepsy in India I: Epidemiology and Public Health. *Ann. Indian Acad. Neurol.* **2015,** *18* (3), 263–277.

Barsakcioglu, D. Y.; Yan L.; Bhunjun, P.; Navajas, J.; Eftekhar, A.; Jackson, A.; Quiroga, R. Q.; Constandinou, T. G. An Analogue Front-End Model for Developing Neural Spike Sorting Systems. *IEEE Trans. Biomed. Circuits Syst.* **2014,** *8* (2), 216–227.

Bihr, U.; Ortmanns, M. In *A Front-End Circuit with Active Spike and LFP Separation Via a Switched Capacitor Filter Structure for Neural Recording Applications*, 2012 IEEE International Symposium on Circuits and System, 2012; pp 2231–2234.

Della Sala, R.; Monsurrò, P.; Scotti, G.; Trifiletti, A. In *Area-Efficient Low-Power Bandpass Gm-C Filter for Epileptic Seizure Detection in 130 nm CMOS*, 2019 26th IEEE International Conference on Electronics Circuits and Systems (ICECS), 2019; pp 298–301.

Devi, S.; Guha, K.; Laskar, N. M.; Nath, S.; Baishnab, K. L.; Iannacci, J.; Krishnaswamy, N. Modelling and Analysis of a Modified Preamplifier for Seizure Detection. *Microsyst. Technol.* **2021,** 1–14.

Gallegos, S. A.; Huq, H. F. In *A 128.7 nW Neural Amplifier and Gm-C Filter for EEG, using gm/ID Methodology and a Current Reference without Resistance*, 2014 IEEE 57th International Midwest Symposium on Circuits and Systems (MWSCAS), 2014; pp 876–880.

Gray, P. R.; Meyer, R. G. MOS Operational Amplifier Design - A Tutorial Overview. *IEEE J. Solid-State Circuits* **1982,** *17* (6), 969–982.

Harrison, R. R.; Charles, C. A Low-Power Low-Noise CMOS Amplifier for Neural Recording Applications. *IEEE J. Solid-State Circuits* **2003,** *38* (6), 958–965.

Kassiri, H.; Tonekaboni, S.; Salam, M. T.; Soltani, N.; Abdelhalim, K.; Velazquez, J. L. P.; Genov, R. Closed Loop Neurostimulators: A Survey and a Seizure Predicting Design Example for Intractable Epilepsy Treatment. *IEEE Trans. Biomed. Circuits Syst.* **2017,** *11* (5), 1026–1040.

Lee, S. Y.; Cheng, C. J. Systematic Design and Modeling of a OTA-C Filter for Portable ECG Detection. *IEEE Trans. Biomed. Circuits Syst.* **2009,** *3* (1), 53–64.

Peng, S. Y.; Lee, Y. H.; Wang, T. Y.; Huang, H. C.; Lai, M. R.; Lee, C. H.; Liu, L. H. A Power-Efficient Reconfigurable OTA-C Filter for Low-Frequency Biomedical Applications. *IEEE Trans. Circuits Syst. I: Reg. Papers* **2017,** *65* (2), 543–555.

Wattanapanitch, W.; Fee, M.; Sarpeshkar, R. An Energy-Efficient Micropower Neural Recording Amplifier. *IEEE Trans. Biomed. Circuits Syst.* **2007,** *1* (2), 136–147.

Yen, C. J.; Chung, W. Y.; Chi, M. C. Micro-Power Low-Offset Instrumentation Amplifier IC Design for Biomedical System Applications. *IEEE Trans. Circuits Syst. I: Reg. Papers* **2004,** *51* (4), 691–699.

Zijlmans, M.; Jacobs, J.; Zelmann, R.; Dubeau, F.; Gotman, J. High Frequency Oscillations and Seizure Frequency in Patients with Focal Epilepsy. *Epilepsy Res.* **2009,** *85* (2-3), 287–292.

CHAPTER 10

A Flow-Induced Piezoelectric Vibration Energy Harvester (PVEH) with Magnetic Force Enhancement in Biomedical and mHealth

SHWETA SINGH[1], SAURABH KUMAR[2], and RAJNISH BHASKER[1]

[1]*Department of Electrical Engineering, Uma Nath Singh Institute of Engineering and Technology, VBSPU, Jaunpur, India*

[2]*Department of Electrical Engineering, Engineering Degree Division, Institute of Engineering and Rural Technology, Prayagraj, India*

ABSTRACT

The advancement of portable electronics and low power consuming devices has evoked the process of micro-level energy harvesting from ambient environment. The basic vibrational energy that is obtained from conventional piezoelectric materials cannot generate the energy that is required by the present electronic devices. In this chapter, a flow-induced piezoelectric vibrational energy harvester with magnetic force enhancement is suited for power generation. For that a device is used in such a way that it can provide increased flow velocity and to transfer the hydrokinetic energy into electrical energy piezoelectric effect is used. Both the processes can be done by using phenomenon called excitation of magnetic force and vortex-induced vibration. The output voltage is increased by using attractive magnetic force, as it reduces the frequency of vibration. The maximization of obtained voltage is done at the point of resonance. This resonance frequency is obtained from the natural frequency of piezoelectric beam and velocity of flow. This chapter

Smart Electronic Devices and Systems for Biomedical and Healthcare Applications. Upendra Kumar, Rajesh K. Kesharwani, & Raj K. Keservani (Ed.)

DOI: 10.1201/9781003560852-10

highlights recent research of the PVEH with magnetic force enhancement and how it can be used in biomedical mhealth care system.

10.1 INTRODUCTION

For assimilating minimal extent of energy from various energy resources of our surroundings, and to collect and accumulate them for future use a process is done, which is defined as energy harvesting or energy scavenging or power harvesting (Gaikwad and Ghosal, 2017). Due to limited life of the battery and very short span when compared to working duration of other devices and with the advent of wireless technology, concept of energy harvesting is introduced (Kumar et al., 2017).

The constant improvement of nano-technologies and micro-technologies allied with innovatory development in wireless connectivity has provided in novel applications of minuscule-sized electronic devices that include wearable and implanted electronic devices, wireless sensing networks (WSNs), micro-electromechanical systems (MEMS), and wireless sensors of standalone type associated with bridges of railways that can be power-derived by using vibrations induced by traffic sounds (Katulka, 2002). These miniature-sized electronic devices along with small sensors require very small amount of power to perform any functions. Due to the issues of power supply their application is quite limited for their effective applications of large-scale devices (Kumar et al., 2013). As frequent charging and replacement of the battery makes the device inefficient, therefore, researchers have provided huge contribution in introducing the concept of energy-harvesting technology (Khanna and Asthana, 2016). The energy obtained from natural resources provides energy in macro range, so to harvest them macro energy-harvesting concept is used. From the energy conversion point of view, energy harvesting is already in use in the form of solar energy, wind energy, geothermal energy, etc. As renewable energy plants generate power in kW or MW range, so to harvest energy of this range a concept of macro energy-harvesting technology is used. And harvesting of energy for reducing the use of conventional battery has introduced a concept called micro energy-harvesting technology (Adhikari and Jackson, 2021). Mechanical vibration, thermal energy from furnace, piezoelectric effect, etc. are the effects on the basis of which micro energy-harvesting technology is used. So, among all the methods the basic method for transduction is piezoelectric vibration energy harvester (PVEH) which works on the principle of piezoelectric effect.

This method is used widely due to its high-power density and also it is easily integrated with MEMS (Suresh et al., 2019). The application of MEMS along with vibration-based piezoelectric energy harvester is restricted in real world and also for commercial application because MEMS devices are not that much robust. So, to overcome this condition a silicon material of poly-crystalline form is introduced in the manufacturing of MEMS devices. As in silicon effects like instants shocks, vibrations of large amplitude and excitations of dynamic behavior obtained from various ambient conditions cause stress of extreme conditions and will make the silicon material very much prone to fracture (Singh et al., 2013). Till now mechanical stoppers are used by the researchers to limit the maximum displacement and also to recover the tendency of shock caused by the harsh use of ambient in PVEH based on MEMS. Due to the existence of complexity in coupling between the devices and external ambient conditions many problems are created in real-world applications. Mechanism of energy dissipation occurs in PVEH based on MEMS. So along with mechanism of linear damping, damping with nonlinearity must be measured for dynamic behavior modeling in case of PVEH based on MEMS mechanism (Deepak and George, 2021). From electromechanical coupling with piezoelectric effect, coulomb friction, drag friction obtained from aerodynamic effect can cause energy dissipation due to nonlinearity. Some other problems of measurement of damping with nonlinearity and complexity are involvement of nonlinear terms in the differential equation of the governing system, because of that it is inappropriate to obtain exact mathematical solution for nonlinear differential equations for the condition of simple geometry and boundaries also (Kumari and Rakotondrabe, 2020). As piezoelectric material provides conversion of mechanical vibration into electrical form, so for a wireless sensor harvesting of energy from piezoelectric effect is considered as self-power source. To generate sustainable energy from micro-range systems like low-power electronic devices and wireless sensor network, vibrational-based energy harvester plays a significant role (Debnath and Kumar et al., 2020). For the generation of such type of power devices that are installed with a fluid environment, a phenomenon called vortex-induced vibration is one of the finest sources. A piezoelectric device with a cantilever cylindrical structure at one end is taken into consideration for the production of energy from cross flows, provided that vibrations resulting from vortex-induced phenomena occur (Bhardwaj and Ravi Teja, 2021). To increase energy's scavenging effectiveness, it is possible to use piezoelectric harvester with external magnetic force enhancement system by creating a nonlinear coupling arrangement. Measurement of harvesting of energy was done by attachment of magnets of tip mass at the face of piezoelectric cantilever beam or at the front of conventional piezoelectric energy harvester and then the extra effect is provided by the addition of external magnetic force effect using movable and fixed magnets of opposite magnetic force of attraction (Suresh et al., 2019). Magnetic holder is there for the attachment of same types of magnets with opposite magnetic force of attraction. A single piezoelectric cantilever beam, connected with mass on its tip, is called conventional piezoelectric energy harvester system. This conventional piezoelectric energy harvester provides oscillations of linear nature with cantilever motion because it has the tendency to scavenge power of single resonance frequency. So, to increase the frequency bandwidth range and to raise the scavenging power of harvester it is required to add external magnetic force called as magnetic force enhancement on piezoelectric energy harvesting (Anand and Kundu, 2019). And due to this enhancement of magnetic force it represents the nonlinear behavior. Significantly, the magnetic force of attraction due to nonlinearity is utilized in such a way that it is buckled by harvester to produce the configurations of mono stability and bistability. However, the performance of such systems could be affected due to uncertainty in the magnitude and direction of the velocity of free stream. Premise of vibrational energy harvesting mainly depends on the relative velocity existing in between the structural base excitation and the mass. Because of this variable and nonlinearity in nature of vortex-induced vibrations, it cannot be described by a single set of governing equations. On the other hand, by vortex shedding phenomenon which is sinusoidal in nature, can model the vortex-induced vibrational cylinder as an oscillator in a linear harmonic form (Kumar et al., 2016). A mathematical formulation of piezoelectric with cantilever beam along with the vibration of end mass is provided under the influence of vortex-induced vibration. According to this, an equivalent mechanical model is developed with the single degree of freedom, and then it is further integrated with the piezosystem electrical model with or without an inductor (Mukherjee et al., 2019). To harvest the energy against the non-dimensional velocity parameter, a deterministic harmonic excitation is required. In this performance analysis, a magnetic force enhancement flow-induced PVEH concept is used and it is experimentally studied, and further it is combined with the piezoelectric effect. In further sections, the impact of flow of velocity on the voltage output is discussed and then merits of the method are presented. And in last conclusion is provided briefly.

10.2 LITERATURE REVIEW

For devices of micro range like power electronic devices of low range it is required to harvest the energy from ambience by using the concept of vibration. To generate the power for devices attached with fluid environment, a phenomenon is performed called vortex-induced vibration phenomenon (Kumar et al., 2016). However, the performance of such systems can get affected by uncertain magnitude and direction of free stream velocity. A complete analysis on piezoelectric energy-harvesting technologies was done by the researchers. But, various methods have been introduced since 2007 so as to augment the properties of materials, architectures of transducers, interfaces in electrical, analytical model, and the use space of piezoelectric energy-harvesting devices. Till now immense amount of work has been done on piezoelectric energy harvesting. Latest improvement in the advancements of small-sized, low-power, handy, and isolated devices has led to the beginning of renewable power sources for the duration of the past two decades. Due to limited life of the battery and very short span when compared to working duration of other devices and with the advent of wireless technology, concept of energy harvesting is introduced (Gaikwad and Ghosal, 2017). But frequent charging and replacement of the battery makes the device inefficient. Therefore, researchers have provided huge contribution in introducing the concept of energy-harvesting technology. The energy obtained from natural resources provides energy in macro range, so to harvest them macro energy-harvesting concept is used. From the energy conversion point of view, energy harvesting is already in use in the form of solar energy, wind energy, geothermal energy, etc. (Khanna and Asthana, 2016). As renewable energy plants generate power in kW or MW range, so to harvest energy of this range a concept of macro energy-harvesting technology is used. And harvesting of energy for reducing the use of conventional battery has introduced a concept called micro energy-harvesting technology. Mechanical vibration, thermal energy from furnace, piezoelectric effect, etc., are the effects on the basis of which micro energy-harvesting technology is used (Rui et al., 2018). The study on harvesting of energy from environmental resources has involved considerable awareness in the past few decades. Along with various sources available in our surroundings, vibration can be taken as one of the most capable sources for harvesting of energy because of its existence in wide range. To learn the capability of electrical energy generators for various sensor nodes, vibrational energy harvesters with so many concepts are introduced. Since the majority of VEHs work mainly on the principle of resonance with linearity under narrow bandwidth region, so they work inappropriately under excitations with broadband (Kim et al., 2018). To get over with the problem of bandwidth of narrow region, researchers have introduced the concept of energy harvesting with nonlinearity by applying compressive load axially and with magnetic coupling. Energy obtained from our surrounding environment like fluid flow of wind and from rotating machinery operations is called vibrational energy. For vibrational-based energy-harvesting technique there are various sources of vibrational energy (Foong et al., 2019). Piezoelectric energy harvesting is the process of conversion of mechanical energy of dynamic form into electrical energy by using a material called piezoelectric material. To harvest low-level energy, from surrounding vibration piezoelectric energy harvesting is used. Piezoelectric material generally works at a very low level of energy, that is, of the order of micro to milli-watts when it is compared with the solar and thermal energy harvesters, as thermal and solar energy harvesters can make hundreds of watts (Fan et al., 2019). There are some merits using piezoelectric transduction principle over harvesting of energy by solar and thermal, which is due to the fact that vibrations of surroundings are frequently importunate because of operational conditions of a method, and hence it does not depend upon uneven and changeable conditions of environment that can vary with time. In addition, piezoelectric harvesters are valuable in situations where there is no availability of solar and thermal energy, and are mainly functional in embedded systems. In reports it is mentioned that worldwide yearly profits of piezoelectric devices have been raised from $22 billion in 2012 to $37 billion in 2017; hence, it represents a drastic boost in the requirement and application of piezoelectric systems (Usharani et al., 2019).

In environments there are several examples of harvesting energy by piezoelectric vibration that can be utilized like bridges and buildings in civil infrastructure, the human body, and system in aerospace. In general, by using piezoelectric, electromagnetic, electrostatics, magneto-strive, and turbo-electric converters, etc., the usable power can be obtained from the vibratory energy (Liu et al., 2016). The inherent transduction capacity is the main advantage of using generators based on piezoelectric above the other methods. In the piezoelectric effect the efficiency of preservation as scale is reduced, power density, and capacity to function in high-frequency applications increases. To obtain energy of vibration from an arrangement, there must be an attachment of piezoelectric harvester with the host structure, so as to transfer the vibrational energy from a host structure effectively (Shukla et al., 2010). With this attachment a mechanical interface is developed in many ways, though, the most favorable option is generally defined by the constraints and characteristics designed for the overall system. Harvesting by piezoelectric systems generally offers a stable, self-sufficient source of power that does not require either maintenance or replacement. When compared with the conventional sources of energy like batteries, cost of replacement of battery reduces when associated with autonomous operation. Energy transducer plays an important role in energy harvesting because energy is present around us everywhere (Sil and Biswas, 2018). As we know that a device which converts any physical quantity into electrical quantity is known as transducer. Among various types of transducers, the energy harvesting is mainly done by using piezoelectric transducer. For the conversion of physical quantity into electrical quantity, if any transducer utilizes the property of piezoelectric materials, then it is called piezoelectric transducer (Anand and Kundu, 2019). The materials that acquire the properties of piezoelectricity are called piezoelectric materials. If any mechanical stress is applied on piezoelectric materials, then it will lead to the generation of voltage in proportional to the stress applied. There are various works that are performed on piezoelectric vibration-based energy harvesting on the basis of base excitation given to it from the vibrational environment (Kumari and Rakotondrabe, 2020). Piezoelectric effect is being utilized in many different applications ranging from doorbells and cigarette lighters to printers and amplifiers. It is also possible to use a piezoelectric material as a transducer to transform strain or movement caused by mechanical action into useable electric voltage or current. It is also possible to generate electricity by applying pressure to the material and thereby harvesting energy from things like floor tiles or shoe bottoms. Such generators are frequently based on a cantilever beam or a membrane that oscillates at the same frequency as the vibration source (Adhikari and Jackson, 2021). IoT applications and low-power wireless sensor networks are primarily powered by piezoelectric energy harvesters. Also, it is known that vibration obtained due to flow-induced fluid consists of huge amount of energies like the energy scavenging due to turbulent effect, vortex-induced energy converter, etc. Flow-induced vibrational energy harvester can be used in several ways like water-induced or wind-induced vibrational energy harvester (Suresh et al., 2019). In case of base excited vibrational energy harvester, for oscillators there is a limitation of frequency bandwidth and uncertainties in sensitivity.

10.3 PIEZOELECTRIC TRANSDUCERS

Transducers are machines that change physical quantities into electrical quantities. Among various types of transducers, the energy harvesting is mainly done by using piezoelectric transducer. Piezoelectric transducer is an active and inverse transducer. Piezoelectric transducer has directional sensitive property because it operates on the idea of piezoelectric effect. Piezoelectric transducer is a dynamic transducer because it works for dynamic inputs only, that is, in piezoelectric transducer we can get output for variable input only and if input is constant means time invariant, then there is no result for us. Basically to convert electric form of energy into mechanical forms and vice versa the transducer used is electromechanical transducer. Among all the main types of such transducers some are earphones, microphones, loudspeakers, vibration pickups; the devices that produce vibrations are the electromechanical transducers. Electromechanical transducers are basically passive in nature because it requires external power supply for its excitation. Networks of electromechanical transducers are generally linear and passive in nature. There are some factors like durability, cost, mechanical and electrical impedances on the basis of which it is decided that which electromechanical transducers are suited for which instruments. Next to the typical electro mechanical methods, piezoelectric appliances have shown to be quite efficient at capturing energy and operating as a supplier of low-power electrical equipment. Typical applications are for portable electronics, wearables, IoTs, systems with sensors that manage and monitor conditions because of their versatility and broad usable frequency range. Notably, piezoelectric harvesters are most cost-effective when utilized for low-power requirements of 1J (1Ws). When compared to modern solar cells/panels and TEGs, this is far lower and should always be considered for design/application feasibility. Vibration or other forms of kinetic energy exist in many forms all around us, such as shock waves, water running, wind, and machine vibration. For many years, corporate and academic research teams have concentrated on employing piezoelectric transducers to capture vibrational energy. Some materials have the ability to generate an electric charge when subjected to mechanical stress. The piezoelectric effect is a phenomenon that is used in many different products, such as doorbells, cigarette lighters, printers, and amplifiers. Piezoelectric materials can also be used as transducers to transform mechanical strain or movement into useful electric voltage or current. However, it is also feasible to generate electricity by applying pressure to the material (and therefore harvesting energy from things like floor tiles or shoe bottoms). Such generators are frequently based on a cantilever beam or a membrane that oscillates at the same frequency as the vibration source. Internet of Things (IoT) applications and low-power wireless sensor systems are the main applications for piezoelectric energy harvesters. The potential for miniaturization and integration of micro-electromechanical systems (MEMS) with piezoelectric energy harvesters is a benefit. Additionally, the design's simplicity and lack of moving components make it simple to create in large quantities at a low cost. A lower power production compared to electromagnetic energy harvesters is the main disadvantage. Additionally, piezoelectric materials are inappropriate for applications requiring high levels of endurance and hostile conditions since they are prone to brittleness and long-term degradation. In comparison to piezoelectric energy harvesters, electromagnetic energy harvesters have a higher power output, which increases the usefulness of the principle in a larger range of applications. Additionally, it is simpler to build electromagnetic energy harvesters for a long service life in challenging settings, which is frequently required for vibration energy-harvesting systems. Piezoelectric energy harvesting has long been considered a viable technique for the IoT and has received a lot of research. The primary alternative for commercial use has been electromagnetic energy harvesting due to the limited power output of piezoelectric energy harvesters.

10.4 ENERGY HARVESTING

Any energy harvesting from environmental vibrations or some form of kinetic energy has experienced a spike in attention over the past few years as a result of people being more aware of alternative energy sources in general. The use of piezoelectric devices as a source of power for low-power electronic devices has proven to be far more efficient than the more traditional electromechanical techniques. Due to the large usable frequency range and versatility, typical applications include portable electronics, wearables, Internet of Things, sensor-controlled, and condition-monitoring systems. Notably, piezoelectric harvesters are most cost-effective when utilized for low-power requirements of 1J (1Ws). This is significantly less when compared to cutting-edge solar cells, panels, and TEGs, and should always be taken into account for design and application practicality. Vibration or other forms of kinetic energy exist in many forms all around us, such as machine vibration, flowing water, wind, human motion, and shock waves. For many decades industrial and academic research units have focused their attention on harvesting energy from vibrations using piezoelectric transducers.

A new kind of electronic device, such as battery-free, autonomous devices relying only on mechanical vibrations harvested by piezoelectric components, is now available because of advancements in ultra-low power electronic components. There are a variety of techniques for converting kinetic and vibrational energy into electricity. Piezoelectric materials and electromagnetic induction are the two of these that are employed in commercial applications most frequently. According to the source, quantity, and kind of energy being converted to electrical energy, the process of energy harvesting takes on numerous shapes. In its most basic form, the energy-harvesting system needs an energy source like heat, light, or vibration. Its primary three components are (1) Transducer/Energy Harvester: This device is an energy harvester, gathering and transforming source energy into electrical energy. Examples of common transducers for different types of energy are photovoltaic for light, thermoelectric for heat, inductive for magnetic, RF for radio frequency, and piezoelectric for vibrations and kinetic energy. (2) Power management: This processes electrical energy into a form that is appropriate for the application. Typical conditioners have regulators and intricate control circuits that can manage the power based on demand and supply. (3) A battery or super capacitor, for example, as a means of energy storage.

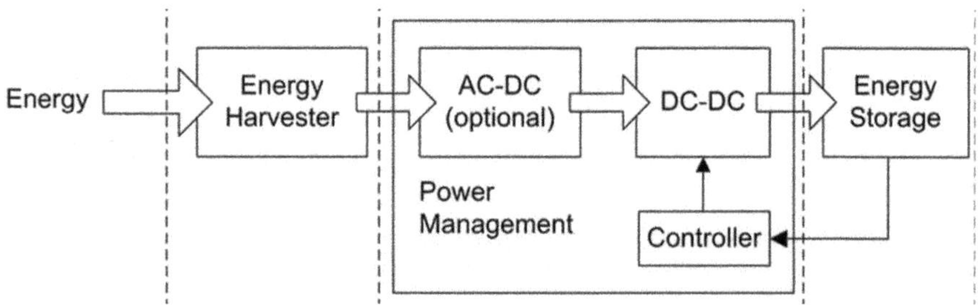

FIGURE 10.1 An energy-harvesting system's block diagram.

10.4.1 PIEZOELECTRIC VIBRATION ENERGY HARVESTING

When under mechanical stress, some materials have the capacity to produce an electric charge. The piezoelectric effect is a phenomenon that is used in many different products, such as doorbells, cigarette lighters, printers, and amplifiers. As depicted in Figure 10.2, it is also feasible to use a piezoelectric substance as a transducer to transform mechanical strain or movement into useful electric voltage or current. However, it is also feasible to generate electricity by applying pressure to the material (and therefore harvesting energy from things like floor tiles or shoe bottoms). Such generators are frequently based on a cantilever beam or a membrane that oscillates at the same frequency as the vibration source. Internet of Things (IoT) applications and low-power wireless sensor systems are the main applications for piezoelectric energy harvesters.

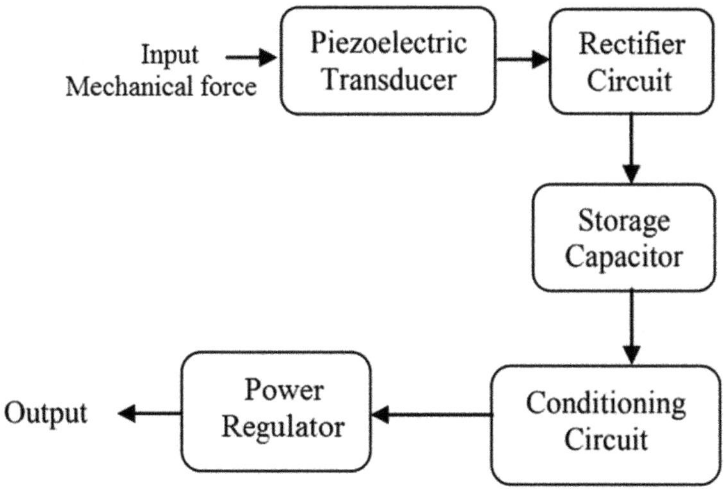

FIGURE 10.2 Block diagram of piezoelectric vibration energy harvesting.

10.4.2 ENERGY HARVESTING FROM ELECTROMAGNETISM

Faraday's law of induction states that when an electric conductor passes through an electric field or when the magnetic field that surrounds the conductor changes, an electric current is produced. Electromagnetic vibration energy harvesters employ a process known as electromagnetic induction to transform vibrations into electricity. The same theory underlies devices like electric motors and transformers, among others. Devices for harvesting electromagnetic energy are frequently made of coils, springs, and magnets that move in relation to one another. As depicted in Figure 10.3, this movement causes a change in the magnetic flux, which generates an electromagnetic force. In wireless sensor networks and Internet of Things applications, electromagnetic energy harvesters—like piezoelectric energy harvesters—are employed as a permanent power source to replace batteries and cables.

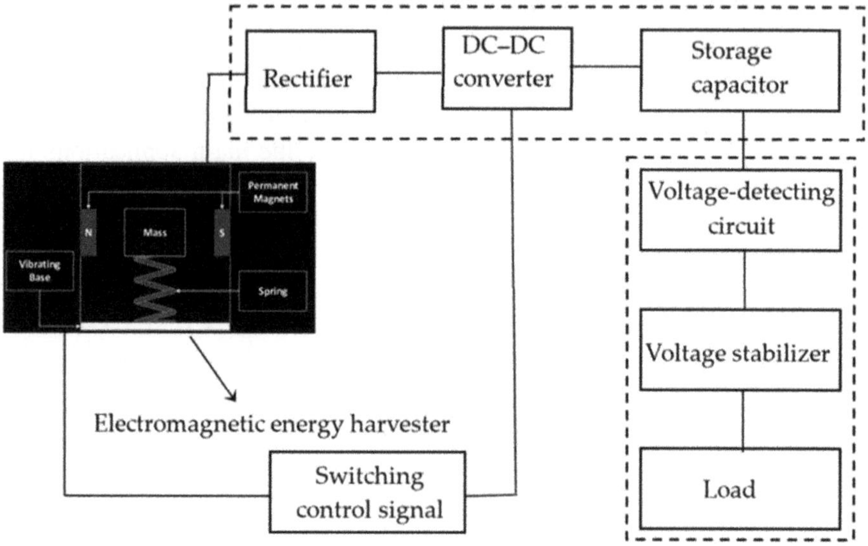

FIGURE 10.3 Block diagram of electromagnetic vibration energy harvesting.

10.4.3 PIEZOELECTRIC VS ELECTROMAGNETIC ENERGY HARVESTING

Resonance is a key component of piezoelectric and electromagnetic energy harvesters. The harvester must therefore be tuned in some fashion to a certain resonance frequency that corresponds to the dominant frequency of the vibration in the intended application. Both the techniques produce an unregulated AC voltage as well, and in order to provide any usable power output, they must be combined with rectification and power management devices. The potential for miniaturization and integration of micro-electromechanical systems (MEMS) with piezoelectric energy harvesters is a benefit. Additionally, the design's simplicity and lack of moving components make it simple to create in large quantities at a low cost. The fundamental flaw is that it produces less power than electromagnetic energy harvesters do. Additionally, piezoelectric materials are unsuitable for severe situations since they tend to be brittle and degenerate over time.

The main benefit with electromagnetic energy harvesters is the high-power output compared to piezoelectric energy harvesters, which makes the principle more suitable for more diverse applications. Additionally, electromagnetic energy harvesters are simpler to build for a long service life in challenging settings, which is frequently necessary for vibration energy-harvesting systems. Piezoelectric energy harvesting has for a long time been seen as a promising future technology to power IoT systems and it has been widely researched. Due to the low-power output of piezoelectric energy harvesters, however, for commercial applications, electromagnetic energy harvesting has been the primary alternative.

10.5 MAGNETIC-FORCE ENHANCEMENT

The rise of use of sensors and transducers incorporated in various systems increases the difficulty in fixing up of wires for supply of power and transmission of data. Because of this rapid growth is observed, in obtaining generators of small size which is equipped with wireless data transmission arrangement along with power transducers. In various cases, maintenance cost increases due to batteries' short life. So, it is better to obtain electrical energy by using ambient energy for sustainable generation of power. Among all available ambient energies, the main focus is provided on mechanical vibrations as it exists all around.

In harvesting of energy by using a concept of vibration the most commonly used device is piezoelectric transducer. Electro-mechanical coupling is the factor on which the efficiency of scavenging depends to a large extent. By using different types of transducers from a level of material to a level of device, it is possible to improve the effect of coupling. Stiffness of cantilever beam and that of the piezoelectric device provides the mechanical stiffness. And the coupling effect of energy harvester based on cantilever beam type structure is inversely proportional to mechanical stiffness at the level of device of piezoelectric transducer. Also the inherent stiffness obtained by the transducer and the inclusion required by the cantilever beam will not provide the easy reduction of mechanical stiffness of the piezoelectric transducer. In this area the mainly used generators are electrostatic and electromagnetic generators. Although, due to solid-state nature facilitated by piezoelectric generators various energy-scavenging approaches have been integrated with piezoelectric generators by using piezoelectric materials. In view of the fact, that piezoelectric element when provided with mechanical vibration can provide an AC voltage across the electrodes. Additionally, in order to store the generated ac voltage, it must first be converted to dc using an AC-DC converter, mostly a bridge rectifier, and then stored using a battery or capacitor.

To improve the performance of energy scavenging it is possible to use piezoelectric harvester with external magnetic force enhancement system by creating a nonlinear coupling arrangement. Measurement of harvesting of energy was done by attachment of magnets of tip mass at the face of piezoelectric cantilever beam or at the front of conventional piezoelectric energy harvester and then the extra effect is provided by the addition of external magnetic force effect using movable and fixed magnets of opposite magnetic force of attraction. Magnetic holder is there for the attachment of same types of magnets with opposite magnetic force of attraction. A single piezoelectric cantilever beam, connected with mass on its tip, is called conventional piezoelectric energy harvester system. This conventional piezoelectric energy harvester provides oscillations of linear nature with cantilever motion because it has the tendency to scavenge power of single resonance frequency. So, to increase the frequency bandwidth range and to raise the scavenging power of harvester it is required to add external magnetic force called enhancing piezoelectric energy harvesting with magnetic forces. And due to this enhancement of magnetic force it represents the nonlinear behavior.

10.6 VORTEX-INDUCED VIBRATION

A layer of boundary develops around the circular cylindrical body, when a fluid flows through a body in such a way that its direction is perpendicular to the direction of axis of the body as shown in Figure 10.4, that is, bluff body. For a large value of Reynolds number, the vortices experience separation from the surface of the body. And if there is further increase in Reynolds number then a phenomenon will occur as shown in Figure 10.5, which is called vortex shedding.

The periodic nature of forces is experienced on the circular cylinder when the shedding of vortices occurs on the surface of the cylinder. As long as the shedding is provided on the surface of circular cylinder, there will be a continuous vibration on the bluff body. When a sufficiently bluff body is kept near a fluid flow that causes occurrence of vortex shedding, then a vortex-induced vibration will occur. So it can be observed that a vortex-induced vibration will occur when frequency of vortex shedding is near or approximately in line with the inherent frequency of the bluff body.

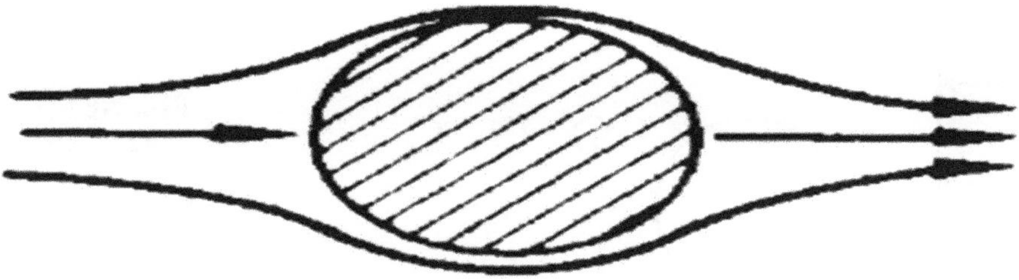

FIGURE 10.4 Flow of fluid through a bluff body.

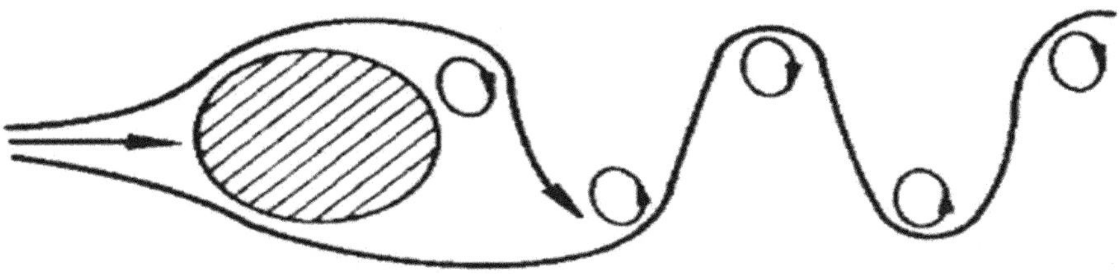

FIGURE 10.5 Vortex shedding.

10.6.1 VARIOUS VORTEX-INDUCED VIBRATION TYPES

1. Vibrations Caused by In-Line Vortices.
2. Cross-flow Vibrations Caused by a Vortex.
3. Vibrations Caused by Hybrid Vortices.

When a bluff body experiences a symmetric shedding then it will cause in-line vortex-induced vibrations as shown in Figure 10.6.

When a vortex shedding on bluff body is not symmetric then there will be vibrations brought on by the cross-flow vortex. Amplitude of motion of vibrations due to cross flow is larger than the amplitude of motion of vibrations due to in-line flow.

Properties of hybrid vortex-induced vibrations lie between cross-flow vortex-induced vibration and in-line vortex-induced vibration. Hybrid vortex-induced vibration is a combination of cross-flow motion and in-line flow motions, that is, the range of velocity is smaller than cross-flow motions and larger than the in-line motions.

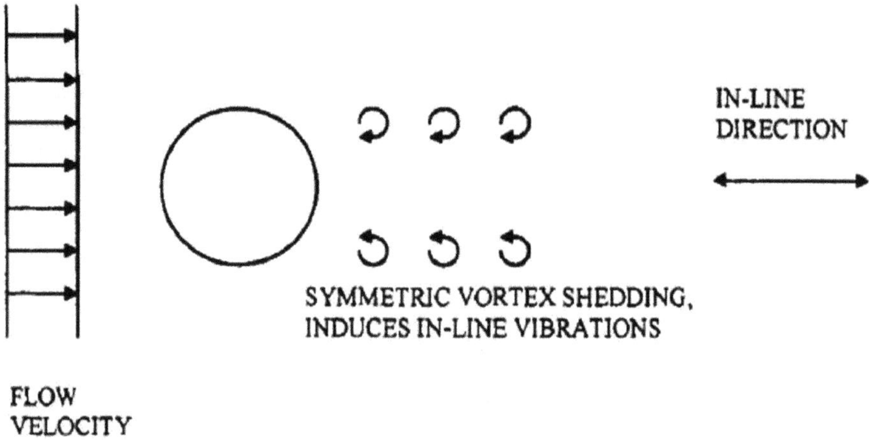

FIGURE 10.6 In-line vortex-induced vibration.

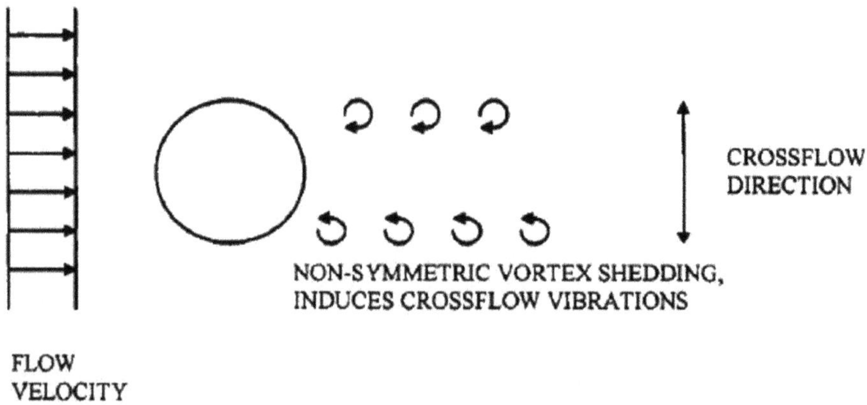

FIGURE 10.7 Cross-flow vortex-induced vibration.

Classification of vortex-induced vibration on the basis of oscillations:

- Self-excited oscillations: when the frequency of vortex shedding is approximately equal to the natural frequency then self-excited vibrations will occur in vortex-induced vibrations. In this type of vortex-induced vibration oscillations are produced by itself naturally.
- Forced oscillations: when the frequency of vortex shedding is not equal to the natural frequency and in this type the amplitude and velocity of oscillations is predefined and can be independently controlled by fluid velocity.

10.7 WORKING PRINCIPLES

A vibrational energy harvesting has engrossed significant consideration because of its utility prospects for powering and charging of micro-electro-mechanical system. Due to abundance availability of a hydro-kinetic energy of water at a very low velocity is enclosed in rivers and oceans, which are fluid environments, which are extensively available in nature like oceans and rivers. In harvesting of energy by using a concept of vibration the most commonly used device is piezoelectric transducer. As we know that a device which converts any physical quantity into electrical quantity is known as a transducer. Among various types of transducers, the energy harvesting is mainly done by using piezoelectric transducer. Piezoelectric transducer is an active and inverse transducer. Piezoelectric transducer has directional sensitive property because it works on the principle of piezoelectric effect. Piezoelectric transducer is a dynamic transducer because it works for dynamic inputs only, that is, in piezoelectric transducer we can get output for variable input only and if input is constant means time invariant, then we do not get any output. Electro-mechanical coupling is the factor on which the efficiency of scavenging depends to a large extent. By using different types of transducers from a level of material to a level of device, it is possible to improve the effect of coupling. Stiffness of cantilever beam and that of the piezoelectric device provides the mechanical stiffness. And the coupling effect of energy harvester based on cantilever beam type structure is inversely proportional to mechanical stiffness at the level of device of piezoelectric transducer. However, the inherent stiffness obtained by the transducer and the inclusion required by the cantilever beam will not provide the easy reduction of mechanical stiffness of the piezoelectric transducer. In this area the mainly used generators are electrostatic and electromagnetic generators. Although, due to solid-state nature facilitated by piezoelectric generators various energy-scavenging approaches have been integrated with piezoelectric generators by using piezoelectric materials. Given that a piezoelectric element can produce an AC voltage across the electrodes when subjected to mechanical vibration, the AC voltage must first be converted into a DC voltage using an AC–DC converter, primarily a bridge rectifier, and then stored using a battery or capacitor. To improve the performance of energy scavenging it is possible to use piezoelectric harvester with external magnetic force enhancement system by creating a nonlinear coupling arrangement. Measurement of harvesting of energy was done by attachment of magnets of tip mass at the face of piezoelectric cantilever beam or at the front of conventional piezoelectric energy harvester and then the extra effect is provided by the addition of external magnetic force effect using movable and fixed magnets of opposite magnetic force of attraction. Magnetic holder is there for attachment of same types of magnet with opposite magnetic force of attraction. A single piezoelectric cantilever beam, connected with mass on its tip, is called conventional piezoelectric energy harvester system. This conventional piezoelectric energy harvester provides oscillations of linear nature with cantilever motion because it has the tendency to scavenge power of single resonance frequency. So, to increase the frequency bandwidth range and to raise the scavenging power of harvester it is required to add external magnetic force called piezoelectric energy harvesting which is strengthened by force. And due to this enhancement of magnetic force it represents the nonlinear behavior. This uses a flow-induced piezoelectric vibration energy harvester (PVEH), with magnetic force enhancement, which is incorporated by piezoelectric beam, circular cylinder magnets. By using both the excitation of the magnetic force and vibration caused by the same, it can convert the hydrokinetic energy into electricity. As the effect of magnetic force on the vibration characteristics is obtained at resonance frequency, so to obtain that resonant frequency the frequency sweep experiment of piezoelectric with cantilever beam is performed.

Most of the researchers mainly focus on the effects of the bluff body for energy harvesting and neglect the mechanical modulation enhancement for the vortex-induced vibrational energy harvesting. The vortex-induced vibrational energy harvesting cannot vibrate easily in case of low level of water flow when this is due to the small vortex-induced force applied to the bluff body and due to large water viscosity and hence the results proved that the vortex-induced vibrational energy harvester has an essential potential application for powering the micro-electronic devices with low flow velocity underwater. A magnetic force-based flow-induced piezoelectric vibration energy harvester (PVEH) with magnetic force enhancement studied it by combining it with piezoelectric cantilever beam, magnets, and a bluff body with circular cylinder. In further sections, the physical model is introduced with experimental platform, and then the impact of velocity flow and the effect of the magnetic poles distance and its interaction properties are discussed in detail and hence the merits of the concept is introduced, that is, a magnetic force increase and a flow-induced PVEH are confirmed and hence in last conclusions are provided.

10.7.1 FLOW-INDUCED PVEH WITH ITS MECHANISM

Flow-induced vibration is a general phenomenon, which explains the physical interaction among the structure and fluid. Vortex-induced vibration is a well-known form of flow-induced vibration. When the bluff body is deformable or elastically mounted, then it can induce vibration and can cause action of fluctuation forces acting on the bluff body, and at the same instant shedding and vortex formation can get altered further due to vibration. In this scavenge the mechanical energy from the flow-induced vibration method is called PVEH with magnetic force enhancement. This mainly provides focus on the design of highly efficient vortex-induced VEH with low velocity of water flow in underwater. This method is generally composed of cantilever beam made of piezoelectric material with bluff of circular cylinder body and the magnets. In the middle of the path of water flow the piezoelectric beam is fixed at the cantilever's open end and the cylindrical circular bluff body is fixed. The bluff body with beam vibration is movable with the permanent magnets. And in the frame the other magnets are fixed. Here, to improve the performance of scavenging of the harvester the concept of excitation with magnet is provided, and to vibrate the PVEH slightly at the low flow velocity further excitation is provided to the bluff body with magnetic force enhancement.

Its experimental system can be represented by its experimental fume and the water supply system is mainly included in the system. The volume and the flow velocity in the fume can be adjusted precisely by controlling the handling door of the outlet gate with the help of stepper motor. To gauge flow velocity, a current meter is employed. The piezoelectric beam, which is completely immersed in water and clamped with the fixed frame, is used as an energy transducer. For the acquisition of output voltage signal the oscilloscope is used. The frame is made up of stainless-steel beam of thin wall so as to decrease the effect of fixed frame on flow field. To provide high insulation to piezoelectric material and to provide easiness to harvesting energy inside the water, water-proof film coating is provided on the surface of piezoelectric beam. With the connection plate lower end of the thin aluminum row is connected and with the clamp the top of the aluminum row is connected and on the surface of the cylindrical, parallel to the magnet on the sheet small magnets are fixed. Due to these connections, there will be an attractive force exertion between the opposite poles and between the same poles the repulsive force is experienced.

10.7.2 MAGNETIC EFFECT

On the basis of the system, the magnetic property effect on the voltage is acquired by the flow-induced PVEH. In this, the pole distance is taken and the velocity of flow is taken. The results of the experiments indicate that the largest amplitude occurs when there is a magnetic force of attraction and repulsion. So it can be concluded that in case of magnetic force of attraction the amplitude of voltage obtained is larger than the amplitude of voltage obtained by harvester with magnetic force of repulsion. Also, it is obvious that the amplitude of output voltage obtained with harvester with no magnetic force is smaller when compared with the other cases. So it can be concluded that the energy obtained by harvesting of piezoelectric voltage is not only provided with vortex-induced force excitation, but it is also get influenced by the magnetic force of attraction and repulsion, due to which there is a change in frequency of vibration and also it causes decrease in vibration deformation. PVEH vibration frequency decreases due to magnetic force of attraction and provides contribution in the increment of output voltage. So, it can be said that by introducing the concept of magnetic force one can improve the performance of energy harvesting. Simultaneously, it can be observed that in case of magnetic force of attraction, the frequency is lesser when compared with other reported methods.

10.7.3 FLOW VELOCITY EFFECT

In piezoelectric voltage-induced effect the main source that causes excitation is flow-induced vibration. Vibration due to flow-induced effect is a very important parameter in this system. It represents the analysis waveform of output voltage and different flow velocities on the basis of FET analysis. Here, it can be observed that with the increase of velocity flow there is an increase in frequency and output voltage. It is already proven that on flow-induced PVEH, velocity flow provides great impact. When there is low flow velocity effect then the effect of power generation is not good. However, it is required to mention that the maximum velocity flow of the experiment fume is 0.7 m/s which is restricted by an experimental arrangement. It is not appropriate to conclude for the larger flow velocity. In general, in piezoelectric beam the resonance frequency is obtained by the natural frequency so acquired voltage is maximized at the point of resonance.

10.7.4 PIEZOELECTRIC BEAM VIBRATIONAL CHARACTERISTICS

For harvesting the energy from piezoelectric vibration, it is important to know the characteristics of frequency of piezoelectric beam. So, to determine natural frequency of arrangement it is important to perform the sweep frequency vibration test. In this chapter, two cases of the magnetic force are considered, the first case is for magnetic force attraction using opposite poles and the second case is for magnetic force repulsion using similar poles.

Two types of magnetic force attraction and repulsion are represented by the amplitude frequency response curve of the output voltage. Additionally, the results of these two methods can be compared with the output obtained by no-magnet case. The poles distance effect on the frequencies for both the cases of attraction and repulsion is represented clearly for distance. Furthermore, in case of magnetic force of attraction, in piezoelectric beam there is a gradual decrease in natural frequency when the distance between the pole decreases. Similarly, in case of magnetic force of repulsion, there is an increase in natural frequency when the distance between the poles decreases. At the same time, it is not obvious to see the difference between the amplitude of output voltage. So, we can say that the distance between the poles provides an important impact on natural frequency, and it also provides less impact on the piezoelectric beam output voltage.

10.8 COMPARATIVE STUDY

A comparative study is done by introducing a concept of nonlinear magnetic force on piezoelectric energy harvesting from vortex-induced vibration. For the convenience of qualitative investigation, a lumped parametric model is considered for piezoelectric with a cantilever cylindrical structure. To harvest the energy from vibration, the most popular structure used for modeling and implementation of energy harvesting devices is piezoelectric cantilever beam. Also, a structure is used which is separated by two sheets with shim in its center and the structure is called bimorphs by virtue of static analysis; evaluation is done for the buckling displacement of harvester in configurations like mono-stable and bi-stable. To determine the coupling frequency and harvester damping with variable electrical load resistance, different values of spacing distance are provided between the magnets. Then, the generated voltage and dynamic behavior of the harvesters are briefly discussed in two configurations, according to environmental wind conditions for significant performance optimization; there is a shift of lock in region by showing the variation in distance of spacing between the magnets. Additionally, the outcome shows that the mono-stable configuration harvester provides hardening behavior and the bi-stable configuration harvester provides softening behavior; to widen the region of synchronization use the harvester with both the configurations. It is not easy to provide a fair comparison of energy harvester technology on the basis of size, material, and different architectural and operational modes. It is difficult to find the comparison of performance, due to the lack of universally available figure of merits. To evaluate the performance of scavenging of energy and evaluate high performance comparison of reported work is provided on the basis of available figure of merits. Power density is the most preferably used parameter for energy harvesting. Power density is nothing but defined as the output power per unit volume.

TABLE 10.1 Comparison of Reported Work.

Reference	Resonance frequency	Estimate size (cm^3)	Voltage (V)	Power density ($\mu W/cm^3$)	Power (μW)	Excitation (m/s^2)
Foong et al. (2019)	17.2	217.5	0.16	34.02	7400	1
Usharani et al. (2018)	21.02	139.95	61.55	47.16	6600	10
Fan et al. (2019)	Random	3.09	21	15.53	48	1
Kim et al. (2018)	30	0.6	5.8	30.93	18.56	17.5
Liu et al. (2016)	21.04	14.4	22.35	29.8	492	3
Rui et al. (2018)	**15.3**	**63.6**	**19.83**	**51.53**	**3277**	**4**

The performance analysis of a flow-induced piezoelectric vibration energy harvester with magnetic force enhancement is compared (as indicated by the bold letters in Table 10.1) along with some published literature. The comparison is conducted in a number of areas, including frequency, size, excitation, power, voltage, and density, which are listed in Table 10.1 as a summary.

10.9 VIBRATIONAL ENERGY HARVESTING IN BIOMEDICAL AND MHEALTH

Numerous energies are continuously available from human beings uninterruptedly and naturally as body heat (W), breathing (mW), locomotion like walking (10–100 mW), hand cranking (W), lifting (10 mW), etc. can be used for energy harvesting. These include: (1) wearable technology that can be recharged through exercise, such as walking or running. (2) A vehicle's driver seat equipped with piezoelectric sensors to track and detect the driver's breathing and heart rate. When a driver's stress level is identified, vibration sensors allow the seat's ventilation and massaging capabilities to instantly turn on. (3) The capacity to power sensors and monitoring equipment in isolated and hazardous locations (pipelines, bridges, etc.). By doing this, the risk that humans face when batteries need to be charged or replaced is eliminated. (4) Lighting for pavements generated by footfalls striking energy-absorbing tiles. (5) A piezoelectric pacemaker that receives its energy from the heartbeat. By doing this, the need for risky and invasive surgery to replace batteries is eliminated.

Some applications of vibrational energy harvesting in biomedical and mHealth are as follows:

10.9.1 FOR BIOMEDICAL DRUG DELIVERY

One of the potential biomedical applications of piezoelectric generators is in implantable drug delivery systems, such as the bio-micro-electro-mechanical systems (BioMEMS). Although such a device has been proposed as a system that can deliver drugs locally to diseased cells and organs, the powering of such devices poses some significant technological challenges. One approach is to use conventional thin-film batteries and power capacitors. These are commercially available and are known to generate sufficient power for bioMEMS. However, they are difficult to recharge within the body. This is attractive because the systems can then be recharged by mechanical energy from moving fluids or solids within the body. The kinetic energy of flowing blood is inadequate, according to basic back-of-the-envelop calculations, to provide the necessary power levels in piezoelectric thin-film generators of the right size for BioMEMS structures. These devices present some major technological problems.

10.9.2 FOR BIOMEDICAL DENTAL CARE

In contrast, the significant mechanical energy associated with occlusal contact (typical maximum loads between 100 and 200 N) may be in the regime that is sufficient for the generation of adequate power for applications in BioMEMS and microelectronics systems. With adequate electrical isolation, the piezoelectric generators may be inserted between the coping of a dental crown and the metal abutment. Such a system could tap the mechanical energy that is provided annually by 1 million contact cycles.

In biomedical sector, continuous power requirement is crucial. Available human energies can be utilized for this purpose. But the small energies available from human body are the hurdles in this area. The main challenge is to enhance this low energy output. Proper electronic circuit designs to capture and enhance the energy output are major issues to be addressed. Further, much more advancement is required to harvest energy from other biological parameters such as body temperature, respiration.

10.9.3 FOR MHEALTH BLOOD PRESSURE MONITORING

The procedure of piezoelectric composite cantilever beams immersed in the flow induced by the manual pumping of the bulb is suggested to power a self-powered blood pressure (BP) monitor. This model for harvesting energy is simple. The vibrations (mechanical) of the beam are harvested by the transducer (piezoelectric) involved in the substrate (metallic). The electrical charge produced by the layer (piezoelectric) is stored to an electrical storage device (capacitor) through a series of conditioning circuits. The size of the transducer is chosen so that when the cuff is fully inflated, an appropriate quantity of electrical energy is stored in the capacitor to facilitate required purposes such as the transduction of pressure, the following processing and analyses are display and storage of BP data.

10.9.4 FOR WEARABLE MHEALTH CARE TECHNOLOGIES

Any electronic device or computer that is incorporated into clothing or small accessories that may be worn on or incorporated into the body is considered to be wearable technology. Many of these gadgets use applications or the devices themselves to wirelessly collect and transmit information. Piezoelectric ceramic is used to convert vibrational energy from the body's various motions, including breathing, walking, and breathing. Piezoelectric components under stress produce an electrical current that can be used right away or saved. The amount of energy generated is still quite low, and the bodily movements needed are frequently irregular and unpredictable. There are a variety of wearable sensors available on the market right now, including activity and fitness bracelets and monitors that track steps, heart rate, distance traveled, and even sleep patterns. Patients' blood pressure is measured using wireless blood pressure cuffs using a smartphone app. Long-established quartz timepieces use the material's inherent piezoelectric property to maintain accurate time. Piezoelectric components are used to transform the vibration into a discernible signal in monitors that track and measure fetal heartbeats.

10.10 CONCLUSION AND FUTURE SCOPE

In this chapter, a flow-induced piezoelectric vibrational energy harvester with magnetic force enhancement is suited for power generation. For this, a tool is employed in a manner that it can provide increased flow velocity and to transfer the hydrokinetic energy into electrical energy piezoelectric effect is used. Both the processes can be done by using phenomenon called excitation of magnetic force and vortex-induced vibration. The output voltage is increased by using attractive magnetic force, as it reduces the frequency of vibration. The maximization of obtained voltage is done at the point of resonance. This resonance frequency is obtained from the natural frequency of piezoelectric beam and velocity of flow. By observing performance analysis, it can be concluded that a flow-induced piezoelectric vibrational energy harvester with magnetic force enhancement has much better performance than a flow-induced piezoelectric energy harvester with no magnet. Additionally, it is observed that in case of low flow velocity effect a flow-induced PVEH with magnetic force of attraction has merits that it has higher voltage output. For better performance of flow-induced PVEH, flow velocity effect plays a significant role, that is, as velocity of flow increases then the output voltage increases. From analysis it can be observed that the harvesting performance improved by providing the concept of magnetic force. The harvesting of acquired output voltage increases for value of flow velocity by considering the magnetic force of attraction. So we can say that the acquired output voltage in case of magnetic force of attraction is more than the output voltage acquired with nonmagnetic force of attraction. And the maximum value occurs at the point of resonance and the results are verified with analysis. Compared with various academic works, it has been demonstrated that magnetic linked structures have a high potential for energy-harvesting density. The harvester has a promising future for use as a result of its simplicity in design and processing. A flow-induced piezoelectric vibrational energy harvester with magnetic force enhancement is used so as to promote the usage of flow-induced vibration for electrical energy generation. To generate an alternative of green energy a viable solution is obtained called vortex-induced vibration energy converter. Factors like properties of fluid, high damping effect, and high amplitude of vortex-induced vibration will determine the efficiency of vortex-induced vibration converter for a wide range of synchronization. In the future it is required to increase the amplitude of vortex-induced vibration converter and provide optimization to lift coefficient and for that more research work is needed. Till now configuration of piezoelectric-based energy harvester with different materials and structural arrangement had been discussed for various vortex-induced vibrational converters. Basic and simplified structure is the main advantage of using a device called piezoelectric energy harvester. In addition, the materials used in these devices do not get influenced by electromagnetic radiations. To adjust the utilization of these harvesting techniques in many ways, many exaggerations are provided by researchers like including the use of materials such as polymers, piezoelectric ceramics. The obtained results are further fabricated for designed optimization of the flow-induced piezoelectric vibrational energy harvesting for case of low rate of velocity flow rate.

In the future this concept of design could be combined with intelligent biomedical monitor controls and devices that detect mHealth care of humans and act as a self-powered system with the biomedical industries.

KEYWORDS

- **energy harvesting**
- **piezoelectric materials**
- **electronic circuits**
- **vortex-induced vibration**
- **piezoelectric property**
- **biomedical and mHealth**
- **magnetic force enhancement**

REFERENCES

Adhikari, R.; Jackson, N. In *Passive Frequency Tuning of Piezoelectric Energy Harvester using Embedded Masses*, 2021 IEEE 20th International Conference on Micro and Nanotechnology for Power Generation and Energy Conversion Applications (PowerMEMS), 2021; pp 176–179. DOI: 10.1109/PowerMEMS54003.2021.9658377.

Anand, A.; Kundu, S. In *Improvement of Output Power in Piezoelectric Energy Harvester under Magnetic Influence*, 2019 Devices for Integrated Circuit (DevIC), 2019; pp 382–385. DOI: 10.1109/DEVIC.2019.8783607.

Bhardwaj, V.; Ravi Teja, A. V. In *Mathematical Modelling and Equivalent Circuit Representation of Bladeless Wind Turbines*, IECON 2021 – 47th Annual Conference of the IEEE Industrial Electronics Society, 2021; pp 1–6. DOI: 10.1109/IECON48115.2021.9589902.

Debnath, B.; Kumar, R. In *Design and Simulation Study of a New Flared-U Shaped Springs Based MEMS Piezoelectric Vibration Energy Harvester*, 2020 IEEE International Conference on Computing, Power and Communication Technologies (GUCON), 2020; pp 101–105. DOI: 10.1109/GUCON48875.2020.9231097.

Deepak, P.; George, B. Piezoelectric Energy Harvesting From a Magnetically Coupled Vibrational Source. *IEEE Sensors J.* **2021,** *21* (3), 3831–3838. DOI: 10.1109/JSEN.2020.3025216.

Fan, K.; Tan, Q.; Liu, H.; Zhang, Y.; Cai, M. Improved Energy Harvesting from Low-Frequency Small Vibrations through a Monostable Piezoelectric Energy Harvester. *Mech. Syst. Signal Proces.* **2019,** *117*, 594–608. DOI: 10.1016/j.ymssp.2018.08.001.

Foong, F. M.; Thein, C. K.; Ooi, B. L.; Yurchenko, D. Increased Power Output of an Electromagnetic Vibration Energy Harvester through Anti-Phase Resonance. *Mech. Syst. Signal Proces.* **2019,** *116*, 129–145. DOI: 10.1016/j.ymssp.2018.06.012.

Gaikwad, S.; Ghosal, M. In *Energy Efficient Storage-Less and Converter-Less Renewable Energy Harvesting System using MPPT*, 2017 2nd International Conference for Convergence in Technology (I2CT), 2017; pp 971–973. DOI: 10.1109/I2CT.2017.8226273.

Katulka, G. L. Micro-electromechanical Systems and Test Results of SiC MEMS for High-g Launch Applications. *Sensors* **2002,** *2*, 1134–1138. DOI: 10.1109/ICSENS.2002.1037273.

Khanna, G.; Asthana, P. In *A Comparative Study of Circuit for Piezo-Electric Energy Harvesting*, 2016 3rd International Conference on Computing for Sustainable Global Development (INDIACom), 2016; pp 1689–1694.

Kumar, K. R.; Morab, S.; Shekar, S.; Mahalingam, A. In *Energy Harvesting from Vortex Induced Vibrations Using Vented Cylinders Mounted on Light Rail Locomotive*, 2016 7th International Conference on Intelligent Systems, Modelling and Simulation (ISMS), 2016; pp 268–275. DOI: 10.1109/ISMS.2016.33.

Kumar, S.; Kumar, R.; Singh, N. In *Comparative Analysis of Battery with Paper Battery for Renewable Energy Storage*, International Conference on Challenges in Sustainable Development from Energy & Environment Perspective, Gorakhpur, India, 2017; pp 222–229.

Kumar, S.; Singh, R.; Singh, D.; Prasad, R.; Yadav, T. Electronics-Waste Management. *Int. J. Environ. Eng. Manag.* **2013,** *4* (4), 389–396.

Kumari, N.; Rakotondrabe, M. In *Development, Presentation and Tests of a Hybrid Thermal Vibrational Energy Harvester based on Lead Free Piezoelectric Material*, 2020 IEEE 2nd International Conference on Electronics, Control, Optimization and Computer Science (ICECOCS), 2020; pp 1–6. DOI: 10.1109/ICECOCS50124.2020.9314542.

Liu, S.; Cheng, Q.; Zhao, D.; Feng, L. Theoretical Modeling and Analysis of Two-Degree-of-Freedom Piezoelectric Energy Harvester with Stopper. *Sens. Actuators A Phys.* **2016,** *245*, 97–105.

Mukherjee, S.; Prasad, D.; Khan, M. I.; Barua, P.; Agarwal, H. In *Hexa Spacer Damper for Vibration Energy Decaying of 765 kV Transmission Line*, 2019 Innovations in Power and Advanced Computing Technologies (i-PACT), 2019; pp 1–6. DOI: 10.1109/i-PACT44901.2019.8960087.

Rui, X.; Li, Y.; Liu, Y.; Zheng, X.; Zeng, Z. Experimental Study and Parameter Optimization of a Magnetic Coupled Piezoelectric Energy Harvester. *Appl. Sci.* **2018,** *8* (12), 2609. DOI: 10.3390/app8122609.

Shukla, R.; Adam Qaisar, S.;Bell, A. J. In *Towards the Development of Efficient Low Frequency Piezoelectric Energy Harvesters*, 2010 IEEE International Symposium on the Applications of Ferroelectrics (ISAF), 2010; pp 1–4. DOI: 10.1109/ISAF.2010.5712228.

Sil, I.; Biswas, K. In *Investigation of Design Parameters in MEMS Based Piezoelectric Vibration Energy Harvester*, 2018 IEEE Electron Devices Kolkata Conference (EDKCON), 2018; pp 64–69. DOI: 10.1109/EDKCON.2018.8770388.

Singh, R.; Kumar, S.; Chauhan, S. Energy:-Conservation, Management, Efficiency & Storage. *Int. J. Appl. Eng. Res.* **2013,** *8* (7), 80–85.

Suresh, K.; Shankar, K.;; Sujatha, C. In *A Lumped Electromechanical Model for Output Power from Partially Covered Piezoelectric Energy Harvester*, 2019 IEEE 2nd International Conference on Power and Energy Applications (ICPEA), 2019; pp 94–98. DOI: 10.1109/ICPEA.2019.8818525.

Usharani, R.; Uma, G.; Umapathy, M.; Choi, S. B. A Novel Piezoelectric Energy Harvester Using a Multi-Stepped Beam with Rectangular Cavities. *Appl. Sci.* **2019,** *8* (11). DOI: 10.3390/app8112091.

CHAPTER 11

Analysis of Integrated Intelligent Methods Toward Medical Decision-Making

CHANDRABHAN SINGH[1], PREM SHANKER YADAV[2], HIMANSHU KUMAR SHUKLA[1], ROHIT SRIVASTAVA[1], and UPENDRA KUMAR[3]

[1]*Department of CSE, F.O.E.T., University of Lucknow, Lucknow, Uttar Pradesh, India*

[2]*Department of Computer Science and Engineering, Malaviya National Institute of Technology, Jaipur, Rajasthan, India*

[3]*Department of CSE, IET, Lucknow, Uttar Pradesh, India*

ABSTRACT

Decision-making in the field of medical to provide treatment is not an easy task. Due to complex nature of disease sign and symptoms, this decision-making process takes a lot of efforts and time. Clinical decision-making process is dependent on practitioner's learning and experience. During the last two decades researchers focus on development of intelligent computing approaches that help the practitioner in correct and fast diagnosis. Various intelligent computing methods are used by researchers to facilitate medical domain. Soft computing method individually or in integration is popularly used by researchers to achieve the goal. To deal with the major problem associated with medical data, that is, classification, soft computing methods like Artificial Neural Network (ANN), Fuzzy Inference System (FIS), Genetic Algorithm (GA) play an important role. This chapter will focus on the study of these methods for the diagnosis of medical problems.

Smart Electronic Devices and Systems for Biomedical and Healthcare Applications. Upendra Kumar, Rajesh K. Kesharwani, & Raj K. Keservani (Ed.)

DOI: 10.1201/9781003560852-11

11.1 INTRODUCTION

Medical data analysis is a delicate topic that requires accuracy for disease analysis, identification, and prediction. Therefore, it is crucial to create and apply appropriate machine learning classification algorithms that accurately and effectively detect and diagnose diseases. Due to the growing use of computationally demanding methodologies, accurate and efficient analysis of such medical data in recent years has sparked a revolution in machine learning. To help doctors correctly diagnose disease, several clinical concerns still need to be resolved, including the requirement for quick, accurate, and trustworthy decision models. The performance of the classification model may be hampered by the noisy, irrelevant, redundant, and missing information present in most medical datasets. The accuracy of the medical data and the models applied throughout the classification process determine how well the classifier performs (disease prediction). Thus, it is crucial to employ classifiers to correctly and accurately interpret sensitive medical data to forecast and detect diseases. Neuro-fuzzy systems (NFS) have an important and essential part in the classification and forecasting of various sorts of medical illnesses (Shihabudheen and Pillai, 2018). To minimize incorrect diagnosis the NFS supports medical practitioners in automating the domain dealing with medical diseases. With time, it has become clear that the NFS strategy can improve accuracy when handling a variety of challenging research issues in medical diagnosis. The author discusses the literature review of the studies conducted to apply NFS in medical diagnostics with a focus on psychiatric diseases in this chapter.

Strength of soft computing approaches plays an important in medical decision-making. In medical decision-making process the major task involves process of diagnosis, prediction, treatment, etc. In clinical decision-making process sign and symptoms also with practitioner observation, knowledge play an important role. Based on knowledge and experience, practitioner assigns level and weight to signs and symptoms. Sometimes, the complex nature of diseases signs and symptoms makes the clinical process lengthy and time taking.

Initially, rule-based systems are designed based on an expert system to facilitate a practitioner. But due to complex nature of diseases sign and symptoms these rule-based systems are not accurate. The complex nature of the medical decision-making process gives birth to introduce soft computing methods for the diagnosis and prediction purpose. Computational intelligence is currently being used to facilitate early disease detection and diagnosis. The size and dimensionality of medical data is huge.

Neural network (NN) came to existence in the 1960s but became popular in the 1980s. Concept of NN is based on biological concept of neurons. Each neuron relates to each other. There are different NN models used to solve complex and uncertain problems associated with medical diagnosis. Neural networks and fuzzy logic are two tools that function well together when creating intelligent systems. While neural networks are a type of low-level computing framework, they are effective when working with unprocessed data. Fuzzy logic uses linguistic data gathered from subject-matter experts to cope with higher-level reasoning. Fuzzy systems are unable to learn new things and cannot adapt to a new environment. NN, on the other hand, is teachable.

The parallel computation and learning capabilities of neural networks can be combined with the fuzzy system's human-like knowledge storage and explanation capabilities to create an integrated neuro-fuzzy system. These two factors work together to make NN more visible and enable learning in fuzzy systems.

The fuzzy inference model's functional counterpart, the neuro-fuzzy system, is a neural network. It can be taught to create If-then fuzzy rules and identify membership functions for the system's input and output variables.

Computational intelligence can be easily adopted in medical field as it can learn, classify, and predict large amount of medical data. Clinical process of decision-making system is dependent on learning capability and experience of practitioner.

The capability of soft computing that deals with imprecise, uncertain, approximation gives birth to build robust and low-cost system. Fuzzy logic (FL), neural network (NN), Support Vector Machine (SVM), Genetic Algorithm (GA), K-Nearest Neighbor (K-NN), Naïve Bayes (NB), and Particle Swarm Optimization (PSO) are commonly used soft computing approaches. Recently, meta-heuristic approaches of optimization are popular based on nature-inspired concepts. These approaches are also used in combination of one or more approaches to overcome the shortcomings of others. In the next section, integrated neuro-fuzzy methods for diagnosis and prediction are analyzed with specific reference to mental health care domain. It is possible to use soft computing techniques to tackle complex computational problems because they give us access to helpful computing methods and algorithms that can be used to handle complex systems and problems in general. For a specific system or issue, it offers a degree of uncertainty, partial truth, and tolerance for imprecision. Computational intelligence is the term used to describe this kind of approach, which enables us to provide and solve problems that the hardware finds either extremely difficult or impossible to solve.

11.2 LITERATURE WORK

Soft computing techniques have been used extensively in the study on medical decision-making systems. This part will concentrate on combined soft computing techniques for informed decision-making. The most practical fuzzy inference system is the Takagi–Sugeno–Kang (TSK) inference system, which is also a potent instrument for modeling nonlinear dynamic systems. The primary benefit of TSK system modeling is that it uses a "multimodal" method that combines linear sub-models to describe the overall behavior of a full complex nonlinear dynamic system (Karaboga, 2016). Adaptive neuro-fuzzy inference system (ANFIS), one of the well-liked neuro-fuzzy methods, has been applied by scholars to regression, modeling, prediction, and control issues (Lin, 2007; Chakravarty, 2012). The ANFIS employs a five-layer network structure with a fuzzy inference system of the TSK type. The two categories of parameters that ANFIS specifies are premise parameters and consequent parameters. The connections between the two groups of parameters are established by fuzzy if-then rules. The primary disadvantage of ANFIS is its computational complexity, which results in complex models being produced for even relatively straightforward problems.

Using characteristics from heart rate variability, Zhang et al. (2011) used a neuro-fuzzy expert system in conjunction with a weighted fuzzy membership function (NEWFM) to diagnose depression. The study's findings were given as ECG signals from 10 depressed people who had been rated depressed using the Zung scale and 10 healthy subjects, each of whom had six characteristics. Fast Fourier transformation was used to analyze the ECG data to calculate the frequency domain features. Standard deviation of RR intervals (SDNN) and root mean square of successive differences (RMSSD) were chosen as time domain features for the experimental study, while very low frequency in the range of 0.0033–0.04 Hz, low frequency in the range of 0.04–0.15 Hz (LF), high frequency in the range of 0.15–0.4 Hz (HF), and ratio of LF and HF (LF/HF) were chosen as frequency domain features. The suggested methodology produced an accuracy of 95% and a sensitivity of 90%.

Using neuro-fuzzy modeling, Chattopadhyay (2012) demonstrated a system for automatically rating depression in elderly individuals (NFM). The author used data from hospitals in Eastern India, which included 270 samples, each with seven attributes in accordance with DSMIV-TR recommendations. Following the use of the t test for feature reduction, the two versions of the NFM model—NFM1 with all seven characteristics as inputs and NFM2 with two features—were created. The Gaussian membership function and the triangular membership function were used to implement both the NFM variants. According to the experimental findings, NFM1 outperformed NFM2 with an accuracy of 94.4%. To simulate the manual process of making a clinical diagnosis of depression after 2 years, Chattopadhyay (2014) used PCA with the Mamdani neuro-fuzzy system.

In the following year, Ekong and Onibere (2015) created a diagnosis method for depression using the Takagi–Sugeno neuro-fuzzy system and case-based reasoning. The study used medical information gathered from a hospital in Nigeria that included 100 cases of depression, each having nine psychological and seven physiological contributing factors. 144 fuzzy rules were produced by the suggested model utilizing a fuzzy inference method. The created rules were then improved by utilizing a neural network to assess local similarities before case-based reasoning was added to create a decision support system. The decision support system was designed using emotional filters to identify mild, moderate, and severe depression in the patient and cognitive filters to compute the cognitive rank between input components. The model's accuracy prediction was 92.4%.

An ANFIS-based model to predict student anxiety was given by Devi et al. (2016). 36 PG students at the G. B. Pant University of Ag. & Technology provided information for the study using the Moudsley Personality Inventory MPI Questionnaire (i.e., the values of neuroticism and extraversion) and Sinha's Comprehensive Anxiety Test SCAT Questioner (i.e., the value of anxiety). This model used Neuroticism and Extraversion as its two input psychological variables, with Anxiety as its output. Back propagation was used to optimize the rules' input parameters, while the least-squares approach was used to optimize the output constants. The suggested model was then assessed for two metrics, MAPE and RMSE errors, and compared with FIS. As the MAPE value was found to be 23.10 in FIS and 21.00 in ANFIS, it was determined that ANFIS was superior to FIS for the prediction of students' anxiety. In FIS and ANFIS, the value of RMSE was found to be 3.381 and 3.280, respectively.

Fuzzy brain structures are used to diagnose Parkinson's illness (Abiyev and Abizade, 2016). We present the FNS structure and learning techniques. The FNS is developed using fuzzy clustering and gradient descent learning methods. Using a 10-fold cross validation data set, learning is conducted.

Several numbers of fuzzy rules are employed in FNS to create the classification system. Using 16 hidden neurons, the classification recognition rate is 100%. Several models are used in the modeling of PD for a comparative study. The findings collected show that FNS performs better than the other models employed for PD classification.

A genetic neuro-fuzzy-based method was proposed by Ashish et al. (2017) for the diagnosis and grading of depression into mild, moderate, and severe. Between 2004 and 2005, the author gathered the data from an Indian mental hospital. It contained 88 samples, each of which had one output that represented depression and seven inputs that represented seven different symptoms of depression, such as feeling down, insomnia, and appetite loss. With the input triangular membership function, the NFS model was created using subtractive clustering and the back-propagation method, yielding 2187 rules. The model was then improved by removing the redundant and unneeded rules using a genetic method.

In the same year, Yoon (2017) described an expert system for diagnosing depression from heart rate variability (HRV) that combined neural networks with the weighted fuzzy membership function (NEWFM) and match count approach. A QRS detection technique was used in the suggested methodology to convert ECG signals into HRV data, which was then followed by the extraction of six features from the time domain, frequency domain, wavelet transform, and Poincare transform. The model was trained to choose the best feature for depression using selected six features as inputs, and four features were retrieved using the Match Count Method in the following stage (MCM). After feature selection, NFS was used to accurately diagnose the condition under investigation with a 97.62% rate of success.

Tejeswinee et al. (2017) produced a novel dataset that contains genetic data related to the two neuro-degenerative diseases Parkinson's and Alzheimer's. The dataset originally comprised 1437 structural and physicochemical protein attributes that were taken from the PROFEAT server and retrieved from the KEGG database. There were 111 instances in which 74 genes contributed specifically to Alzheimer's disease and 37 genes specifically to Parkinson's disease. This investigation's goal was to evaluate the effectiveness of various classification algorithms on the created dataset, including Support Vector Machine (SVM), Random Forest, Decision Tree, Naive Bayes, Adaboost, and K-NN. These algorithms' classification accuracy was evaluated by tabulating the outcomes of two measurements: accuracy and Matthew's Correlation Coefficient (MCC). The study showed that, compared to other classification methods, Random forest and K-NN classifiers predicted the diagnostic classes with high accuracy (82%) before feature selection. SVM has the highest accuracy (around 94%) while evaluating the CFS subset. Random Forest demonstrated remarkable results (85%) in the Gain Ratio technique. SVM and decision tree classifiers came after it. From the study, choosing the best features using CFS and then putting them into practice using DFS, producing a hybrid feature selection technique, increases classification accuracy and helps in early disease diagnosis. These findings can be used to develop suitable medications that will relieve the symptoms of the patients.

A model for the detection of Postpartum Depression (PDD) in nursing mothers was proposed by Osubor and Egwali (2018) using a neuro fuzzy method. The study's data collection came from the Federal Neuro-Psychiatric Hospital in Uselu, Benin City, Edo State, Nigeria. The Postpartum Depression Screening Scale was used to develop a questionnaire with 35 questions to determine the likelihood that women will experience postpartum depression. Using Pearson product-moment correlation feature selection approaches, redundant features were removed from the model and only 6 features remained to be used as inputs into the ANFIS model to forecast PDD.

A neuro-fuzzy system based on fuzzy c-means was created for the diagnosis of diabetes by Othman and Yan in 2007. On a dataset of 6172 records, the FCM classification technique's efficacy and efficiency as a classifier in the neural fuzzy model ANFIS for the diagnosis of diabetes were compared to those of the conventional sub-clustering method. Weka was used to further enhance the diagnosis system, producing results that were more precise than those obtained using the traditional method.

Kour et al. (2019) examined the efficacy of subtractive clustering-based ANFIS hybrid system (SCANFIS) and fuzzy C-means (FCM)-based ANFIS system (FCMANFIS) in predicting Alzheimer's disease (AD). Three factors are used to compare the performance of these two algorithms: forecast accuracy, precision, and root mean square error (RMSE). In a forecast analysis of Alzheimer's disease, experimental data showed that the FCMANFIS model works better than the SCANFIS model.

By using a chaotic tent map for the initialization, El-Hasnony et al. (2020) modify the parameters of the adaptive neuro-fuzzy inference system (ANFIS). In order to achieve this, they make use of the advantages of the grey wolf optimization (GWO) and the particle swarm optimization (PSO). Fog processing at the edge of gateways gathers data, analyzes it, and instantly transmits it to the neighborhood. When compared to other optimization techniques, the proposed model outperformed them using a variety of evaluation metrics, including the root mean square error (RMSE), the mean square error (MSE), the standard deviation (SD), and the accuracy. These metrics included the mean square error (MSE), the root mean square error (DE), the genetic algorithm (GA), and the ant colony optimization (ACO). ANFIS + PSOGWO, which had an accuracy of 87.5% when used to predict Parkinson's disease, was also advised. PSO, GWO, GA, ACO, DE, and some novel literature were compared to PSO, GWO, and the proposed ANFIS + PSOGWO, which produced more encouraging results for Parkinson's disease prediction.

A neuro-fuzzy system (NFS) with feature selection is used in Lee's (2022) approach to detect Parkinson's disease. The five most precise reduced features from all 22 features were selected using neural networks with weighted fuzzy membership functions (NEWFMs), thereby confirming the effectiveness of the nonoverlapping region approach (NORM). By eliminating the weakest features, NORM can select the condensed traits that make up each interpretable fuzzy membership. The performance sensitivity, specificity, and accuracy of the 22 traits as an input to the NEWFMs were all 87.43%, 96.43%, and 88.72%, respectively. At least five NEWFM features had performance sensitivity, precision, and accuracy of 95.24%, 85.42%, and 92.8%, respectively.

To preprocess medical records and conduct disease prediction, Nagarajan and Thirunavukarasu (2022) suggest a neuro-fuzzy-based healthcare framework. The framework develops a multi-layered method for carrying out the task, including preprocessing of healthcare data, normalization through the fuzzification process, disease prediction by using the appropriate rules, and de-fuzzification of output values to obtain information pertaining to predicted disease. The fuzzy rule foundation is skilfully created to improve the decision-making process. The effectiveness of the suggested system is tested in an experimental setting and contrasted with fuzzy-based and language neuro-fuzzy models that include feature extraction. The accuracy value of the suggested neuro-fuzzy-based method is 87.7%, which is higher than that of the existing methods.

11.3 WORKING OF NEURO-FUZZY SYSTEM

The following layers in Figure 11.1 explain how the neuro-fuzzy system functions:

Layer 1: The incoming layer is layer 1. In this layer, every neuron transmits external crisp signals straight to the layer below.

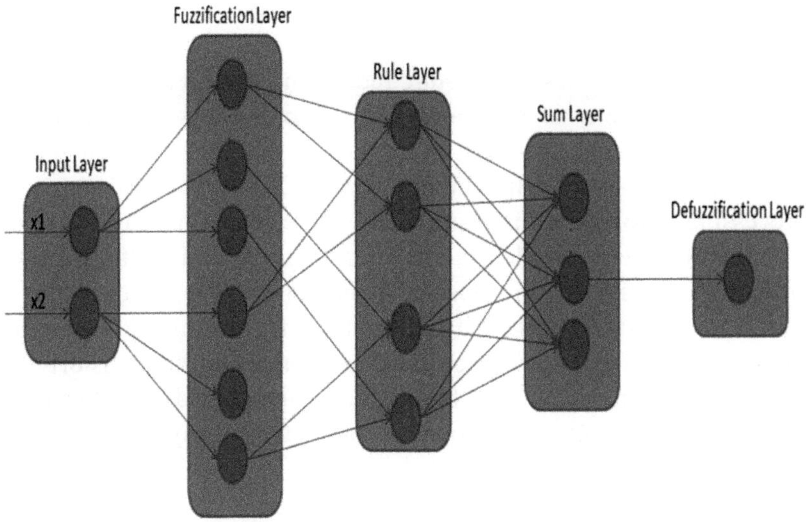

FIGURE 11.1 Block diagram of neuro-fuzzy system.

Layer 2: The second layer is the layer of fuzzification. Fuzzy sets used in the predecessors of fuzzy rules are represented by neurons in this layer. When a crisp input is given to a fuzzification neuron, it decides whether this input is part of the fuzzy set. The set to the function that identifies the imprecise set of the neuron is the activation function of the membership neuron.

Layer 3: The fuzzy rule layer is the third layer. In this layer, each neuron correlates to a singular fuzzy rule. Fuzzification neurons, which stand in for fuzzy sets in the rule antecedents, send information to a fuzzy rule neuron.

Layer 4: The layer responsible for output membership. In this stratum, neurons stand in for the fuzzy sets that fuzzy rules produce. An output membership neuron uses the fuzzy algorithm union to combine all its inputs.

Layer 5: Defuzzification is done at layer 5. Each neuron in this layer corresponds to a unique neuro-fuzzy system output. It creates a fuzzy system set by combining the result fuzzy sets that have been reduced by the reparative integrated firing strength. Neuro-fuzzy systems can use common defuzzification techniques, such as the centroid approach.

Being basically a multi-layer neural network, a neuro-fuzzy system can use learning algorithms created for neural networks, such as the back-propagation algorithm.

11.4 WORKING OF CLASSIFICATION SYSTEM

Data collection, preprocessing, feature reduction, classification, and prediction are the key stages in integrated soft computing-based techniques for diagnosing and decision-making. Raw data is gathered through the date collection process from different hospitals or medical data repositories. Preprocessing involves normalization and noise reduction. Scaling can be done in this stage as well. Medical data is ambiguous and multidimensional by design. A crucial stage is feature reduction, which removes the less crucial characteristic from the data. Algorithms for optimization are crucial in this stage. The high accuracy of the classification technique is driven by learning with better feature sets. In the end, an integrated classifier is used for categorization. Figure 11.2 shows the basic steps of any learning system.

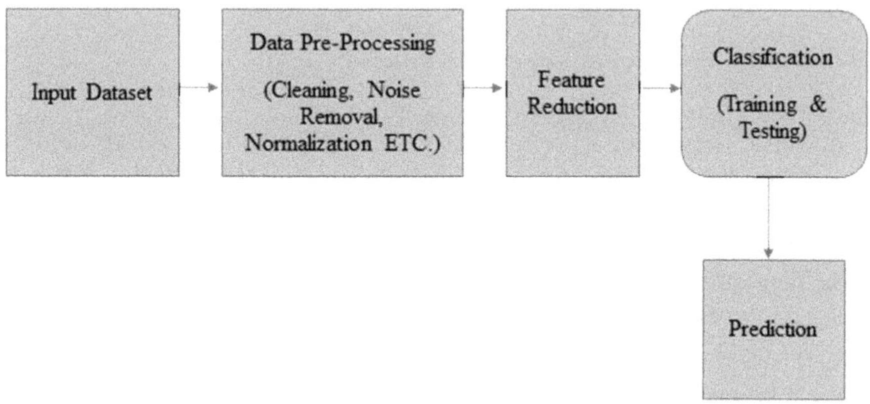

FIGURE 11.2 Basic steps of learning system.

11.4.1 PREPROCESSING

Data preprocessing is a crucial step in machine learning because the quality of the data and the information that can be extracted from it directly influence how well our model can learn. It is crucial that we preprocess our data before introducing it to any model. Few null numbers can ever be found in a real-world dataset. No model can handle these NULL or NaN values on its own, so we need to step in regardless of whether it is a regression, classification, or other types of issue. Imputation is just the replacement of our dataset's missing numbers. This preparation stage is essential in and of itself. When standardizing, we change our numbers so that their mean is 0 and their standard deviation is 1. Another crucial component of machine learning is handling categorical factors. Discrete and noncontinuous variables are what categorical variables are in general.

11.4.2 FEATURE REDUCTION

The process of reducing the number of features in a resource-intensive computation without losing crucial data is referred to as feature reduction, also known as dimensionality reduction. Reducing the number of features results in fewer variables, which makes the computer's job simpler and more rapid. Feature selection and feature extraction are the two steps that make up feature reduction. Feature minimization can be done using a variety of methods. The most well-liked ones include principal component analysis, generalized discriminant analysis, autoencoders, and non-negative matrix factorization.

The goal of feature reduction is to lessen the quantity of features (or factors) that the computer processes to carry out its function. Less resources are required to finish calculations or tasks because of feature reduction. The machine can perform more work because it requires less processing time and storage space. Feature reduction improves the machine learning model being used by removing multicollinearity during the learning process. Another advantage of feature reduction is that it makes data simpler for humans to visualize, especially when the data is reduced to two or three dimensions that can be graphically represented with ease. The burden of dimensionality is a fascinating issue that feature reduction can assist with.

11.4.3 CLASSIFICATION

Here ANFIS enters in the picture, as it can autonomously choose appropriate parameters for the membership functions in certain situations. When we already have a collection of input and associated output variables and values, for example, this is particularly true. The ANFIS system's nodes and connections between them can adjust "automatically," much like in an artificial neural network.

Certainly, it can be applied to the classification of diseases. Since the ANFIS concept combines an imprecise system with ANN design, in this situation, ANFIS has two primary advantages. First, you can use fuzzy variables, which are linguistically supported and suitable for disease signs that are frequently used as system inputs. ANFIS can perform training to produce results that are more precise because the design is mapped to ANN layers.

11.5 ISSUES AND CHALLENGES IN THE DESIGN OF LEARNING SYSTEM MODEL

The general issues that come in the design of any learning system are as follows:

i. Quality of data.
To reach the intended outcome, high data quality is more important than having a lot of data, though. Noisy, ambiguous, and uncorrelated data heavily affects classification results. In many cases, variations may be outside the expected range. Output wrongly fluctuates at various probabilities and predicts error-prone classification. Diagnostic reports and photographs are both included in medical data; hence, medical dataset quality must be guaranteed. It is quite difficult to apply this to integrated architecture, as well as to do sound research and make informed decisions.
ii. Ratio of training and testing data.
Proper distribution of data in teaching–learning procedure reflects an effective classification. The ratio 80:20, which designates 80% of the data as training and 20% as testing, is frequently employed. In practise, other ratios like 70:30, 60:40, and even 50:50 are also employed. High data ratio is assumed for effective training in deep learning. If handling for medical decision-making demands 100% accuracy, exercising for this is necessary. We occasionally have to build datasets according to our needs, while dealing with a small dataset that yields terrible results. Various approaches are to be involved in the reduction of learning overhead. Data ratio balancing is an acceptable task for significant installation prior to training or testing in an integrated environment.
iii. Selection of methods
Support vector machines, neural networks, logistic regression, KNN, k- means, and deep learning techniques are a few of the popular classification methods. Recognition, comprehension, the categorizing of concepts and things into predetermined groups or "sub-populations" are all parts of the classification process. The chance that new data will fall into one of the established categories is predicted by classification algorithms using input training data.

Emails that are classified as "spam" or "non-spam" are one of the most often used categorization techniques. Naive Bayes and SVM are used to classify documents, Naive Bayes and SVM are used to classify emails to identify whether they are spam, and Bag of Words and time series approaches are used to do text-based sentiment analysis. So it is very critical and challenging to choose and apply an appropriate method to execute classification.

iv. Curse of Dimensionality

Many aspects are consistently investigated in conjunction with observations in any integrated method. So, the appropriate concept must be used at the moment of feature extraction to fulfil this difficulty. Images have a significant part in clinical data, which is now the most fascinating field of study. Without neglecting the minute particulars of diagnostic samples, feature variety is an interesting but difficult topic of research.

If the data used to train the system contains noise and it is unclear, there must be chance that it causes overfitting or underfitting. Adoptable handling must be used rigorously to address these pressing problems. Learning systems have also some challenges that must be handled to build a better system. These challenges are as follows:

a. Handling with imbalanced dataset.
b. Dimensionality reduction.
c. Reducing learning overhead.

A dataset's dimensionality is defined as the quantity of input characteristics or factors. Techniques to lower the quantity of incoming factors in a dataset are known as dimensionality reduction. The curse of dimensionality, which is more widely known, describes how adding more input features frequently makes a predictive modeling job more difficult to model. Data illustration frequently makes use of high-dimensional statistics and dimensionality reduction methods. To reduce a classification or regression dataset, however, to better build a predictive model, these methods can be used in applied machine learning. In order to improve the classifier overhead, these issues and challenges must be addressed.

11.6 CONCLUSION

The study clearly indicates that the integrated method shows a better accuracy and overcomes the weakness of each other. It has both the learning ability of neural network and the explanation ability of fuzzy system. This combination works complementary to each other and overcomes the shortcoming of each other. Furthermore, various optimization algorithms specially based on nature-inspired approaches can also be combined with these approaches to build more accurate and reduce the computation time. Since neural networks can recognize patterns and link networks, the integrated methods function as a complement to one another. However, neural networks are not good at explaining how decisions are made. While fuzzy systems are excellent at making decisions (using reasoning like that of a human), they are unable to automatically acquire the rules required for making those decisions. GA can only be used to optimize searches effectively and to determine the best course of action. Based on the various nature or domain of the application area, the integration of soft computing techniques can be advantageous. Meta-heuristic optimization techniques founded on ideas borrowed from nature have recently gained popularity. These methods are also used in conjunction with one or more other methods to surmount the drawbacks of those other methods. Neuro-fuzzy methods for illness detection can be combined with these optimization techniques, and it will also lower classifier overhead.

KEYWORDS

- **artificial neural network**
- **fuzzy inference system**
- **genetic algorithm**
- **neuro fuzzy systems**
- **support vector machine**
- **K-nearest neighbor**
- **adaptive neuro-fuzzy inference system**

REFERENCES

Abiyev, R. H.; Abizade, S. Diagnosing Parkinson's Diseases Using Fuzzy Neural System. *Comput. Math. Methods Med.* **2016,** *2016* (3), 1–9.

Ashish, K.; Dasari, A.; Chattopadhyay, S.; Hui, N. B. Genetic-neuro-fuzzy system for grading depression. *Appl. Comput. Inform.* **2017,** *14* (1).

Chakravarty, S. A PSO based integrated functional link net and interval type-2 fuzzy logic system for predicting stock market indices. *Appl. Soft Comput.* **2012,** *12* (2), 931–941.

Chattopadhyay, S. Neurofuzzy Models to Automate the Grading of Old-Age Depression. *Expert Syst.* **2012,** *31* (1), 48–55.

Chattopadhyay, S. A Neuro-Fuzzy Approach for the Diagnosis of Depression. *Appl. Comput. Inform.* **2014,** *13* (1), 10–18.

Devi, S.; Kushwaha, G. S.; Kumar, S. In *An Adaptive Neuro Fuzzy Inference System for Prediction of Anxiety of Students*, 8th International Conference on Advanced Computational Intelligence Chiang Mai, Thailand, February 14–16, 2016.

Ekong, V. E.; Onibere, E. A. A Soft Computing Model for Depression Prediction. *Egypt Comput. Sci. J.* **2015,** *39* (4).

El-Hasnony, I. M.; Barakat, S. I.; Mostafa, R. R. Optimized ANFIS Model using Hybrid Metaheuristic Algorithms for Parkinson's Disease Prediction in IoT Environment. *IEEE Access* **2020,** *8*, 119252–119270.

Karaboga, D. An Adaptive and Hybrid Artificial Bee Colony Algorithm (aABC) for ANFIS Training. *Appl. Soft Comput.* **2016,** *49*, 423–436.

Kour, H.; Manhas, J.; Sharma, V. Brief Paper: Evaluation of Subtractive Clustering Based Adaptive Neuro-Fuzzy Inference System with Fuzzy c-Means Based ANFIS System in Diagnosis of Alzheimer. *J. Multimedia Inf. Syst.* **2019,** *6* (2), 87–90.

Lee, S. H. Minimized Feature Selection for Detection of Parkinson's Disease Using Neuro-Fuzzy System. *J. Mech. Med. Biol.* **2022,** *22* (03), 2240004.

Lin, C. The Design of Neuro-Fuzzy Networks using Particle Swarm Optimization and Recursive Singular Value Decomposition. *Neurocomputing* **2007,** *71*, 297–310.

Nagarajan, R.; Thirunavukarasu, R. A Neuro-Fuzzy Based Healthcare Framework for Disease Analysis and Prediction. *Multimed. Tools Appl.* **2022,** *81* (5), 1–17.

Osubor, V. I.; Egwali, A. O. A Neuro Fuzzy Approach for the Diagnosis of Postpartum Depression Disorder. *Iran J. Comput. Sci.* **2018,** *1* (4).

Othman, M. F.; Yau, T. M. In *Neuro Fuzzy Classification and Detection Technique for Bioinformatics Problems*, Proceedings of the IEEE First Asia International Conference on Modelling & Simulation, 2007.

Shihabudheen, K. V.; Pillai, G. N. Recent Advances in Neuro-Fuzzy System: A Survey. *Knowl. Based Syst.* **2018,** *152*, 136–162.

Tejeswinee, K.; Shomona, G. J.; Athilakshmi, R. Feature Selection Techniques for Prediction of Neuro-Degenerative Disorders a Case Study with Alzheimer's and Parkinson's Disease. *Procedia Comput. Sci.* **2017,** *115*, 188–194.

Yoon, H. Depression Diagnosis using Feature Selection of Neuro-Fuzzy System-Based Match Count Method. *J. Adv. Res. Dyn. Control Syst.* **2017,** 200–204.

Zhang, Z. X.; Tian, X. W.; Lim, J. S. In *New Algorithm for the Depression Diagnosis using HRV: A Neuro-Fuzzy Approach*, International Symposium on Bioelectronics and Bioinformations, 2011; pp 283–286.

CHAPTER 12

Image Description Based Medical Image Retrieval

PREM SHANKER YADAV[1], HIMANSHU KUMAR SHUKLA[2], CHANDRABHAN SINGH[2], SHUDHA SHUKLA[2], and UPENDRA KUMAR[3]

[1]*Department of Computer Science and Engineering, Malaviya National Institute of Technology, Jaipur, Rajasthan, India*

[2]*Department of CSE, F.O.E.T., University of Lucknow, Lucknow, India*

[3]*Department of CSE, IET, Lucknow, India*

ABSTRACT

Recent developments in medical technology have led to a massive increase in the availability of medical diagnostic photographs. Designing a precise retrieval model to return the most relevant photos in response to the query is, therefore, a crucial area of urgent study. The complex content of medical pictures, however, cannot be sufficiently described by the visual descriptors. A novel image retrieval model based on picture captions is created to address this issue. The objective is to produce textual visuals for clearly identifying the relationship between things with captions. Here, the pre-training deep learning models are used to achieve picture captioning. Text length of caption, frequency-inverse document frequency (TF-IDF), and Bag of Words are hybrid to get the best match between picture captions and the query text. Bidirectional Long Short-Term Memory (Bi-LSTM), Convolutional Neural Network, Recurrent Neural Network (CNN+RNN), and Deep Hierarchical Encoder-Decoder Network (DHEDN) are used for image captioning on dataset created from Kvasir (a multi-class image dataset for gastrointestinal disease detection). Experiments show that DHEDN is outperforming compared to Bi-LSTM and CNN+RNN.

Smart Electronic Devices and Systems for Biomedical and Healthcare Applications. Upendra Kumar, Rajesh K. Kesharwani, & Raj K. Keservani (Ed.)

DOI: 10.1201/9781003560852-12

12.1 INTRODUCTION

The use of image captioning benefits self-driving cars retrieves specific photos, and helps the blind. It is obvious that automating the process of turning an image into text reduces the workload associated with captioning. To get the necessary photos, typical image retrieval systems rely significantly on manual tagging. Academics have paid close attention to classic large-scale image retrieval methodologies. These operations are carried out with the help of hashing learning methods.

Medical image retrieval is a critical component of modern medicine. Advanced medical imaging technologies have enabled healthcare providers to collect extremely detailed pictures of many regions of the human body to help in illness diagnosis, treatment planning, and monitoring. Medical image retrieval helps healthcare practitioners to swiftly and efficiently retrieve a large number of medical pictures, allowing them to make more accurate and fast diagnoses. Medical image retrieval is critical in the creation of a therapy plan. Doctors can choose the best course of action to pursue by analyzing a patient's medical photos. Medical image retrieval allows healthcare providers to track illness development and therapy efficacy. Doctors can decide if a specific treatment is effective or if it needs to be amended by comparing current medical imaging to earlier ones. Medical researchers may examine and comprehend a variety of medical disorders with the help of medical image retrieval. Researchers may analyze and detect patterns and trends in vast databases of medical pictures, allowing them to design novel treatments and cures. Medical image retrieval is crucial for advancing medical research, better patient outcomes, and raising the standard of healthcare.

In content-based image retrieval (CBIR), a lot of work is being done to retrieve similar images for queries. CBIR relies on two steps: the first is image description, which is the same as the query image and evaluates similarity among the features, and the second is image retrieval, which is the same as the query image and evaluates similarity among the features. For the purpose of determining the semantic picture contents, the conventional content-based technique relies on handcrafted qualities. To minimize feature-searching levels, classic algorithms take handcrafted features as input and map the high-dimensional feature to a lower-dimensional feature (Zhang et al., 2017). The goal was to extract useful material from huge volumes of accessible data in an efficient manner. Because retrieval efficiency is based on representative features, substantial work was previously devoted to mining strong feature models. The spatial resolution and quality of the medical pictures have significantly improved. With the increased availability of large photo collections in recent years, medical captioning is becoming useful for managing vast image libraries by giving relevant captions. It also discovers applications for material retrieval (Srivastava and Srivastava, 2018). The objective is to generate and use textual descriptions based on picture captions in the context of medical images. The pre-training approaches are used to encode visual characteristics and decode them with a text dataset to generate the caption. Captions are used to extract features like as TF-IDF, Bag of Words, and text length. Meanwhile, comparable characteristics are mined using the query text.

The rest of the section is organized as follows: Section 12.2 discusses previously developed image retrieval methods. Section 12.3 presents the dataset and experimental result. Section 12.4 discusses the challenges and Section 12.5 concludes and future scope.

12.2 PREVIOUS WORK

Image captioning can be categorized into three types: template-based, encoder-decoder, and retrieval based. Fixed templates with numerous empty spaces are used to generate captions in template-based systems. Template-based medical captioning has also been explored by Onita et al. (2020) and Wang et al. (2021). To execute more accurate reporting, Ayesha et al. (2021) and Beddiar et al. (2022) used grammar literature and language rules. Li et al. (2018) gather phrases referring to detected objects, attributes, and their linkages in order to map efficient caption. Ayesha et al. (2021) and Beddiar et al. (2022) adopted grammar literature and language rules to perform more accurate reporting.

Retrieval concept based on the approach of the same images will be the same descriptions. For query image, a set of similar images are identified from the huge database. Ayesha et al. (2021) discussed two viewpoints to generate new caption for novel image. First, map the most similar candidate description to the same description. Later, exploit the candidate description and merge them based on pre-specified rules and logic to generate a description for the novel image. Similarity comparison plays a golden factor in the retrieval approach. Finding comparable photos in big datasets or archives and creating a new sentence. The most similar sentence is used in place of the new description, or the retrieved sentences are combined to generate an entirely new sentence. Various distance matrices like Hamming distance, Hellinger, kendell tau, and Euclidean may be used as per requirement. Wang et al. (2019) applied Lucene Image Retrieval (LIRE) approach and used a classifier along with a topic modeling technique to retrieve medical-based images, making the clusters of various classes of images, the content of the images, and formulating the captions. The suggested method could be enhanced to extract the input image's most pertinent captions from captions of photos in its class.

Syedo-Mohmood et al. (2020) proposed a domain-centric retrieval system for achieving similar chest X-ray radiology report document from the available huge dataset. Feature extraction is the main phase as it leads to fine detailed descriptions of the scene. Labeling is properly integrated into features and carried out using concept detection. In the proposed algorithm, a group of phrases is critical factor, responsible for effective labeling and conversion into textual descriptions of the scene. The final report is retrieved using the labeling of similar images from the archive dataset. Charalampakos et al. (2021) proposed a model integrated with classical model to consolidate retrieval approach effective. The author found that KNN and the previous work (Kougia et al., 2021) impose a more realistic impact on the retrieval concept. Additionally, when creating new words, the authors used a visual attention method (Pavlopoulos et al., 2021) to concentrate on certain vectors of the input text encoding.

To retrieve medical investigation is also very critical involvement for better health care system. Numerous supervised learning algorithms appear to be applicable to the classification problem. Unfortunately, it ran into major issues. The limitation is that it only contains a few annotated examples. The labels are accompanied by questions and relevant remarks. However, LSTM, Bi-LSTM (Bidirectional LSTM), and GRU (Gated recurrent units), among others, may be utilized on the decoder side. Deep learning requires the construction of convolution, pooling, and flattening phases. This is the most often used attentive notion in picture-to-text conversion since deep learning is now a solid performer. These methods include encoder-decoder topologies, fully linked networks, and CNNs. A deep-learning model, developed by J. Pavlopoulos et al. (2021), Vinyals et al. (2015), is primarily to blame for the strategies now being used in the literature.

Poor classification may arise due to a lack of training data. Another problem is learning dimensionality, because high-dimensional data can be complicated, requiring selection and dimensionality reduction (Dharani et al., 2013). A report was produced by Alsharid et al. (2020) utilizing deep learning. RNNs are used to translate newly produced words into understandable English after gathering traits from images. Attention-based encoder-decoder architectural model was addressed (Yin et al., 2019; Tsuneda et al., 2021; Zeng et al., 2020) to discriminate between normal and pathological results in generated medical reports. A two-LSTM model was suggested by Harzig et al. (2020) to explain both positive and negative captioning values. An attention-based network was developed by Yang et al. in 2021 to identify and describe important fine qualities in ultrasound medical pictures.

A multi-task approach is provided by Chelaramani S. et al. (2020) to recognize and diagnose fundus-based illnesses. In order to extract features from photos and feed them to an LSTM in order to provide medical descriptions, Zeng et al. (2020) utilized an encoder, then focused on features to do so. Hoxha et al. (2020) developed a hybrid image retrieval model that incorporates a convolutional neural network with a recurrent neural network (CNN+RNN). The similarity of vectors with the query picture was created for image retrieval. It causes problems when modeling the primitives present in photos.

In B. Wang et al. (2019), mapped texts and images are created in the semantic space, and distance measure is used to discover the similarity between words and images. In general, retrieval-based strategies are used to create sentences that are exact in terms of grammaticality and syntax. The representation of pictures is described for scalable image searching and retrieval. In Zhang et al. (2018), a deep hashing neural network that utilized the cross entropy loss is employed to generate binary codes, while a metric learning-based deep hashing network that leveraged the triplet loss function is used.

Ye et al. (2018) developed and applied a retrieval approach based on a weighted distance measure to visual features provided by CNN. The re-ranking strategy is developed, which recognizes each RS image using CNN features and then modifies image-to-class distance metrics to address retrieval issues (F. Ye et al., 2019). Deep neural network advancements (Wang et al., 2016) have resulted in key efficiency in CBIR. The deep model-based models optimized both image feature learning and retrieval.

To model the primitives, Cornia et al. (2020) developed an unsupervised technique for picture retrieval in RS pictures. Visual Big Transfer (BiT) Models were used to represent the picture, and text descriptions were encoded using a Bi-LSTM network. The approach has a high level of computational complexity. Iyer et al. (2019) created VGG16 CNN to retrieve images based on image captioning while taking query images into account to reduce complexity. To index the photographs available from the server, the elastic search methodology was employed, and for searching and retrieval, an intermediate captioning method was applied. However, this method was unable to discover latent traits. In order to reveal latent features, Xiao et al. (2019) developed a DHEDN based on picture captioning for image retrieval. In this, the deep hierarchical model was employed to divide the functions of the decoder and encoder. The deep network modeling capabilities of the model were used efficiently for semantic fusion. It did not, however, evoke additional inputs such as features or visual attention in DHEDN. To add more qualities, Wang et al. (2018) developed Bidirectional LSTM (Bi-LSTM) for image retrieval based on picture captioning. The following figure shows the structure of image retrieval using image captioning.

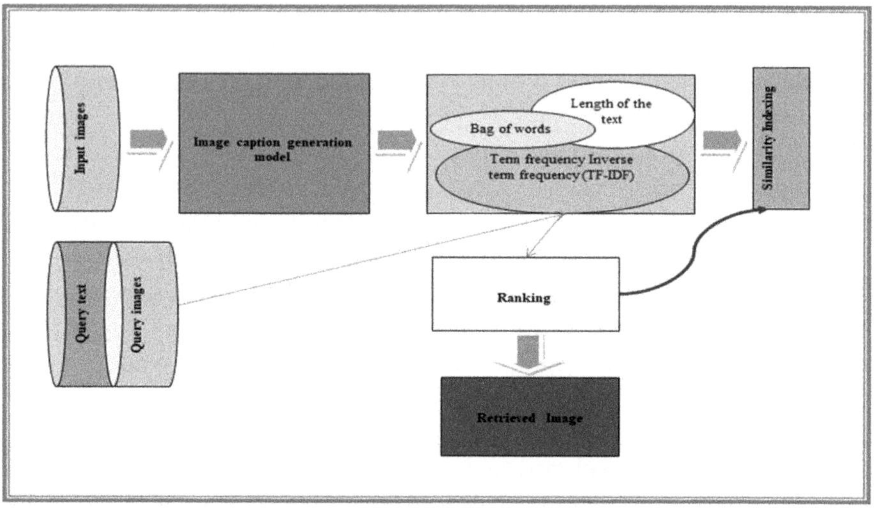

FIGURE 12.1 Structure of Image captioning-based Image retrieval.

12.3 IMAGE RETRIEVAL STRUCTURE FOR MEDICAL IMAGES

The image caption is generated using the encoder-decoder architecture of CNN and LSTM.

This model is useful for performing fine-tuning models. Our image retrieval model architecture is presented here, consisting of caption encoder-decoder for generating captions, using TF-IDF, text length, and word of bags features for indexing, ranking is performed by calculating the similarity between query and archive images using metrics Hellinger distance, Kendall tau and Hamming distance.

To develop image descriptions, we have employed CNN+RNN, Bi-LSTM, and DHEDN. Once the image caption is created, properties such as TF-IDF, text length, and Bag of Words are extracted from the captions in the input picture. Following that, the picture is indexed using the obtained characteristics. TF-IDF, text length, and Bag of Words are just a few of the characteristics that are extracted from the query text on the query side. Finally, feature vectors from the text query are taken into consideration while doing similarity matching utilizing hybrid similarity measures including Kendall tau, Hamming distance, and Hellinger distance.

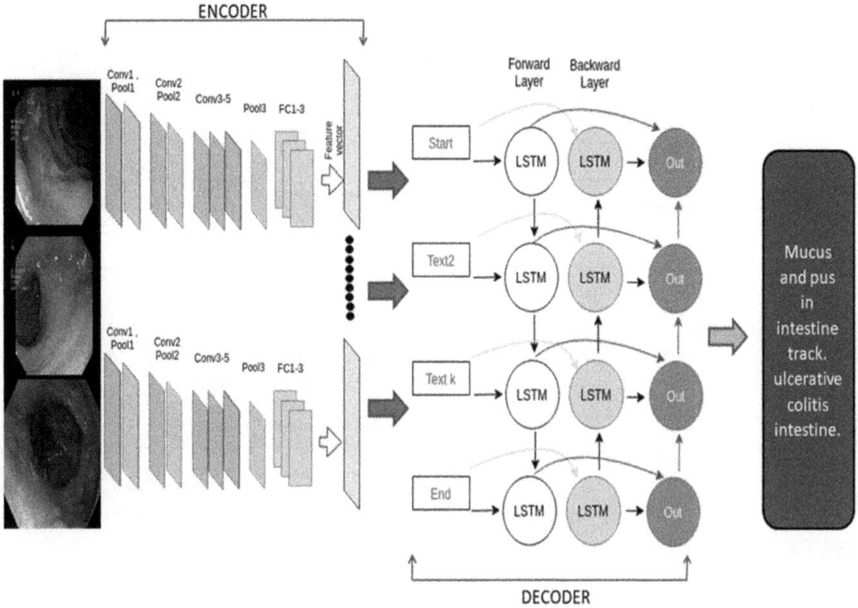

FIGURE 12.2 Encoder-decoder architecture.

12.3.1 COLLECTING FEATURES

After generating the caption we are calculating features to proper indexing. These features are the length of text, Bag of Words, and TF-IDF.

12.3.1.1 TF-IDF

The TF-IDF (Liu et al., 2018) is used to gauge the significance of strings in data, such as words and phrases. It also aids in determining how relevant a word is to the sequence under consideration. In statistics, a phrase's frequency tells us how frequently it happens.

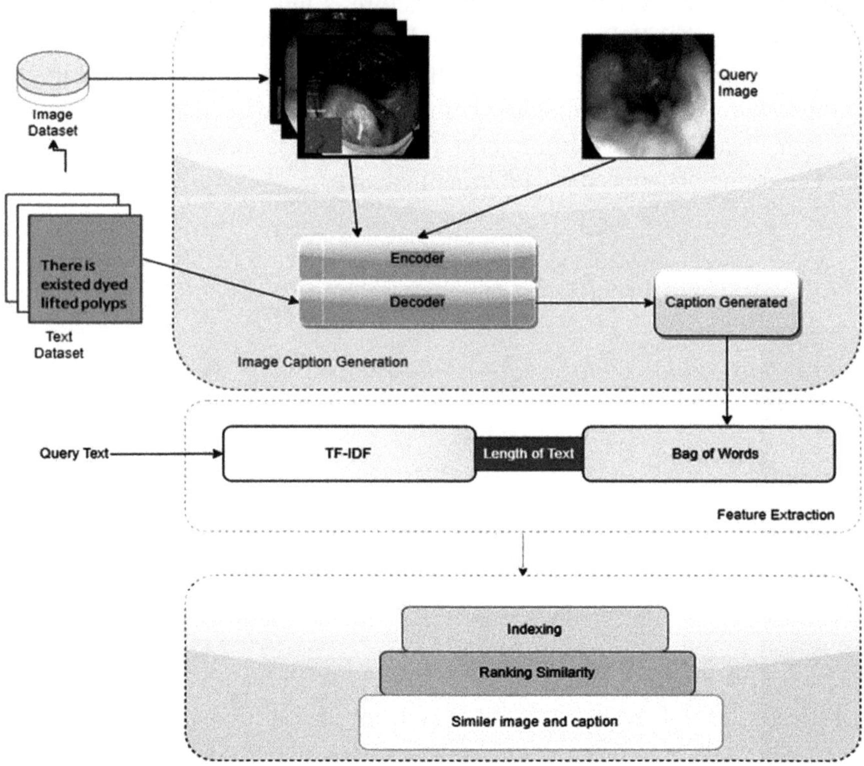

FIGURE 12.3 Architecture for image retrieval used in this work.

12.3.1.2 LENGTH OF THE TEXT

It refers to the total number of characters in an image caption.

12.3.1.3 BAG OF WORDS

The Bag of terms denotes text modeling that defines the recurrence of particular terms in an image caption.

12.3.2 INDEXING OF IMAGES

When photos are retrieved using mining characteristics, image indexing reduces the search space. The fundamental strategy is to use the qualities as indexes and to build a data structure to accelerate indexing. The index enables the computer to find photos relevant to a query without having to scan the entire database.

12.3.3 HYBRID SIMILARITY MEASURE

The recovered features are compared with the indexed picture and the query while taking into account the Hellinger, Kendal, and Hamming distances using the hybrid similarity measure.

12.4 DATASET

The automatic identification of illnesses through the use of computers is a significant yet uncharted area of research. Such advancements could enhance medical procedures and global healthcare systems. Kvasir contains photographs of the gastrointestinal (GI) tract. Three key anatomical landmarks and three clinically relevant findings are divided into the collection of photos. Two types of photos pertaining to the excision of endoscopic polyps are also included. The dataset is sorted and annotated by medical professionals. The dataset consists of hundreds of photos for each class that illustrate anatomical landmarks, pathological findings, or endoscopic operations in the GI tract. There are enough photos to employ for a variety of applications, including image retrieval, machine learning, deep learning, transfer learning, etc. The clinical findings include esophagitis, polyps, ulcerative colitis, etc. whereas the anatomical markers include Z-line, pylorus, cecum, etc. The excision of lesions is also covered in a number of sets of photos, such as "dyed and lifted polyp" and "dyed resection margins." The photos in the dataset range in resolution from 720×576 to 1920×1072 pixels, and they are arranged in the dataset in discrete folders with names that are appropriate for the content, with each folder including a group of photographs. The position and configuration of the endoscope inside the intestine are shown in some of the image classes that are presented as a green picture in picture using an electromagnetic imaging system. Some of the recognizable features within the GI tract are shown as:

a) The Z-line designates the point where the esophagus and stomach meet. The border between the white esophageal mucosa and the red stomach mucosa may be seen endoscopically as a distinct line. To ascertain whether or not illness is present, the Z-line must be identified and evaluated.

b) The region around the passage from the stomach into the first portion of the small bowel is known as the pylorus. Circumferential muscles around the entrance control how food leaves the stomach.
c) The largest bowel's cecum is its closest section. The appendiceal orifice may be seen as a crescent-shaped slit in the image on the left, and the scope configuration for the appendiceal opposition can be seen in the green image inside the image.

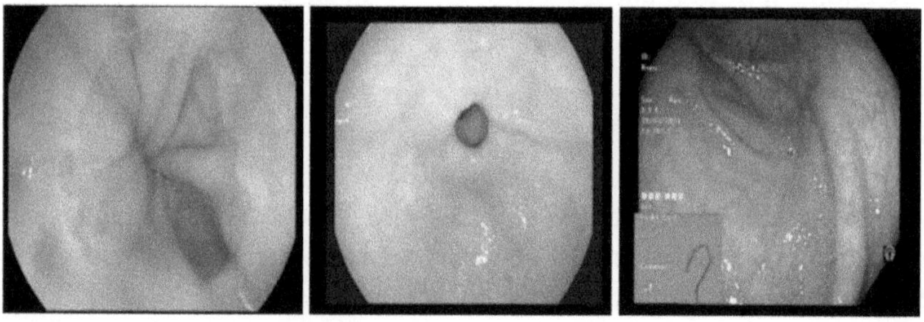

FIGURE 12.4 The recognizable features within the GI tract are (a) Z-line, (b) pylorus, and (c) cecum.

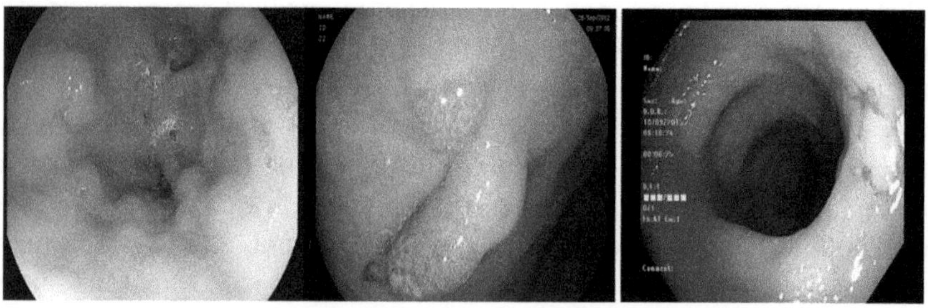

FIGURE 12.5 (a) Esophagitis (b) Polyps and (c) Ulcerative colitis. (a) Esophagitis is an esophageal Figure 12.4. Figure 12.5. Abnormal features (a) Esophagitis (b) Polyps and c) Ulcerative colitis (5a) Esophagitis is an esophageal inflammation that manifests as a Z-line-related rupture in the esophageal mucosa. (5.b) Bowel polyps are lesions that can be seen as mucosal outgrows. The color and surface pattern of the polyps can be used to identify them from normal mucosa. They can be flat, raised, or pedunculated. However, certain intestinal polyps have the potential to develop into cancer. Therefore, it is crucial to find and remove polyps in order to stop the growth of colorectal cancer. The image's green boxes provide a visual representation of the endoscope arrangement. (5.c) The big bowel is affected by the chronic inflammatory condition known as ulcerative colitis. The diagnosis is mostly based on colonoscopic findings, and the condition may have a significant impact on quality of life. There are four levels of inflammation: none, mild, moderate, and severe, each having a unique endoscopic profile.

In our work, we have collected in ratio of 30 out of 100 images from various labels of Kvasir datasets. A two- to three-sentence text explanation that carefully considers medical sense is created for each image.

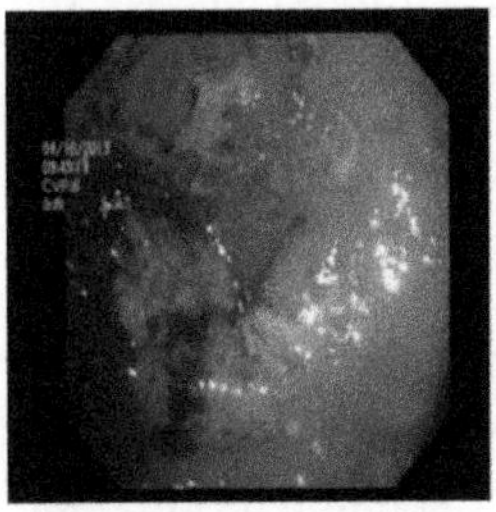

Z-line is not clearly visible. It is damaged as white mucus and pus cells are covering densely. There is clear ulcerative-colitis in intestine

FIGURE 12.6 Image of ulcerative colitis in intestine.

12.5 RESULT ANALYSIS

The finest analytical metrics for result analysis are precision, recall, and F1-score.

12.5.1 PRECISION

It refers to proximity of several data amidst each other, and is modeled as,

$$P_r = \frac{y_p}{y_p + v_p}$$

where, y_p is true positive, and v_p express false positive.

12.5.2 RECALL

It evaluates actual positives in which technique attains true positive label and formulated as,

$$R_l = \frac{y_p}{y_p + n_f}$$

where, n_f refers to false negative.

12.5.3 F1-SCORE

The accuracy of a model is assessed using the F1 score, a statistic for machine learning evaluation. Scores for a model's recall and accuracy are included. The accuracy statistic shows the percentage of times a model across the whole dataset made an accurate prediction.

$$F_1 = 2 * \frac{precision * recall}{precision + recall}$$

Precision, recall, and F1-score are calculated for CNN+RNN, Bi-LSTM, and DHEDN. In the first experiment, we applied CNN for encoding and LSTM for decoding. For the next experiment, Bi-LSTM is taken as encoder. Finally, a deep hierarchal encoder and decoder are applied. We have used VGG16 for our setup and performed training for 50 epochs. Experiment results are tabulated as given.

TABLE 12.1 Precision, Recall, and F1-score for Various Deep Learning Models.

Datasets	Metrics	CNN+RNN	Bi-LSTM	DHEDN
Kvasir-based dataset	Precision	0.853	0.886	0.904
	Recall	0.831	0.873	0.902
	F-measure	0.842	0.879	0.903

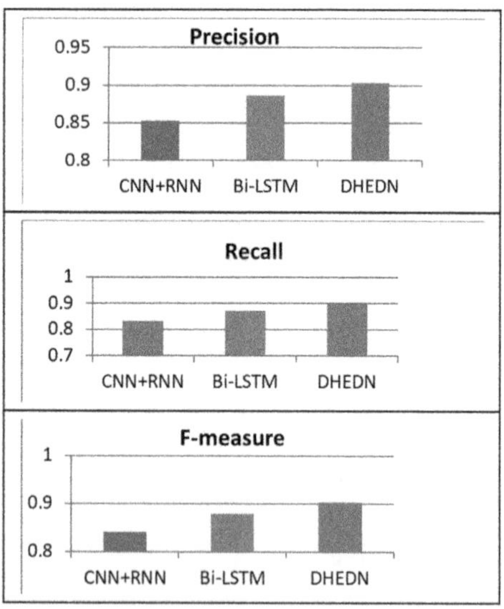

FIGURE 12.7 Precision, recall, and F-measure of various architectures

12.5.4 OUTPUT

Some of the outputs are presented here in various cases. In this, retrieved images are presented with text output for a given query.

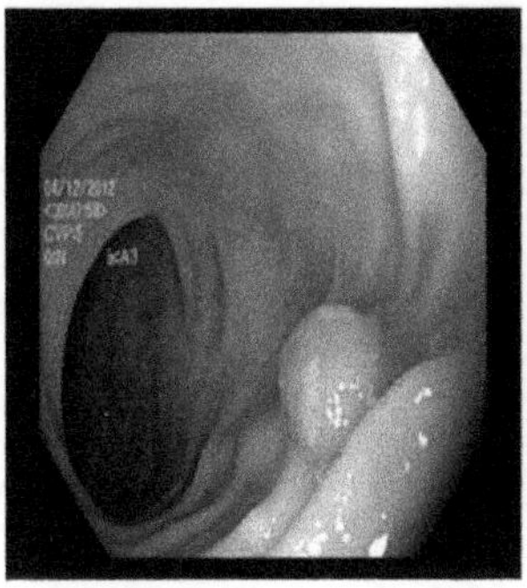

Query: Intestine with clear edges. A polyp is clearly visible. Intestine is damaged

Retrieved: Intestine with distinct with polyp. Intestine is not proper.

FIGURE 12.8(a) Retrieved image and text for given query related to polyp.

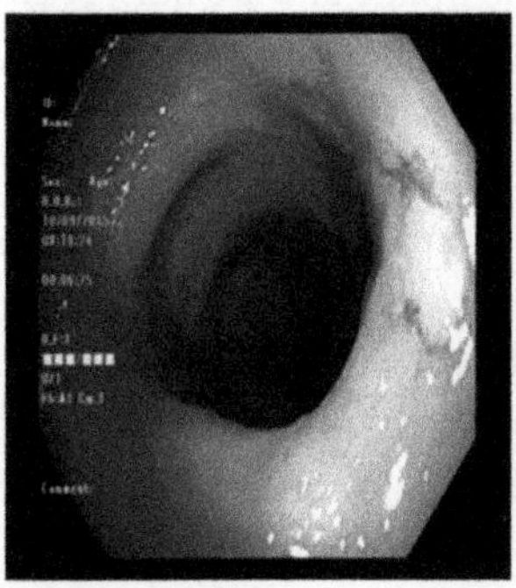

Query: ulcerative colitis with bleeding, swelling and ulceration of the mucosa

Retrieved: swelled and bleeding part is seen. Colitis is found.

FIGURE 12.8(b) Retrieved image and text for given query related to colitis.

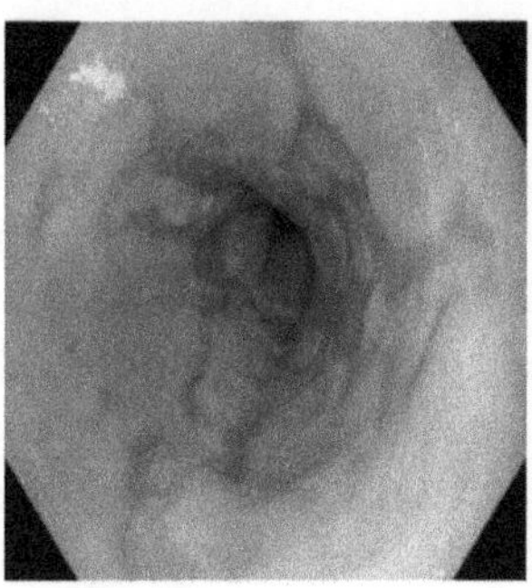

Query: red mucosal tongues projecting up in the white esophageal lining,

Retrieved: inflammation of the esophagus visible in the esophageal mucosa.

FIGURE 12.8(c) Retrieved image and text for given query related to esophagus.

12.6 LIMITATIONS

Although numerous approaches for describing medico-images have been offered in the literature, they are still limited by a number of difficulties. Some of these restrictions are listed below:

- Because a limited dataset is used for analysis in this case, a larger dataset is necessary for a more efficient outcome.
- One text caption phrase is written in the gathered dataset to explain the picture content or image scenario. Several description reports may result in improved retrieval; therefore, instead of one, several text caption phrases for each image may be used.
- Because no medico-expert was engaged in the dataset's gathering, the inclusion of a medico-expert in the dataset's development may vary the final results slightly.
- Although just one sentence is used per image to characterize disease behavior, there is still the issue of incoherent or long comments that do not sufficiently express the content of the image and are not yet clinically suitable since they may be extremely deceiving (Harzig et al., 2020).
- Although existing models perform better, human assessment is still crucial (Pavlopoulos et al., 2021). We require human experts to check the results, but obtaining such assessors remains challenging and expensive.
- Model and disease generalization is still very difficult and challenging.
- While it is beneficial to use pre-defined and trained models to handle the problem of little data, there is a large amount of domain transfer from classical natural photography to medical content images, which may result in various degrees of error.

12.7 CONCLUSION AND FUTURE SCOPE

The medical picture is obtained here by generating image captions. The goal of this technology is to effectively retrieve photos by merging natural language processing with medical diagnostics. The use of created captions to describe picture information allows for the expression of complex medical image material. The medical picture is obtained from the image captions generated here. The goal of this technology is to retrieve pictures effectively by merging natural language processing with medical diagnostics. The use of created captions to describe picture content allows for the expression of sophisticated medical image material. Features such as TF-IDF, Bag of Words, and length of text are derived from image captions and query strings. The query text features and picture caption features are utilized for similarity matching, which is done using a hybrid similarity measure. The image index is used for picture retrieval and indexing. Future studies will include a discussion of other databases to test the model's viability. Other deep-learning models might potentially be used to create better captions.

KEYWORDS

- **bidirectional long short-term memory**
- **convolutional neural network**
- **recurrent neural network**
- **deep hierarchical encoder-decoder network**
- **gated recurrent units**
- **bag of words**

REFERENCES

Al Rahhal, M. M.; Bazi, Y.; Abdullah, T.; Mekhalfi, M. L.; Zuair, M. Deep Unsupervised Embedding for Remote Sensing Image Retrieval Using Textual Cues. *Appl. Sci.* **2020,** 10 (24), 8931.

Allaouzi, I.; Ben Ahmed, M.; Benamrou, B. In *Automatic Caption Generation for Medical Images*, Proceedings of the 3rd International Conference on Smart City Applications (SCA'18), 2018.

Alsharid, M.; El-Bouri, R; Sharma. A Curriculum Learning Based Approach to Captioning Ultrasound Images. In *Medical Ultrasound, and Preterm, Perinatal and Paediatric Image Analysis*; 2020, p 12437.

Askari, Q.; Younas, I.; Saeed, M. Political Optimizer: A Novel Socio-inspired Meta-heuristic for Global Optimization. *Knowl. Based Syst.* **2020,** *195,* 105709.

Aswiga, R. V.; Shanthi, A. P. A Multilevel Transfer Learning Technique and LSTM Framework for Generating Medical Captions for Limited CT and DBT Images. *J. Digit. Imaging* **2022,** *35* (3), 564–580.

Ayesha, H.; Iqbal, S.; Tariq, M. Automatic Medical Image Interpretation: State of the Art and Future Directions. In *Pattern Recognition*; 2021; p 107856.

Banerjee, S.; Lavie, A. In *Meteor: An Automatic Metric for MT Evaluation with Improved Correlation with Human Judgments*, Proceedings of the ACL Workshop on Intrinsic and Extrinsic Evaluation Measures for Machine Translation and/or Summarization, 2005; pp 65–72.

Chaudhuri, U.; Banerjee, B.; Bhattacharya, A. Siamese Graph Convolutional Network for Content Based Remote Sensing Image Retrieval. *Comput. Vis. Image Understand.* **2019,** *184,* 22–30.

Cornia, M.; Baraldi, L.; Tavakoli, H. R.; Cucchiara, R. A Unified Cycle-Consistent Neural Model for Text and Image Retrieval. *Multimed. Tools Appl.* **2020,** *79* (35), 25697–25721.

Dharani, T.; Aroquiaraj, I. L. A Survey on Content Based Image Retrieval. In *Pattern Recognition, Informatics and Mobile Engineering*; 2013; pp 485–490.

Hashim, F. A.; Houssein, E. H.; Hussain, K.; Mabrouk, M. S.; Al-Atabany, W. Honey Badger Algorithm: New Metaheuristic Algorithm for Solving Optimization Problems, *Math. Comput. Simulat.* **2022,** *192,* 84–110.

Hoxha, G.; Melgani, F.; Demir, B. Toward Remote Sensing Image Retrieval Under a Deep Image Captioning Perspective. *IEEE J. Sel. Top. Appl. Earth Obs. Remote Sens.* **2020,** *13,* 4462–4475.

Irvin, J.; Rajpurkar, P.; Ko, M.; et al. In *Chexpert: a large Large-scale Remote Sensing Image Retrieval by Deep Hashing Neural Networks Expert Comparison*, Proceedings of the AAAI Conference on Artifcial Intelligence, 2019; pp 590–597.

Iyer, S.; Chaturvedi, S.; Dash, T. Image Captioning-Based Image Search Engine: An Alternative to Retrieval by Metadata. In *Soft Computing for Problem Solving*; 2019; pp 181–191.

Li, C. Y.; Liang, X.; Hu, Z.; Xing, E. P. In *Knowledge-Driven Encode, Retrieve, Paraphrase for Medical Image Report Generation*, Proceedings of the AAAI Conference on Artificial Intelligence, Atlanta, GA, USA, 8–12 October 2019; Volume 33, pp 6666–6673.

Li, C.; Liang, X.; Hu, Z.; Xing, E.; Li, C. Y.; Liang, X.; Hu, Z.; Xing, E. P. In *Hybrid Retrieval-Generation Reinforced Agent for Medical Image Report Generation. In Advances in Neural Information Processing Systems 31 (NIPS 2018)*, Proceedings of the Annual Conference on Neural Information Processing Systems 2018, NeurIPS 2018, Montreal, QC, Canada, 3–8 December 2018; Curran Associates, Inc.: New York, NY, USA, 2018; Volume 31.

Li, Y.; Zhang, Y.; Huang, X.; Zhu, H.; Ma, J. Large-scale Remote Sensing Image Retrieval by Deep Hashing Neural Networks. *IEEE Transac. Geosci. Remote Sens.* **2017,** *56* (2), 950–965.

Lin, C. Y. Rouge: a Package for Automatic Evaluation of Summaries. In *Text Summarization Branches Out*, 2004; pp 74–81.

Liu, Q.; Wang, J.; Zhang, D.; Yang, Y.; Wang, N. In *Text Features Extraction Based on TF-IDF Associating Semantic*, IEEE 4th International Conference on Computer and Communications (ICCC), December 2018; pp 2338–2343.

Park, H.; Kim, K.; Yoon, J. In *Feature Difference Makes Sense: A Medical Image Captioning model Exploiting Feature Difference and tag information*, 58TH Annual Meeting of the Association for Computational Linguistics (ACL 2020): Student Research Workshop, pp. 95–102.

Pavlopoulos, J; Kougia, V.; Androutsopoulos, I. Diagnostic Captioning: A Survey. arXiv preprint arXiv:2101.07299, 2021.

Srivastava, G.; Srivastava, R. In *A Survey on Automatic Image Captioning*, International Conference on Mathematics and Computing, 2018; pp 74–83.

Syeda-Mahmood, T.; Wong, K.; Gur, Y. In *Chest X-ray Report Generation Through Fne-grained Label Learning*, 23rd International Conference on Medical Image Computing and Computer-Assisted Intervention, MICCAI 2020; vol. 12262, pp 561–571.

Tomar, N. K.; Debesh, J.; Riegler, M. A.; Johansen, H. D.; Johansen, D.; Rittscher, J.; Halvorsen, P.; Ali, S. FANet: A Feedback Attention Network for Improved Biomedical Image Segmentation. *IEEE Transac. Neural Networks and Learning Syst.* **2022.**

Wang, B.; Lu, X.; Zheng, X.; Li, X. Semantic Descriptions of Highresolution Remote Sensing Images. *IEEE Geosci. Remote Sens. Lett.* **2019,** *16* (8), 1274–1278.

Wang, C.; Yang, H.; Meinel, C. Image Captioning with Deep Bidirectional LSTMs and Multi-task Learning. In *ACM Transactions on Multimedia Computing, Communications, and Applications*; 2018; pp 1–20.

Wang, X.; Zhang, Y.; Guo, Z.; Li, J. A Computational Framework Towards Medical Image Explanation. In *Artificial Intelligence in Medicine: Knowledge Representation and Transparent and Explainable Systems*; Springer: Cham, Switzerland, 2019; Vol 11979, pp. 120–131.

Xiao, X.; Wang, L.; Ding, K.; Xiang, S.; Pan, C. Deep Hierarchical Encoder–Decoder Network for Image Captioning. *IEEE Trans. Multimedia* **2019,** *21* (11), 2942–2956.

Xu, N.; Liu, A. A.; Liu, J.; Nie, W.; Su, Y. Scene Graph Captioner: Image Captioning based on Structural Visual Representation. *J. Visual Commun. Image Represent.* **2019,** *58,* 477–485.

Xu, N.; Zhang, H.; Liu, A. A.; Nie, W.; Su, Y.; Nie, J.; Zhang, Y. Multi-level Policy and Reward-based Deep Reinforcement Learning Framework for Image Captioning. *IEEE Transac. Multimedia* **2019,** *22* (5), 1372–1383.

Ye, F.; Dong, M.; Luo, W.; Chen, X.; Min, W. A New Re-ranking Method Based on Convolutional Neural Network and Two Image-to-Classdistances for Remote Sensing Image Retrieval. *IEEE Access* **2019,** *7,* 141498–141507.

Ye, F.; Xiao, H.; Zhao, X.; Dong, M.; Luo, W.; Min, W. Remote Sensing Image Retrieval Using Convolutional Neural Network Features and Weighted Distance. *IEEE Geosci. Remote Sens. Lett.* **2018,** *15* (10), 1535–1539.

Zhou, L.; Palangi, H.; Zhang, L.; Hu, H.; Corso, J. J.; Gao, J. In *Unified Vision-Language Pre-Training for Image Captioning and VQA*, In Proceedings of the AAAI Conference on Artificial Intelligence, vol. 34, no. 07, pp 13041–13049, April 2020.

Zhou, W.; Newsam, S.; Li, C.; Shao, Z. PatternNet: A Benchmark Dataset for Performance Evaluation of Remote Sensing Image Retrieval. *ISPRS J. Photogramm. Remote Sens.* **2018,** *145,* 197–209.

CHAPTER 13

Deep Learning, Extreme Learning Machine, and Kernel-Based Elm for the Diagnosis of Alzheimer's Disease

ASHUTOSH MISHRA

Computer Science and Engineering Department, Thapar Institute of Engineering and Technology, Patiala, India

ABSTRACT

The diagnosis of Alzheimer's disease (AD) has recently become a priority in the area of biomedical research which has led to advancements in classification techniques for the diagnosis of diseases. A series of three different cases consisting of classification models including deep learning, extreme learning machine (ELM), and kernel-based extreme learning machine (KELM) is proposed in this chapter to classify subjects as infected or fit using AD Neuroimaging Initiative (ADNI*) data. The three cases are shuffling of presence or absence of Principal Component Analysis (PCA), and k-fold cross-validation in our operation carried out for the diagnosis. Then, a comparative study of accuracy and error obtained by these methods is also done. For further establishing the correctness of this work, the classification models are also applied to three individual files of the dataset (set1, set2, and set3). The dataset is taken by random selection of 30,600 records for validating the homogeneity of available data and evaluating the performance of these computing models for different sets.

Smart Electronic Devices and Systems for Biomedical and Healthcare Applications. Upendra Kumar, Rajesh K. Kesharwani, & Raj K. Keservani (Ed.)

DOI: 10.1201/9781003560852-13

13.1 INTRODUCTION

The term AD was coined after a German psychiatrist Alois Alzheimer who first identified the case in the year 1901 in a woman named Auguste D (Alzheimer, 1907, 1987). AD is a fatal neurodegenerative disease that causes deterioration of mental capacity and memory and has caused 60–70% dementia cases (Salman and Mordechai, 2017; Sarvabowma, 2015). According to World Health Organization (WHO), nearly 47 million people have AD or AD-related dementia worldwide which is expected to increase to 75 million by 2030. Common symptoms of AD include short-term memory loss, disorientation, problems with language, loss of interest, self-care and motivation, mood swings, and withdrawal from family and society (Salman and Mordechai, 2017; Sarvabowma, 2015). Severe AD patients cannot even recognize their own family members and so become incapable of caring for themselves (Weiner et al., 2012). This disease gradually leads to losing bodily functions (Sarvabowma, 2015) and ultimately leads to death (Salman and Mordechai, 2017). The exact or actual cause of AD is not known for a large number of cases, but a few are exceptions where genetic differences have been detected Reitz and Mayeux, 2014). Also, there are several hypotheses that explain the cause of AD among which Cholinergic, Tau, and Amyloid are the major ones.

The classification of cases of AD has been a well-looked-after problem for a while because of the difficulty in diagnosing the early stages of AD and the lack of any particular effective prevention measures.

The Cerebral Blood Flow (CBF) values from Single Photon Emission Computed Tomography (SPECT) that support ANN training are proposed (Liu et al., 2014) as a tool to aid in the diagnosis of AD. To avoid having to start the learning phase from scratch each time new input-output correlations are introduced, an incremental learning technique is described. We propose a two-stage process: first, using the patient's CBF values from the lab, offline learn the global features that are constant regardless of implementation; and second, in situ learn the features specific to each location.

Multiple sclerosis (MS) and AD are complex neurodegenerative diseases that might benefit greatly from combining protein-protein interaction research with gene expression studies, as is done by Goñi et al., 2008.

With this study, the researchers aimed to determine if the degree and betweenness parameters, two essential measurements in network theory, are characteristics that distinguish implicated (seed-proteins) and non-implicated (neighbors) nodes in MS and AD. Their research reveals that proteins whose gene expression is disrupted in neurodegenerative disorders have varied qualities of centrality.

Integrating disparate data sets using a kernel technique for AD prediction is proposed (Ye et al., 2008). To better identify characteristics (biomarkers) from disparate data sets, the kernel structure was further developed. Sixty-nine healthy controls and sixty-nine people diagnosed with AD each had their Magnetic Resonance Imaging (MRI) data analyzed using the proposed method. Tensor factorization is used for MRI data pre-processing. In this research, voxel-based data and MRI ROI data are both considered independent data sources, and an approach is proposed to combine the two sets of data. The experimental findings demonstrate that by merging data from several sources, prediction accuracy is much improved.

Akgül et al., 2009 present a learning framework relying on the nearest neighbor (NN) approach for diagnosing AD at an early stage from Magnetic Resonance Imaging (MRI) using visual similarity and user feedback in 2009. The dataset comprised an MRI report of 175 elderly patients from the OASIS neuroimaging dataset. The proposed similarity learning-based NN approach is effective and has a high accuracy of 92% in AD prediction even with global visual descriptors with frail indications of the presence of AD.

There is new interest in using graph theory for neuroscientific data. Using graph theoretical analysis of resting-state electroencephalogram (EEG) recordings, de Haan et al., 2009 examine the topological alterations in patients with AD and frontotemporal lobar degeneration (FTLD) who use functional brain imaging. Twenty patients with mild to moderate AD, fifteen patients with FTLD, and twenty-three controls all had resting-state EEGs recorded with their eyes closed. The large-scale functional brain network architecture in AD deviates from the ideal small-world network structure in favor of a more random form, as measured by a reduction in both local and global connection metrics. This lends credence to the disconnection hypothesis of AD because there is a correlation between this and inefficient communication between different parts of the brain. Curiously, FTLD patients display alterations toward an overly organized network structure, the reverse of what is seen in AD patients, which may indicate a distinct underlying pathophysiological process.

Plant et al., 2011 introduced a bootstrapping-based feature extraction technique for identifying AD at an early stage using resting-state functional resonance images in 2011. The proposed method used SVM on the extracted features and showed an accuracy of 79%.

Zou et al., 2012 propose a new strategy for the genome-wide association (GWA) analysis of AD that combines random forests with enrichment analysis. Firstly, the random forest was used for performing backward feature selection on a GWA dataset of AD patients carrying the apolipoprotein gene (APOEε4). Among the 1058 risky single nucleotide polymorphisms (SNPs) identified, several are already known to be related to AD. Then, the investigation of the susceptible SNPs was done by enrichment analysis. Significantly associated gene functional annotations, such as alternative splicing, glycoprotein, and neuron development, were successfully discovered which indicated that these biological mechanisms play important roles in the development of AD in APOEε4 carriers.

Mo et al., 2013 propose an approach to classify subjects using an ensemble of classifiers. The classifiers were trained using ADNI data and validated using 13-fold validation. The performance was enhanced using the PCA augmentation approach. The accuracy of classification in this paper was 89%.

Multiple types of data, including neuroimaging data, gene/protein expression data, genetic information, etc. are collected and analyzed together to improve the predictive power of AD research. The biggest obstacle is figuring out how to efficiently combine data from various disparate sources. In a single cohesive form, the models presented by Xiang et al., 2013 deal with feature-level and source-level analysis. By not directly imputing the missing values, this incomplete data model achieves better results. Effective optimization strategies for both the full and the missing models are also provided.

An automated image processing-based strategy is proposed (Farhan et al., 2014) for detecting AD using brain MRI scans. The proposed method in this research is unusual since it achieves a high accuracy value of 93.75 while using a more condensed feature set than is typical.

Decades of research have established neuroimaging as a promising approach for the early detection of AD and its precursor, Mild Cognitive Impairment (MCI).

The primary topics are feature representation and merging of multimodal information from MRI and PET scans (Suk et al., 2014). (PET). In this work, we propose a deep learning-based approach to extracting latent and shared features from neuroimaging modalities at a high level. Experiments on the ADNI dataset, as well as comparisons to state-of-the-art methodologies, confirmed its accuracy. Maximum accuracies of 95.35 percent, 85.7 percent, and 74.58 percent were achieved, respectively, in three binary classification problems consisting of AD versus healthy Normal Control (NC), MCI versus NC, and MCI converter versus MCI non-converter, thus outperforming the competing methods.

To quantify the AD progression of participants, Jedynak et al., 2015 verify an AD progression score model that employs numerous biomarkers. There is a sigmoidal relationship between disease progression and all biomarkers, subjects' ages at the disease beginning, and rates of disease progression, and thirdly, all patients experience a unique disease progression.

Using a pattern-classification lens, Zheng et al., 2016 provide a review of the research on automated dementia identification techniques. In this research, we examine four distinct types of classifiers—linear discriminant analysis, Bayesian classifiers, support vector machines, and artificial neural networks—and three classes of feature extraction methods—voxel-, vertex-, and ROI-based approaches. A comparison of these two sets of results demonstrates that many algorithms can distinguish between AD and MCI with satisfactory accuracy.

The ADNI data population includes 317 females and 342 males and is classified by a total of 243 biological and clinical characteristics (Gamberger et al., 2016). To help find groups of patients that share similar clinical and molecular characteristics, researchers used a novel approach to clustering.

In Krashenyi et al., 2016, automatic diagnosis of AD using MRI and PET image features is achieved through the use of a fuzzy inference classification system. The best classification result, at 94% accuracy, was achieved by combining data from 7 MRI areas and 39 PET regions.

A handheld point-of-care (PoC) device for AD diagnosis is being developed, and (Jammeh et al., 2016) aims to determine the optimal number of blood biomarkers (with respect to some biomarkers and sensitivity/specificity) for such a device. Blood biomarkers for AD were found using the ADNI database. Six different biomarkers (A1Micro, A2Macro, AAT, ApoE, complement C3, and PPP) were found. When age was included as a covariate, it was able to distinguish between AD patients and normal people with a sensitivity of 85.4% and a specificity of 78.6%.

Classifier systems developed using Machine Learning (ML) methods can aid physicians in the early diagnosis and prediction of disease, which is crucial in reducing healthcare costs. Because of their effectiveness and low cost in resolving bioinformatics issues, these technologies may be employed for analyzing massive amounts of complex biological data.

It is a rapidly expanding topic that is becoming increasingly important in the medical community for teaching professionals how to recognize and categorize illnesses. Diagnostics in the medical field increasingly rely on classification schemes. Big data have become a viable analytics tool because of technological advancements that make its recording and storage very painless (Gamberger et al., 2016). There are already proposals in the literature for deterministic ML algorithms that can be applied to clinical data. While most of these approaches get decent predictive accuracy, comparing them is tricky because they are tested on distinct pathologically unverified data sets.

In our work, we have employed three classification models which include deep learning, ELM, and KELM for the diagnosis of AD. Genetic data of 30,600 individuals from ADNI were used to check their efficiency for this purpose. Their comparison was done using four types of error and five types of accuracy measures. After this simple evaluation, the process was again repeated thrice on the same sample by taking into account PCA and k-fold cross-validation to further prov the correctness of our method. As our sample was made using three different files of genetic data, there was a need for validation of these files from which we took 30,600 random individuals' information for given features, this was done using the Kruskal–Wallis test.

The remaining sections of this chapter are laid out as follows: In Section 13.2, we describe the existing work in illness diagnostics using ML and deep learning approaches. Section 13.3 talks about the problem statement. The dataset, its features, and its approach are briefly described in Section 13.4. In Section 3.5, we provide an overall summary of the ML approaches that were employed in this study. Section 13.6 explains the proposed approach. Section 13.7 discusses evaluation parameters and analysis, while Section 13.8 wraps up the chapter.

13.2 COMPUTING MODELS

13.2.1 DEEP LEARNING

Recently, deep learning has emerged as a popular technique for modeling data's extremely non-linear forms in ML. Data representations at different levels of abstraction can be learned by computational models with numerous processing layers (LeCun et al., 2015; LeCun, 2015). Deep neural network (DNN) is a type of artificial neural network having the input and output layer units with multiple hidden layers between them (Bengio, 2009; Schmidhuber, 2015). The extra layers help in the modeling of complex data with fewer units as compared to a shallow network with similar performance (Bengio, 2009) by enabling layer composition which is a hierarchical notion wherein one layer uses the output of a lower layer as input to a higher layer.

Deep learning can be both supervised as well as unsupervised. Multilayer Perceptron Networks, Convolutional Neural Networks, and Long Short-Term

Memory Recurrent Neural Networks are the most popular techniques for deep learning. In 1965, Ivakhnenko and Lapa published the first working learning algorithm for supervised deep feedforward multilayer perceptrons called the Group Method of Data Handling (GMDH) (Ivakhnenko et al., 1965; Ivakhnenko and Lapa, 1967; Ivakhnenko, 1968). In 1980, Kunihiko Fukushima introduced Neocognitron, deep learning working architectures built from artificial neural networks (Fukushima and Miyake, 1982)

In our work, we have applied a convolutional neural network (CNN) for deep learning. Convolutional networks, a type of multilayer perceptron, are optimized to require as little pre-processing as possible (LeCun et al., 2015; LeCun, 2015). CNN is a sort of feedforward artificial neural network with connections between neurons modeled after the structure of the visual cortex in animals. In CNN, individual cortical neurons respond to inputs within a very small area called a receptive field. Different neurons' receptive fields partially overlap, creating a tiling effect over the visual field. CNN is also called shift or space invariant artificial neural network (SIANN), based on its translation invariance characteristics and shared weights architecture. The response of a single neuron to an external stimulus can be approximated mathematically via a convolution operation.

To take advantage of spatially local correlations, CNNs impose a local connection pattern between neurons of neighboring layers, with the inputs of hidden units in layer l coming from a subset of units in layer l-1 with spatially contiguous receptive fields.

Graphically, a basic architecture of CNN is illustrated in Figure 13.1. It depicts a four-layer CNN with two inputs, three groups are there in the first hidden layer, the second hidden layer is divided into two groups, and the final output is a single real number.

In CNNs, each filter is repeated across the entire image. Each of these duplicated units contributes to the feature map with the same weight vector and bias. Applying a function several times to different regions produces a feature map, or it can be said, by convolution of the input with a linear filter, then adding a bias term before applying a non-linear function.

To obtain the k-th feature map at a given layer, let us call it h^k, whose filters are set by the bias b^k and weights W^k, then the feature map is h^k obtained as:

$$h_{ij}^k = \tanh\left((W^k * x)_{ij} + b_k\right) \tag{13.1}$$

Each hidden layer is constructed from several feature maps to get a more complete picture of the data $\{h^{(k)}, k = 0 \ldots K\}$. All possible permutations of the source feature map, destination feature map, vertical position, and horizontal position are represented in the weights W of a hidden layer, which is a

4D tensor. Each feature map at the final destination contributes one element to the vector that represents the biases b. In a network having a cost function $J(W, b; x, y)$, where (W, b) are the parameters and (x, y) are the training data and label pairs, the error term for the l+1th layer would be $\delta^{(l+1)}$. If layer l is strongly connected to layer l+1, then the layer l's error can be calculated as

$$\delta^{(l)} = \left((W^{(l)})^T \delta^{(l+1)}\right) * f'(z^{(l)}) \tag{13.2}$$

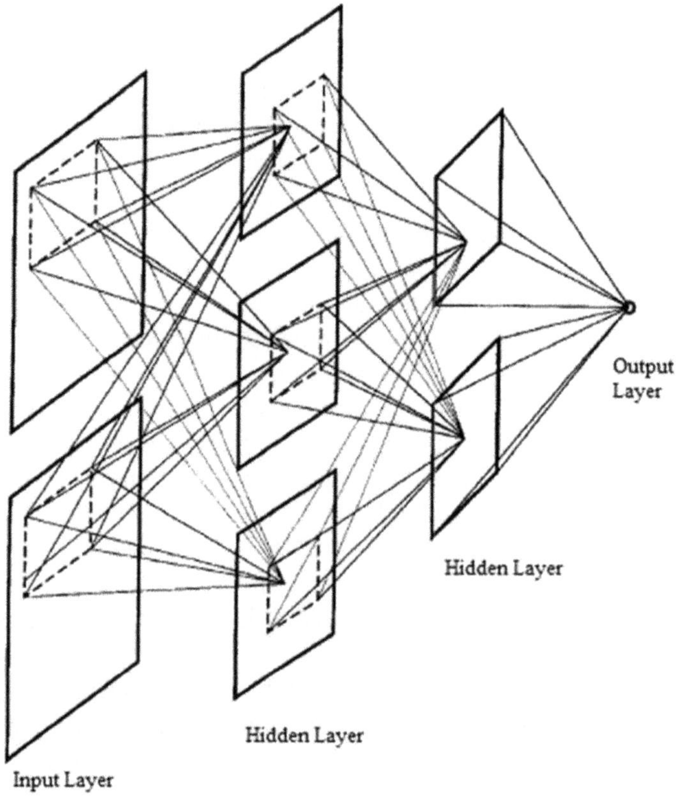

FIGURE 13.1 An illustration for CNN.

If the lth layer is a subsampling convolutional layer, the error will be passed down the stack as

$$\delta_k^{(l)} = upsample\left((W_k^{(l)})^T \delta_k^{l+1}\right) * f'(z_k^{(l)}) \tag{13.3}$$

Here, $f'(z_k^{(l)})$ is the derivative of the activation function, and k is the index of the number of filters to use. The upsampling process is responsible for spreading the error across the pooling layer by determining the error relative to each unit feeding into the pool.

13.2.2 ELM

Unlike traditional gradient-based learning algorithms (like backpropagation), the ELM algorithm for training SLFN (Single Hidden Layer Feedforward Networks) does not necessitate iterative tuning or setting parameters like learning rate, momentum, etc. As the "hidden neurons," ELM supports a wide variety of non-linear piecewise continuous functions (Li et al., 2014). Therefore, for N arbitrary distinct samples $\{(x_i, t_i) \mid x_i \in R^n, t_i \in R^m, i = 1,..., N\}$ the output function in ELM with L hidden neurons can be written as:

$$f_L(x) = \sum_{i=1}^{L} \beta_i h_i(x) = h(x) \tag{13.4}$$

The data are transformed from input space to ELM feature space as follows: (Huang et al., 2004, 2006, 2012) where $\beta = [\beta_1, \beta_2, \ldots, \beta_L]$ is the vector of the output weights between the hidden layer of L neurons and the output neuron, and $h(x) = [h_1(x), h_2(x), \ldots h_L(x)]$ is the output vector of the hidden layer with respect to the input i.

In ELM, both the training error and the output weights must be minimized simultaneously in order to reduce the training error and improve the generalization performance of neural networks.

$$\text{Minimize}: \|H\beta - T\|, \|\beta\| \tag{13.5}$$

The least-square solution of the above equation can be written as:

$$\beta = H^T \left(\frac{1}{C} + HH^T\right)^{-1} T \tag{13.6}$$

where, H, C, and T are the hidden layer output matrix, regulation coefficient, and sample expected output matrix, respectively.

The output function of ELM is:

$$f(x) = h(x)H^T \left(\frac{1}{C} + HH^T\right)^{-1} T \tag{13.7}$$

13.2.3 KELM

The kernel implementation of ELM is said as KELM. If the feature mapping $h(x)$is not known and based on Mercer's conditions, the kernel matrix of ELM can be defined as:

$$M = HH^T : m_{ij} = h(x_i)h(x_j) = k(x_i, x_j) \tag{13.8}$$

KELM's output function is:

$$f(x) = [k(x, x_1), \ldots, k(x, x_N)](\frac{1}{C} + HH^T)^{-1} T \tag{13.9}$$

where, $M = HH^T$ and $k(x, y)$ are single hidden layer feedforward neural network kernel functions.

Many kernel functions such as linear kernel, polynomial kernel, Gaussian kernel, and exponential kernel can satisfy the Mercer condition. In this paper, the Gaussian kernel is used for the performance analysis which is as given below:

$$k(x, y) = \exp(-a\|x - y\|) \tag{13.10}$$

13.3 PRELIMINARIES

13.3.1 DATASET DESCRIPTION

The source of data used for initial evaluation is the second release of ADNI array data. A total of 30,600 records were selected at random from the entire set ensuring that no chosen record had any missing or unsupported values. Appendix A below explains the attributes of the data that we have used, out of which Sample Index, Sample ID, SNP Index, and SNP Name are omitted due to them being irrelevant for evaluation. The rest of the data was homogenized.

13.3.2 METHODOLOGY OF PERFORMANCE MEASURE

13.3.2.1 ACCURACY PARAMETERS

AUC: The term "area under the curve," or "AUC," is an acronym. In classification analysis, it is used to compare different models and choose the one that provides the most accurate predictions.

Sensitivity: Sensitivity is the rate at which positives are accurately classified as such.

$$Sensitivity = \frac{number\ of\ truepositives}{number\ of\ truepositives + number\ of\ falsenegatives} \tag{13.11}$$

Specificity: The percentage of false negatives that are accurately identified as false negatives is the measure of specificity.

$$Specificity = \frac{number\ of\ truenegatives}{number\ of\ truenegatives + number\ of\ falsepositives} \quad (13.12)$$

13.3.2.2 ERROR PARAMETERS

Mean Absolute Error (MAE): The mean absolute error is a statistic for gauging how accurately predictions are made. Mathematically, the MAE is given as

$$MAE = \frac{1}{n}\sum_{i=1}^{n} | a_i - p_i | = \frac{1}{n}\sum_{i=1}^{n} | e_i | \quad (13.13)$$

Where, absolute error = $|e_i| = |a_i - p_i|$, actual = a_i and predicted = p_i

Relative Absolute Error: It is nothing but the total absolute error, normalized by dividing by the total absolute error of the simple predictor.

$$RAE = \frac{\sum_{i=1}^{n} | a_i - p_i |}{\sum_{i=1}^{n} | p_i - \overline{a} |} \quad (13.14)$$

Where, $\overline{y} = \frac{1}{n}\sum_{i=1}^{n} a_i$

Root Mean Square Error (RMSE): RMSE is the measure of differences between values that are predicted by a model and the values observed. The mathematical representation of RMSE is given as:

$$RMSE = \sqrt{\frac{\sum_{i=1}^{n}(p_i - a_i)^2}{n}} \quad (13.15)$$

Root Mean Squared Logarithmic Error (RMSLE): The root mean squared logarithmic error is similar to RMSE but instead, logis taken of the prediction and actual values.

$$RMLSE = \sqrt{\frac{1}{n}\sum_{i=1}^{n} (\log(p_i + 1) - \log(a_i + 1))^2} \quad (13.16)$$

Root Relative Squared Error (RRSE): The RRSE is the total squared error normalized by dividing by the total squared error of the simple predictor.

$$RMSE = \sqrt{\frac{\sum_{i=1}^{n}(a_i - p_i)^2}{\sum_{i=1}^{n}(p_i - \overline{a})^2}} \quad (13.17)$$

13.3.2.3 VALIDATION TESTS

Kruskal–Wallis Test: Whether or not there is a statistically significant difference between two or more groups of an independent variable on a continuous or ordinal dependent variable can be determined with the help of the Kruskal–Wallis test, which is a rank-based non-parametric test (Mckight and Najab, 2010). It is a non-parametric alternative to one-way analysis of variance (ANOVA), and it is an extension of the Mann–Whitney U test that allows for the comparison of more than two distinct groups.

Dunn's Multiple Comparison Test: Following the failure of a Kruskal–Wallis test, Olive Dunn (1964) developed a post hoc pairwise multiple comparisons approach. Dunn's test, as described by (Dinno, 2016), is a non-parametric pairwise multiple comparisons approach based on rank sums. As a result, it is a non-parametric alternative to multiple pairwise tests once the null hypothesis of an ANOVA has been rejected.

13.3.3 IMPLEMENTATION

The dataset containing 30,600 records was split into a ratio of 2:1 for training and testing data, that is, 20,400 records were used to train the models, and 10,200 records were used for purpose of testing. The computation was done for three different cases. First, the models were trained and tested using the above-described datasets. Second, the models were trained but then subjected to 10-fold cross-validation as well. Third, PCA is applied to the dataset at a pre-processing stage reducing the number of attributes from 20 to 6. Then this pre-processed data was used for training and 10-fold cross-validation. The process is clearly shown in Figure 13.2.

The computing models that are used for classification purpose include deep learning, ELM, and KELM. The evaluations were implemented on a 64-bit system with 4GB RAM and a Windows7 operating system using R Studio. The implementation of these models is given below:

13.3.3.1 DEEP LEARNING

In this study, we employ MXNetR, an R package that provides cutting-edge deep learning capabilities along with lightning-fast GPU processing. The procedure is outlined in detail below:

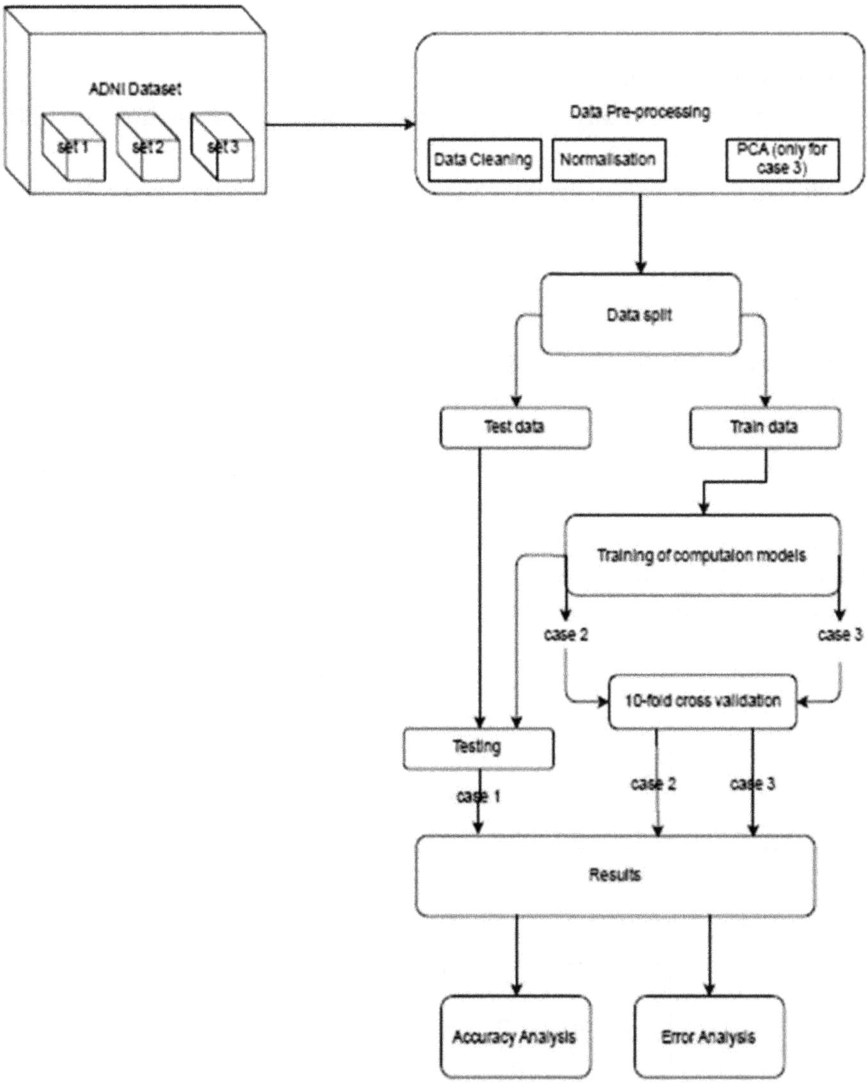

FIGURE 13.2 Flowchart for evaluation of ADNI data.

Step 1. To ensure compatibility with mxnet and the R standard, it is necessary to transpose the input matrix to column-major format.

Step 2. After the system has the data, the next action is to set up the framework for the network.

Step 3. In mxnet, its own data type symbols are used to configure the network. data <-mx.symbol. Variable ("data") uses data to represent the input data, or it can be said as the input layer. The argument is the name of the result symbol which is of data type string.

Step 4. Then the first hidden layer is set by:

fc1 <-mx.symbol.FullyConnected (data, name="fc1", num_hidden=128).

Input data, the layer's label, and the number of hidden neurons are the parameters of this layer. This function can also have arrays of weight and bias, or a no_bias of type boolean (which is false by default) as other parameters.

The activation is set by:

act1 <-mx.symbol.Activation (fc1, name="relu1", act_type="relu").

The activation function takes the output from fc1, the first hidden layer.

Step 5. Using the output of act1 as its input, the second hidden layer (labeled "fc2") consists of 64 hidden neurons. The second activation is very similar to the first (act1), with the exception of the input source and the name.

Step 6. Now, the output layer. Since there are only two classes (whether the person has AD or not), the number of neurons is set to 2.

Step 7. In order to arrive at a probabilistic forecast, we switch the activation to softmax in the end.

Figure 13.3 shows the implementation steps for deep learning using CNN.

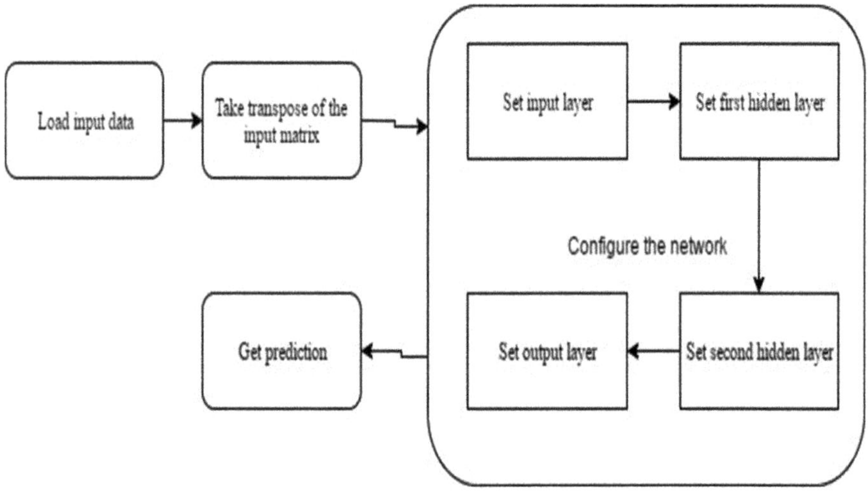

FIGURE 13.3 Flow diagram showing implementation of deep learning.

13.3.3.2 ELM

A package in R, called elmNNis used for the evaluation of ELM in RStudio.

Training of an SLFN with ELM is a three-step learning model:

Step 1. Assign randomly hidden node parameters (a_i, b_i), $i = 1,..., L$. This indicates that both the hidden layer bias and the arc weights connecting the input and hidden layers are produced at random.

Step 2. Use any of the accessible activation functions to derive H.

Step 3. Calculate the output weights β: β = ginv(H)*T (matrix multiplication), where ginv(H) is the Moore-Penrose generalized inverse of hidden layer output matrix H.

Once the SLFN has been trained, the output of a generic test set is simply Y = H*B (matrix multiplication) [11]. Salient features:

- ELM learns at a lightning-fast rate.
- ELM applies to any bounded non-constant piecewise continuous activation functions, whereas conventional gradient-based learning algorithms are limited to differentiable activation functions.
- ELM has a tendency to arrive at the answers directly without such trifling difficulties as local minima, incorrect learning rate, overfitting, etc. which plague conventional gradient-based learning methods.
- Compared to other well-known learning algorithms, including neural networks and support vector machines, the ELM learning process appears to have fewer moving parts.

As described in Gosso and Martinez-de-Pison, 2012 to fit a neural network, the function to use is elmtrain (default version is elmtrain.formula). It begins by randomly generating input weights and a bias for the hidden layer and then determines the hidden layer's output based on a specified activation function. Finally, it determines the neural network's output weights. The trained neural network is represented by an ELM model (an object of class elmNN) that is returned by the elmtrain. The parameters used were x for training dataset, *y* for labels of the training dataset, nhid for number of hidden neurons, actfun for type of activation function, and seed for setting the number of seeds. n.blockswas also used for 10-fold cross-validation which was set to 10 in our case.

To predict values, the function to use is predicted. It takes as parameters a trained neural network and a new dataset to calculate the output from the model.

The training and prediction functions rely on other functions for internal use.

13.3.3.3 KELM

The same package and functions which were used by ELM are used for KELM in R.A parameter in elmtrain, 'kernel' was introduced and set to

Guassian kernel and the same process is repeated. The performances of these methods were analyzed using accuracies and errors obtained for all three cases. Further, the computing models were applied to the individual data-sets which were merged to make the final dataset used above. Each of the three sets—set 1, set 2, and set 3, went through the same three-step process described above and the results were analyzed.

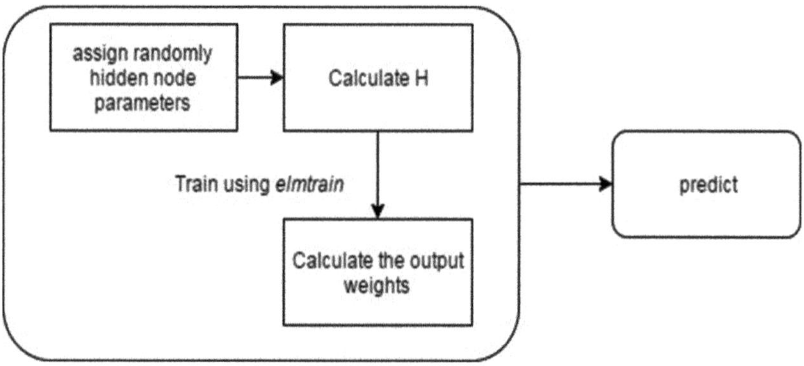

FIGURE 13.4 Flow diagram showing implementation of ELM.

13.4 RESULT ANALYSIS

The performances of various computing models were analyzed using accuracy and error obtained for all three cases named case 1, case 2, and case 3, where case 1 is validation using test and train data, case 2 is validation using 10-fold cross-validation, and case 3 is pre-processing with PCA and validation using 10-fold cross-validation.

13.4.1 ACCURACY ANALYSIS

The accuracy of various computing models was compared based on three parameters—AUC, sensitivity, and specificity. The following observations were made for all the various cases:

13.4.1.1 DEEP LEARNING

Table 13.1 and Figure 13.5 show the values of accuracy parameters obtained for deep learning for all three cases.

TABLE 13.1 Accuracy of Deep Learning.

	AUC	Sensitivity	Specificity
Case 1	0.6204	0.585098	0.368628
Case 2	0.69982	0.741961	0.486078
Case 3	0.6984	0.756667	0.513726

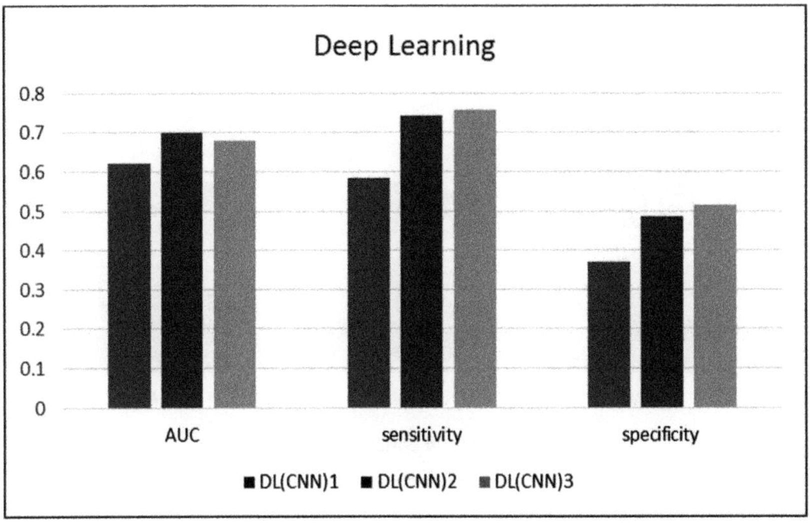

FIGURE 13.5 Accuracy of deep learning.

As can be seen from Table 13.1 and Figure 13.3, the AUC is maximum for case 2, but sensitivity and specificity are maximum for case 3. Case 1 comes out to be the worst for all three parameters. Also, the difference between AUC for case 2 and case 3 is not much but the difference between their sensitivity and specificity values is quite significant, therefore, it can be inferred that deep learning works best for case 3.

13.4.1.2 ELM

Table 13.2 and Figure 13.6 show the values of accuracy parameters obtained for ELM for all three cases.

Unlike deep learning, the value of AUC for ELM is maximum for case 1 and minimum for case 3, and the same is the case for specificity. For sensitivity, case 3 is the maximum, and case 1 is the minimum.

TABLE 13.2 Accuracy of ELM.

	AUC	Sensitivity	Specificity
Case 1	0.61615	0.410196	0.431569
Case 2	0.61483	0.431756	0.382157
Case 3	0.58068	0.587843	0.370784

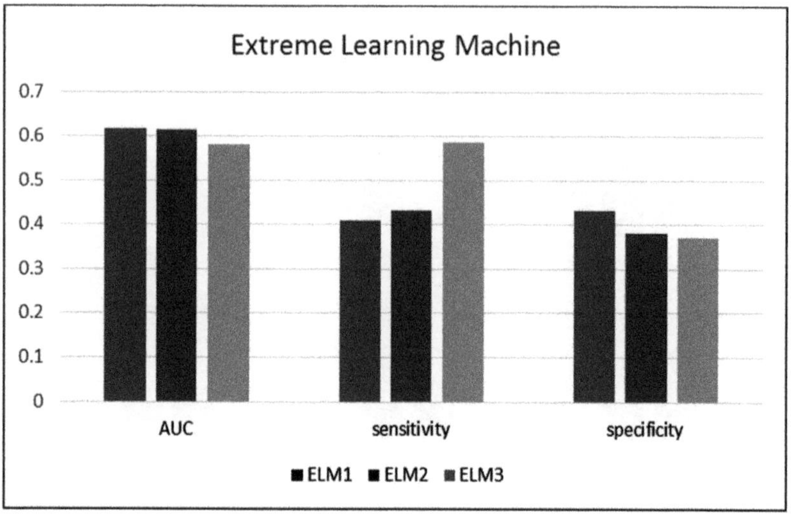

FIGURE 13.6 Accuracy of ELM.

13.4.1.3 KELM

Table 13.3 and Figure 13.7 show the values of accuracy parameters obtained for KELM for all three cases.

TABLE 13.3 Accuracy of KELM.

	AUC	Sensitivity	Specificity
Case 1	0.5963	0.543529	0.381373
Case 2	0.60855	0.48902	0.39
Case 3	0.55992	0.612941	0.373137

For KELM, case 2 shows the maximum value for AUC while case 3 shows the minimum, and specificity follows similar ordering. Sensitivity is maximum for case 3 and worst for case 2, which is just the opposite of the order for the other two parameters.

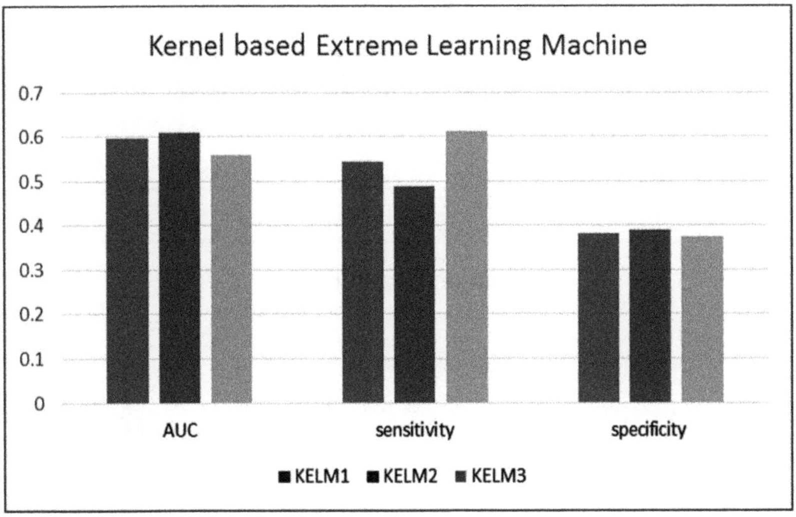

FIGURE 13.7 Accuracy for KELM.

If there is a requirement to find out the overall best technique among all the cases, then the best according to AUC is case 2 of deep learning with AUC = 0.69982. Case 3 of deep learning is also very close with AUC = 0.6984. If sensitivity is taken into consideration for finding the best technique, then case 3 of deep learning rules over the others with a value of 0.756667, and the same is true for specificity where its value is 0.513726. Therefore, it can be said that the best model for diagnosis of AD according to accuracy parameters is case 3 of deep learning.

13.4.2 ERROR ANALYSIS

The errors of various computing models were compared based on five parameters— (MAE), Relative Absolute Error (RAE), RMSE, Root Mean Logarithmic Squared Error (RMLSE), and RRSE. The following observations were made for the various cases:

13.4.2.1 DEEP LEARNING

Table 13.4 and Figure 13.8 show the values of five parameters of error for deep learning.

TABLE 13.4 Error for Deep Learning.

	MAE	RAE	RMSE	RMLSE	RRSE
Case 1	0.52314	1.04628	0.72328	0.501341	1.4466
Case 2	0.38598	0.77196	0.62127	0.43063	1.2425
Case 3	0.3648	0.72961	0.60399	0.418654	1.2079

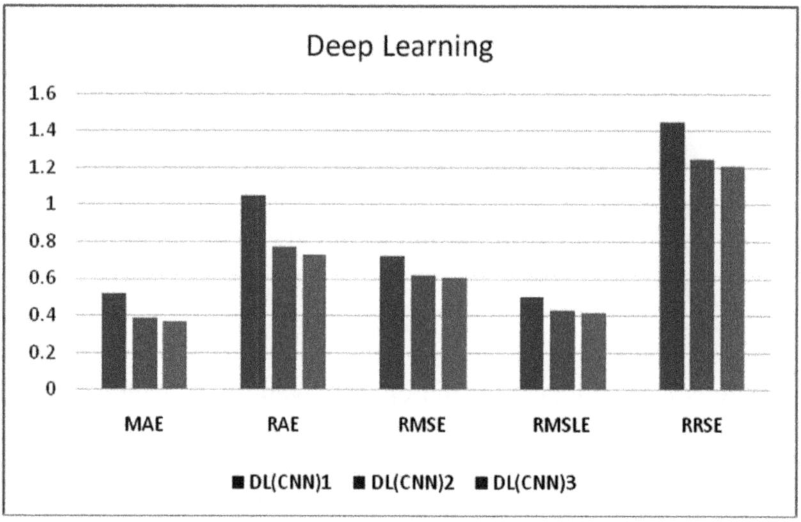

FIGURE 13.8 Error for deep learning.

The error for case 3 is the least and the error for case 1 is the most for all five parameters in case of deep learning.

13.4.2.2 ELM

Table 13.5 and Figure 13.9 show the values of five parameters of error for deep learning.

TABLE 13.5 Error for ELM.

	MAE	RAE	RMSE	RMLSE	RRSE
Case 1	0.57912	1.15824	0.761	0.52748	1.522
Case 2	0.59304	1.18608	0.77009	0.533786	1.54018
Case 3	0.52069	1.04137	0.72159	0.500165	1.44317

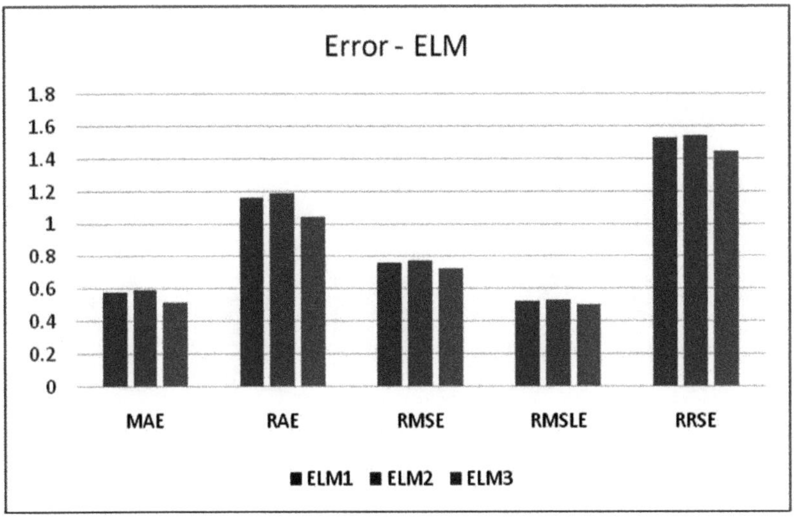

FIGURE 13.9 Error for ELM.

The error is maximum for case 2 and minimum for case 3 for all the five parameters in case of ELM.

13.4.2.3 KELM

Table 13.6 and Figure 13.10 show the values of five parameters of error for deep learning.

TABLE 13.6 Error for KELM.

	MAE	RAE	RMSE	RMLSE	RRSE
Case 1	0.57912	1.15824	0.761	0.52748	1.522
Case 2	0.59304	1.18608	0.77009	0.533786	1.54018
Case 3	0.52069	1.04137	0.72159	0.500165	1.44317

In case of KELM, the error is maximum for case 2 and minimum for case 3 for all the five parameters.

According to error analysis, the overall best technique would be the one with minimum error. In all the three models discussed here, case 3 has having least error for all the five parameters. For MAE, case 3 of deep learning has the least value, same is the case with the other four parameters as well.

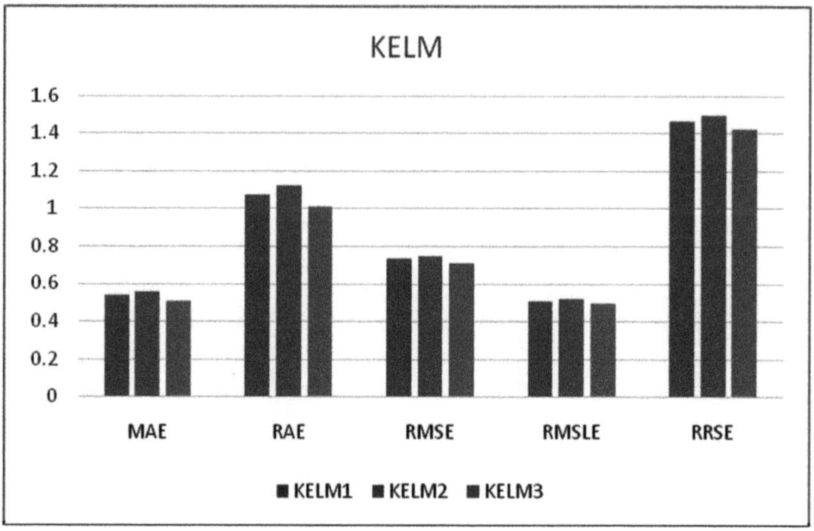

FIGURE 13.10 Error for KELM.

After both accuracy and error analyses, it can be said that deep learning using 10-fold cross-validation on genetic data which is pre-processed by PCA is the best option for diagnosis of AD.

13.4.3 SET WISE ANALYSIS

The dataset used in this paper has data gathered from three different sets—adni_gwas_v2_set1, adni_gwas_v2_set2, and adni_gwas_v2_set3. The data from these sets were analyzed to check for homogeneity across the sets. 6800 and 3400 records for training and testing purposes, respectively, were used from each of the three sets. The computing models were applied individually to these three different sets of data, with each of them having 10,200 records. Accuracy and error analysis was done on all of them.

13.4.3.1 ACCURACY ANALYSIS

Deep Learning. The accuracies for all the three sets are given in Table 13.7 and Figure 13.11 for deep learning.

Table 13.8 shows the best model according to each parameter for all the sets.

TABLE 13.7 Accuracy of all Sets for Deep Learning.

		AUC	Sensitivity	Specificity
Set 1	DL(CNN)1	0.5227	0.5158824	0.5417647
	DL(CNN)2	0.65337	0.54	0.5494118
	DL(CNN)3	0.59353	0.7288235	0.4670588
Set 2	DL(CNN)1	0.86759	0.8035294	0.7623529
	DL(CNN)2	0.89441	0.8970588	0.7482353
	DL(CNN)3	0.62986	0.5517647	0.6282353
Set 3	DL(CNN)1	0.84296	0.9323529	0.1523529
	DL(CNN)2	0.92037	0.9588235	0.1635294
	DL(CNN)3	0.88962	0.9276471	0.1529412

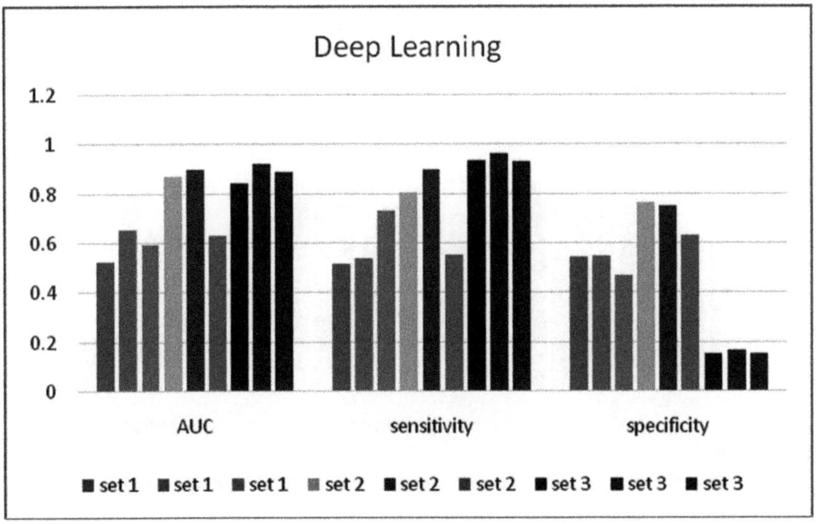

FIGURE 13.11 Accuracy of all sets for deep learning.

TABLE 13.8 Best Model for Each Set for Deep Learning.

	AUC	Sensitivity	Specificity
Set 1	Case 2	Case 3	Case 2
Set 2	Case 2	Case 2	Case 1
Set 3	Case 2	Case 2	Case 2

ELM. The accuracies for all the three sets are given in Table 13.9 and Figure 13.12 for ELM.

TABLE 13.9 Accuracy of all Sets for ELM.

		AUC	Sensitivity	Specificity
Set 1	ELM1	0.53283	0.4017647	0.5788235
	ELM2	0.5181	0.5958824	0.5264706
	ELM3	0.59155	0.6411765	0.52
Set 2	ELM1	0.64234	0.6805882	0.55644706
	ELM2	0.62083	0.6211765	0.54
	ELM3	0.63014	0.5552941	0.6341176
Set 3	ELM1	0.6501	0.6823529	0.1923529
	ELM2	0.67482	0.6894118	0.1805882
	ELM3	0.70709	0.9129412	0.1670588

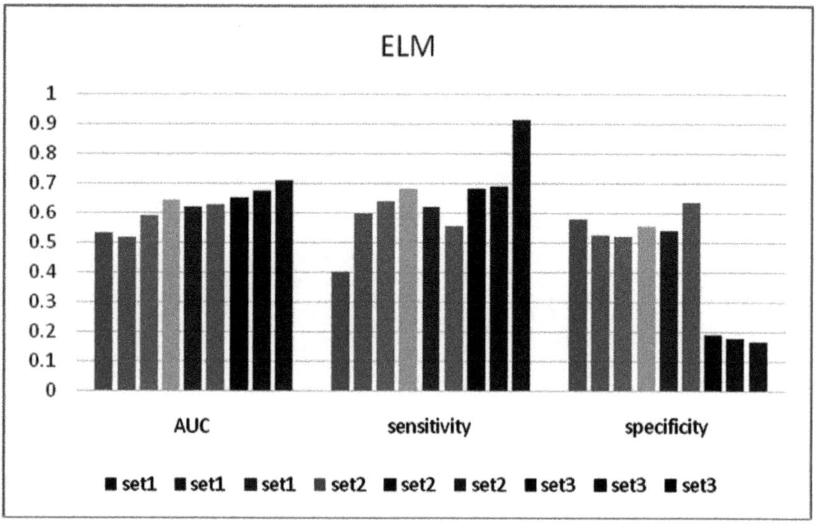

FIGURE 13.12 Accuracy of all sets for ELM.

Table 13.10 shows the best model according to each parameter for all the sets.

TABLE 13.10 Best Model for Each Set for ELM.

	AUC	Sensitivity	Specificity
Set 1	Case 3	Case 3	Case 1
Set 2	Case 1	Case 1	Case 3
Set 3	Case 3	Case 3	Case 1

KELM: The accuracies for all the three sets are given in Table 13.11 and Figure 13.13 for KELM.

TABLE 13.11 Accuracy of All Sets for KELM.

		AUC	Sensitivity	Specificity
Set 1	KELM1	0.53059	0.49176	0.56941
	KELM2	0.52912	0.50529	0.55294
	KELM3	0.56559	0.61	0.52118
Set 2	KELM1	0.60529	0.6241176	0.5864706
	KELM2	0.60824	0.6894118	0.5270588
	KELM3	0.60853	0.59	0.6270588
Set 3	KELM1	0.57265	0.6682353	0.1864706
	KELM2	0.59853	0.5847059	0.2182353
	KELM3	0.79169	0.89	0.1723529

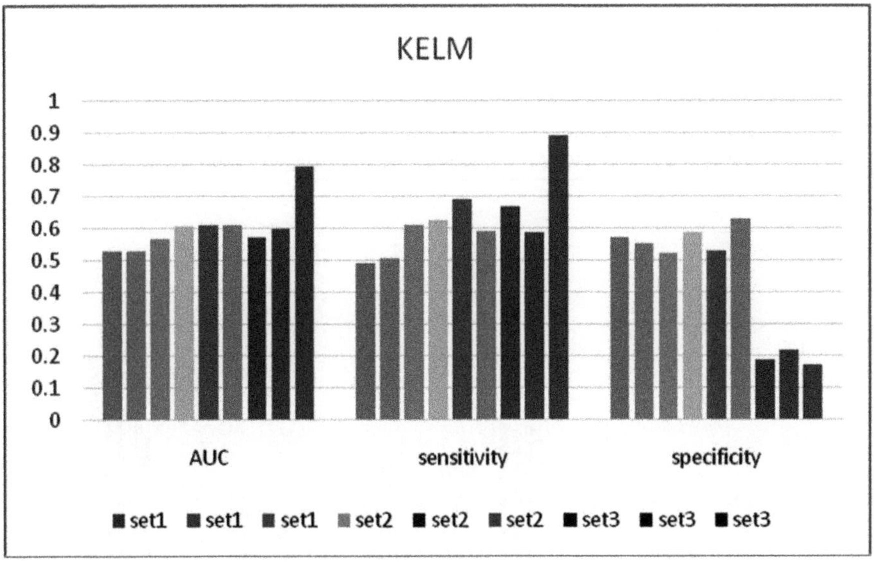

FIGURE 13.13 Accuracy of all sets for ELM.

Table 13.12 shows the best model according to each parameter for all the sets.

TABLE 13.12 Best Model for Each Set for KELM.

	AUC	Sensitivity	Specificity
Set 1	Case 3	Case 3	Case 1
Set 2	Case 3	Case 2	Case 3
Set 3	Case 3	Case 3	Case 2

13.4.3.2 ERROR ANALYSIS

Deep Learning: The value of error parameters are shown in Table 13.13 and Figure 13.14.

TABLE 13.13 Error of All Sets for Deep Learning.

		MAE	RAE	RMSE	RMLSE	RRSE
Set 1	DL(CNN)1	0.47118	0.94235	0.68642	0.47579	1.37285
	DL(CNN)2	0.45529	0.91059	0.67475	0.4677	1.34951
	DL(CNN)3	0.40206	0.80412	0.63408	0.43951	1.26816
Set 2	DL(CNN)1	0.21706	0.43412	0.4659	0.32293	0.93179
	DL(CNN)2	0.17735	0.35471	0.42113	0.29191	0.84227
	DL(CNN)3	0.41	0.82	0.64031	0.44383	1.28063
Set 3	DL(CNN)1	0.45765	0.91529	0.6765	0.46891	1.35299
	DL(CNN)2	0.43882	0.87765	0.66244	0.45917	1.32488
	DL(CNN)3	0.45971	0.91941	0.67802	0.46997	1.35603

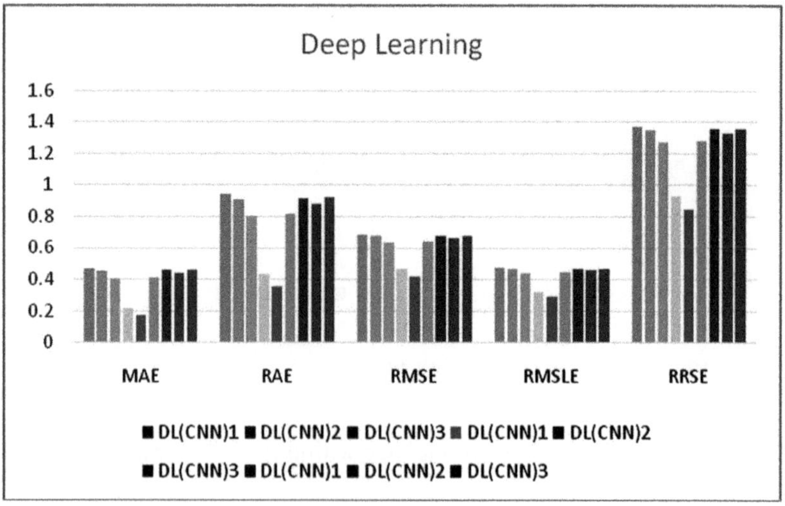

FIGURE 13.14 Error of all sets for deep learning.

Table 13.14 shows the best model (minimum error) according to each parameter for all the sets.

TABLE 13.14 Best Model for Each Set for Deep Learning.

	MAE	RAE	RMSE	RMLSE	RRSE
Set 1	Case 3	Case 3	Case 3	Case 3	Case 3
Set 2	Case 2	Case 2	Case 2	Case 2	Case 2
Set 3	Case 2	Case 2	Case 2	Case 2	Case2

ELM: The value of error parameters are shown in Table 13.15 and Figure 13.15.

TABLE 13.15 Error of All Sets for ELM.

		MAE	RAE	RMSE	RMLSE	RRSE
Set 1	ELM1	0.50971	1.01941	0.71394	0.49486	1.42787
	ELM2	0.43882	0.87765	0.66244	0.45917	1.32488
	ELM3	0.41941	0.83882	0.64762	0.4489	1.29524
Set 2	ELM1	0.38147	0.76294	0.61763	0.42811	1.23527
	ELM2	0.41941	0.83882	0.64762	0.4489	1.29524
	ELM3	0.40529	0.81059	0.63663	0.44128	1.27325
Set 3	ELM1	0.56265	1.12529	0.7501	0.51993	1.5002
	ELM2	0.565	1.13	0.75166	0.52101	1.50333
	ELM3	0.46	0.92	0.67823	0.47012	1.35647

Table 13.16 shows the best model (minimum error) according to each parameter for all the sets.

TABLE 13.16 Best Model for Each Set for ELM.

	MAE	RAE	RMSE	RMLSE	RRSE
Set 1	Case 3	Case 3	Case 3	Case 3	Case 3
Set 2	Case 1	Case 1	Case 1	Case 1	Case 1
Set 3	Case 3	Case 3	Case 3	Case 3	Case 3

KELM: The value of error parameters is shown in Table 13.17 and Figure 13.16.

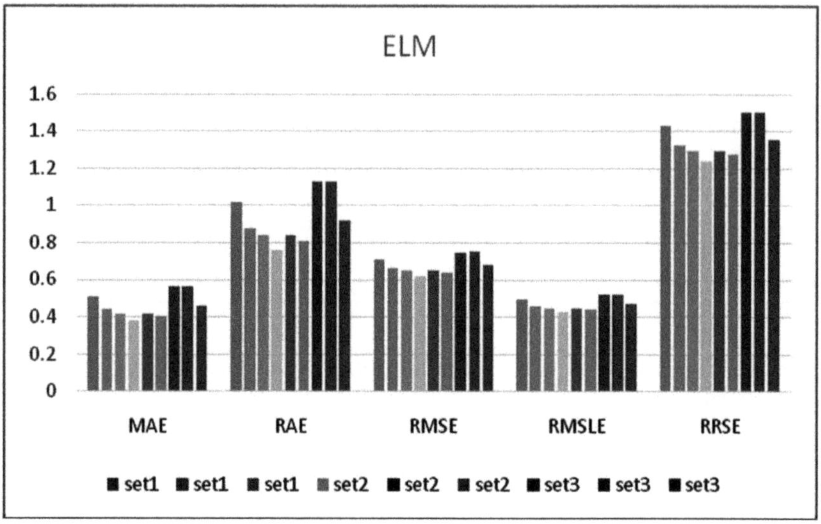

FIGURE 13.15 Error of all sets for ELM.

TABLE 13.17 Error of All Sets for KELM.

		MAE	RAE	RMSE	RMLSE	RRSE
Set 1	KELM1	0.46941	0.93882	0.68514	0.4749	1.37027
	KELM2	0.47088	0.94176	0.68621	0.47564	1.37242
	KELM3	0.43441	0.86882	0.6591	0.45685	1.3182
Set 2	KELM1	0.39471	0.78941	0.62826	0.43547	1.25651
	KELM2	0.39176	0.78353	0.62591	0.43385	1.25182
	KELM3	0.39147	0.78294	0.62568	0.43369	1.25135
Set 3	KELM1	0.57265	1.14529	0.75673	0.52453	1.51347
	KELM2	0.59853	1.19706	0.77365	0.53625	1.54729
	KELM3	0.46882	0.93765	0.68471	0.4746	1.36941

Table 13.18 shows the best model (minimum error) according to each parameter for all the sets.

TABLE 13.18 Best Model for Each Set for KELM.

	MAE	RAE	RMSE	RMLSE	RRSE
Set 1	Case 3	Case 3	Case 3	Case 3	Case 3
Set 2	Case 3	Case 3	Case 3	Case 3	Case 3
Set 3	Case 3	Case 3	Case 3	Case 3	Case 3

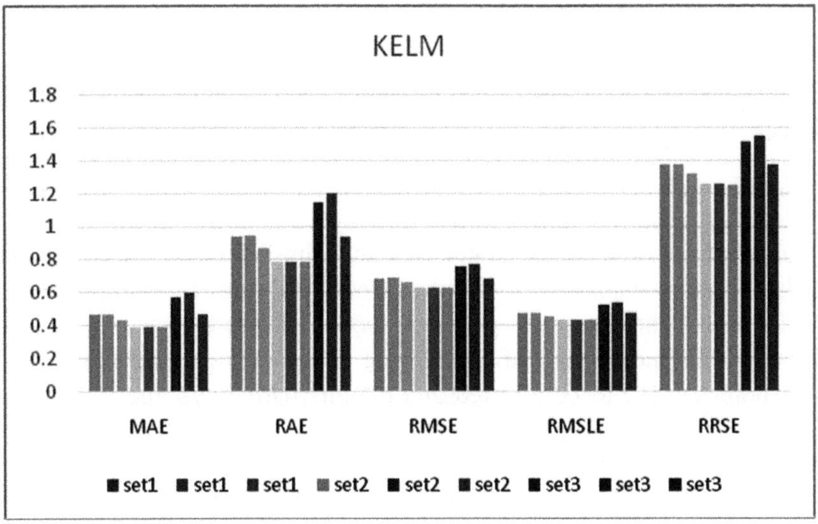

FIGURE 13.16 Error of all sets for KELM.

13.4.4 KRUSKAL–WALLIS TEST

To analyze the homogeneity of data present in the three sets, the Kruskal–Wallis test was performed on the computing models under consideration like it is done by Rathore and Kumar, 2016Since, the predicted value of computing methods followed non-normal distribution, Kruskal– Wallis test was used for the analysis of data using AUC, sensitivity and specificity. The results of Kruskal–Wallis test are given in Tables 13.19–13.21.

TABLE 13.19 Kruskal–Wallis Test for AUC.

Chi-squared	5.4222
DF	2
p-value	0.06646
Alpha	0.05

TABLE 13.20 Kruskal–Wallis Test for Sensitivity.

Chi-squared	5.6
DF	2
p-value	0.06081
Alpha	0.05

TABLE 13.21 Kruskal–Wallis Test for Specificity.

Chi-squared	5.9556
DF	2
p-value	0.05091
Alpha	0.05

For the tables above, H_0 states that the samples come from the same population and H_a states that the samples do not come from the same population. At least one of the computed p-values should be less than the significance value $\alpha = 0.05$ (indicating that 5% of false positives are permissible) for the null hypothesis, H_0, to be rejected and the alternative hypothesis, H_a, to be accepted. As all are greater than α in our case, the null hypothesis is accepted and the alternative hypothesis is rejected.

The results of the Kruskal–Wallis test indicate that the data in the three sets are not significantly different from each other and the three used computing models do not have much statistically significant performance difference from each other for the samples from these sets.

13.4.5 DUNN'S MULTIPLE COMPARISON TEST

To further validate the absence of statistically significant differences among the sets, Dunn's multiple comparison test was performed. The results of multiple comparison test for AUC, sensitivity, and specificity are given in Tables 13.22–13.24.

TABLE 13.22 Dunn's Test for AUC.

Sl. No	Comparison	Z	P. unadj	P.adj
1	Set 1 verses Set 2	–2.0869968	0.03688843	0.11066528
2	Set 1 verses Set 3	–1.9379256	0.05263230	0.07894845
3	Set 2 verses Set 3	0.1490712	0.88149745	0.88149745

TABLE 13.23 Dunn's Test for Sensitivity.

Sl. No	Comparison	Z	P. unadj	P.adj
1	Set 1 verses Set 2	–1.7888544	0.07363827	0.11045741
2	Set 1 verses Set 3	–2.2360680	0.02534732	0.07604196
3	Set 2 verses Set 3	–0.4472136	0.65472085	0.65472085

TABLE 13.24 Dunn's Test for Specificity.

Sl. No	Comparison	Z	P. unadj	P.adj
1	Set 1 verses Set 2	–0.745356	0.45605654	0.45605654
2	Set 1 verses Set 3	1.639783	0.10105026	0.15157538
3	Set 2 verses Set 3	2.385139	0.01707266	0.05121798

Here, Z represents a vector of all m of Dunn z test statistics, P.unadj represents a vector of p-values corresponding to Z and P.adj represents a vector of adjusted p-values corresponding to Z.

Comparing the p.adj values with $\alpha/2$, with $\alpha = 0.05$, we see that all the tests give positive results of similarity.

13.5 CONCLUSION AND FUTURE WORK

The performance of three efficient computing models was evaluated and compared for the classification of records of AD patients. The computing models included deep learning, ELM, and KELM. The evaluation was done on a dataset obtained from ADNI.

Three different cases including validation using a test dataset, 10-fold cross-validation, and pre-processing with PCA followed by 10-fold cross-validation were taken and the performance of the computing models was evaluated based on accuracy and error.

A set-wise analysis was also performed to check the homogeneity of the available dataset and Kruskal–Wallis Test and Dunn's Multiple Comparison Test were done to check the dependency of the data from different sets.

Different models performed differently under different conditions. After both accuracy and error analyses, it was seen that deep learning using 10-fold cross-validation on genetic data which is pre-processed by PCA is the best option for diagnosis of AD.

Apart from the work done in this paper, various other classical as well as modern classification approaches can be applied to heterogeneous data in the future. Also, this study can be extended to other similar neurodegenerative diseases like Parkinson's disease.

KEYWORDS

- **AD**
- **machine**
- **kernel**
- **diagnosis**

REFERENCES

Alzheimer, A. Uber Eineeigenartigeerkrankung der Hirnrinde. *Allgemeine Zeitschrife Psychiatrie* **1907**, *64*, 146–148.

Alzheimer, A. *A Characteristic Disease of the Cerebral Cortex: Meeting of South-West Germany Psychiatrists Held in Tubingen on November 3rd and 4th, 1906,* 1987.

Akgül, C. B.; Ünay, D.; Ekin, A. In *Automated Diagnosis of Alzheimer's Disease using Image Similarity and User Feedback*, Proceedings of the ACM International Conference on Image and Video Retrieval (p. 34). ACM, 2009.

Bengio, Y. Learning Deep Architectures for AI. *Foundations and trends® in Machine Learning* **2009,** *2* (1), 1–127.

de Haan, W.; Pijnenburg, Y. A.; Strijers, R. L.; van der Made, Y.; van der Flier, W. M.; Scheltens, P.; Stam, C. J. Functional Neural Network Analysis in Frontotemporal Dementia and Alzheimer's Disease Using EEG and Graph Theory. *BMC Neurosci.* **2009,** *10* (1), 101.

Dinno, A. Dunn's Test of Multiple Comparisons Using Rank Sums. R package version, 132.

Gamberger, D.; Ženko, B.; Mitelpunkt, A.; Shachar, N.; Lavrač, N. **2016,** Clusters of Male and Female Alzheimer's Disease Patients in the Alzheimer's Disease Neuroimaging Initiative (ADNI) Database. *Brain Inform.* **2016,** *3* (3), 169–179.

Goñi, J.; Esteban, F. J.; de Mendizábal, N. V.; Sepulcre, J.; Ardanza-Trevijano, S.; Agirrezabal, I.; Villoslada, P. A computational Analysis of Protein-protein Interaction Networks in Neurodegenerative Diseases. *BMC Syst. Biol.* **2008,** *2* (1), 52.

Farhan, S.; Fahiem, M. A.; Tauseef, H. An ensemble-of-classifiers Based Approach for Early Diagnosis of Alzheimer's Disease: Classification Using Structural Features of Brain Images. *Comput. Math. Methods Med.* **2014,** 2014.

Fukushima, K.; Miyake, S. In *Neocognitron: A Self-Organizing Neural Network Model for a Mechanism of Visual Pattern Recognition*, Competition and Cooperation in Neural Nets (pp. 267–285). Springer Berlin Heidelberg, 1982.

Gosso, A.; Martinez-de-Pison, F. J. elmNN: Implementation of ELM (Extreme Learning Machine) Algorithm for SLFN (Single Hidden Layer Feedforward Neural Networks). R package version, 1, 2012.

Huang, G. B.; Zhu, Q. Y.; Siew, C. K. In *Extreme Learning Machine: A New Learning Scheme of Feedforward Neural Networks*, Neural Networks, 2004. Proceedings. 2004 IEEE International Joint Conference on (Vol. 2, pp. 985–990). IEEE, 2004.

Huang, G. B.; Zhu, Q. Y.; Siew, C. K. Extreme Learning Machine: Theory and Applications. *Neurocomputing* **2006,** *70* (1), 489–501.

Huang, G. B.; Zhou, H.; Ding, X.; Zhang, R. Extreme Learning Machine for Regression and Multiclass Classification. *IEEE Trans. Syst. Man Cybern. Part B (Cybern.)* **2012,** *42* (2), 513–529.

Ivakhnenko, A. G.; Lapa, V. G.; United States. Cybernetic Predicting Devices. CCM Information Corp: New York, 1965.

Ivakhnenko, A. G.; Lapa, V. G. Cybernetics and Forecasting Techniques 1967.

Ivakhnenko, A. G. The Group Method of Data Handling-a rival of the Method of Stochastic Approximation. *Soviet Automatic Control* **1968,** *13* (3), 43–55.

Jammeh, E.; Zhao, P.; Carroll, C.; Pearson, S.; Ifeachor, E. In *Identification of Blood Biomarkers for Use in Point of Care Diagnosis Tool for Alzheimer's Disease*, Engineering in Medicine and Biology Society (EMBC), 2016 IEEE 38th Annual International Conference of the. IEEE, 2016; pp 2415–2418.

Jedynak, B. M.; Liu, B.; Lang, A.; Gel, Y.; Prince, J. L.; Alzheimer's Disease Neuroimaging Initiative. A Computational Method for Computing an Alzheimer's Disease Progression score; experiments and validation with the ADNI data set. *Neurobiol. Aging* **2015,** 36, S178-S184.

Krashenyi, I.; Popov, A.; Ramirez, J.; Gorriz, J. M. In *Fuzzy computer-aided diagnosis of Alzheimer's disease using MRI and PET statistical features*, Electronics and Nanotechnology (ELNANO), 2016 IEEE 36th International Conference on. IEEE, 2016; pp 187–191.

LeCun, Y.; Bengio, Y.; Hinton, G. Deep Learning. *Nature* **2015,** *521* (7553), 436–444.

LeCun, Y. LeNet-5, Convolutional Neural Networks 2015 [Online]. http://yann. lecun. com/exdb/lenet.

Li, B.; Rong, X.; Li, Y. An Improved Kernel Based Extreme Learning Machine for Robot Execution Failures. *Sci. World J.* **2014,** 2014.

Liu, S.; Liu, S.; Cai, W.; Pujol, S.; Kikinis, R.; Feng, D. In *Early Diagnosis of Alzheimer's Disease with Deep Learning*, Biomedical Imaging (ISBI), 2014 IEEE 11th International Symposium on. IEEE, 2014; pp 1015–1018.

McKight, P. E.; Najab, J. Kruskal-Wallis Test. *CorsiniEncyclopedia Psychol.* 2010.

Mo, J.; Siddiqui, S.; Maudsley, S.; Cheung, H.; Martin, B.; Johnson, C. A. In *Classification of Alzheimer Diagnosis from ADNI Plasma Biomarker Data*, Proceedings of the International Conference on Bioinformatics, Computational Biology and Biomedical Informatics, ACM, 2013; p 569.

Pant, R.; Mukhopadhyay, S.; Lakshmayya, B. Review on Alzheimer's Disease: A Neurodegenerative Disorder.

Plant, C.; Sorg, C.; Riedl, V.; Wohlschläger, A. In *Homogeneity-based Feature Extraction for Classification of Early-stage Alzheimer's Disease From Functional Magnetic Resonance images*, Proceedings of the 2011 Workshop on Data Mining for Medicine and Healthcare, ACM, 2011; pp 33–41.

Rathore, S. S.; Kumar, S. An Empirical Study of Some Software Fault Prediction Techniques for the Number of Faults Prediction. *Soft Comput.* **2016,** 1–18.

Reitz, C.; Mayeux, R. Alzheimer Disease: Epidemiology, Diagnostic Criteria, Risk Factors and Biomarkers. *Biochem. Pharmacol.* **2014,** 88 (4), 640–51.

Salman, A.; Mordechai, S. The Potential of Infrared Spectroscopy and Multivariate Analysis of Peripheral Blood Components as a Validated Clinical Test for Early Diagnosis of Alzheimer's Disease. *Cell. Mol. Med.* **2017,** *2,* 1.

Sarvabowma, S. Sexual Disinhibition in An Elderly Man-Could It Be Alzheimer's Dementia: A Case Report. *J. Evol. Med. Dent. Sci.* **2015,** *4* (92), 15809–15810.

Schmidhuber, J. Deep Learning in Neural Networks: An Overview. *Neural Netw.* **2015,** *61,* 85–117. arXiv:1404.7828 Freely accessible. doi:10.1016/j.neunet.2014.09.003

Suk, H. I.; Lee, S. W.; Shen, D.; Alzheimer's Disease Neuroimaging Initiative. Hierarchical Feature Representation and Multimodal Fusion with Deep Learning for AD/MCI Diagnosis. *NeuroImage* **2014,** *101,* 569–582.

Weiner, M. W.; et aI. The Alzheimer's Disease Neuroimaging Initiative: A Review of Papers Published Since its Inception. *Alzheimers Dement.* **2012,** *8* (1 Suppl), 61–68.

Xiang, S.; Yuan, L.; Fan, W.; Wang, Y.; Thompson, P. M.; Ye, J. In *Multi-source Learning with Block-wise Missing data for Alzheimer's Disease Prediction*, Proceedings of the 19th ACM SIGKDD International Conference on Knowledge Discovery and Data Mining, ACM, 2013; pp 185–193.

Ye, J.; Chen, K.; Wu, T.; Li, J.; Zhao, Z.; Patel, R.; Reiman, E. In *Heterogeneous Data Fusion for Alzheimer's Disease Study*, Proceedings of the 14th ACM SIGKDD International Conference on Knowledge Discovery and Data Mining, ACM, 2008; pp 1025–1033.

Zheng, C.; Xia, Y.; Pan, Y.; Chen, J. Automated Identification of Dementia Using Medical Imaging: A Survey from a Pattern Classification Perspective. *Brain Inform.* **2016,** *3* (1), 17–27.

Zou, L.; Huang, Q.; Li, A.; Wang, M. A Genome-wide Association Study of Alzheimer's Disease Using Random Forests and Enrichment Analysis. *Sci. China Life Sci.* **2012,** *55* (7), 618–625.

Index

C

D

E

F

G

I

K

M

For Product Safety Concerns and Information please contact our EU representative GPSR@taylorandfrancis.com
Taylor & Francis Verlag GmbH, Kaufingerstraße 24, 80331 München, Germany

www.ingramcontent.com/pod-product-compliance
Ingram Content Group UK Ltd.
Pitfield, Milton Keynes, MK11 3LW, UK
UKHW022209190926
475702UK00003B/276
9781774918166